Residential Property Appraisal

Residential Property Appraisal, Volumes 1 and *2*, are handbooks not only for students studying residential surveying but also for those involved in the appraisal of residential property.

Volume 1 has been updated and covers the valuation process as it relates to residential properties, particularly when valuation is undertaken for secured lending purposes. It addresses the basic skills required, the risks posed in a valuation, the key drivers of value, emerging issues that impact valuation and the key legal and RICS regulatory considerations that a valuer needs to understand.

Volume 2 of the series goes on to consider the practical aspects of the survey and inspection of residential properties in more detail. Not only does this include updated sections on the most common defects (for example, building movement, moisture problems, wood rot and wood-boring insects), it also covers emerging challenges, including assessing personal safety hazards, modern construction technologies and materials and invasive plants. The volume also takes account of the Home Survey Standard recently published by the RICS and the changes resulting from climate change, the energy crisis and concerns about fire safety.

Building services in domestic residential properties is another area of rapid change, especially with the development of low-carbon and renewable technologies. To ensure that this aspect is covered in sufficient detail, the content is to be included in *Volume 3: Assessing Building Services*.

An essential book for students studying to enter the residential survey and valuation profession and for existing practitioners who wish to improve their knowledge of current practices.

Phil Parnham is a Chartered Building Surveyor and was a director at BlueBox partners and has over 40 years' experience in the property sector. He spent his first ten years working in local government and for housing associations where he developed an interest in residential defects. He took up an academic post at Sheffield City Polytechnic (latterly Sheffield Hallam University) where he helped develop their building surveying course at all levels. He left academia and joined up with Chris Rispin in 2010 to form BlueBox partners where he jointly organised and delivered a range of training courses and other services for those working in the residential property sector. In September 2021, BlueBox partners was acquired by Sava and Phil set up his own training and development consultancy creatively named Phil Parnham Ltd.

During this time Phil has also written a number of books, journal articles and conference papers. The most notable examples include: the *RICS Survey Writer Sample Phrases for the RICS Condition and HomeBuyer Reports, The Surveyor's Guide to RICS Home Surveys*, the *Domestic Energy Assessor's Handbook*, and *Assessing Building Services* (all for the RICS) and *Residential Property Appraisal*. In addition, he was the technical author for RICS's information paper on Japanese knotweed (first edition) and the RICS professional statement on the Home Survey Standard.

Phil is aiming to retire in the near future but like many of those in the residential sector he is unlikely to achieve this goal. Although he has given up rushing around the conference and seminar circuit, he can often be found in his office in Sharrow, Sheffield engaging in what he calls 'deep thinking'.

Larry Russen is a chartered building surveyor with over 40 years' practical experience in residential, industrial and commercial property. He is a chartered building engineer and party wall surveyor and established Russen & Turner Chartered Surveyors in King's Lynn in 1981. Over the last 20+ years, he has enjoyed training thousands of surveyors, engineers and energy assessors with RICS, Sava and BlueBox partners. His mantra for any surveyor, aspiring or practicing, is 'it's all about your attitude and attention to detail'.

Residential Property Appraisal

Volume 2 – Inspections, defects and reports

Phil Parnham and Larry Russen

LONDON AND NEW YORK

Designed cover image: © Phil Parnham; Larry Russen

First published 2023
by Routledge
4 Park Square, Milton Park, Abingdon, Oxon OX14 4RN

and by Routledge
605 Third Avenue, New York, NY 10158

Routledge is an imprint of the Taylor & Francis Group, an informa business

British Library Cataloguing-in-Publication Data
A catalogue record for this book is available from the British Library

Library of Congress Cataloging-in-Publication Data
Names: Parnham, Phil, author. | Russen, Larry, author.
Title: Residential property appraisal / Phil Parnham and Larry Russen.
Description: New York, NY : Routledge, 2023. | Includes bibliographical
    references and index. |
Identifiers: LCCN 2022014368 (print) | LCCN 2022014369 (ebook) |
    ISBN 9781032181509 (hbk) | ISBN 9781032159911 (pbk) |
    ISBN 9781003253105 (ebk)
Subjects: LCSH: Residential real estate—Valuation.
Classification: LCC HD1390.5 .P37 2023 (print) | LCC HD1390.5 (ebook) |
    DDC 333.33/8—dc23/eng/20220607
LC record available at https://lccn.loc.gov/2022014368
LC ebook record available at https://lccn.loc.gov/2022014369

ISBN: 978-1-032-18150-9 (hbk)
ISBN: 978-1-032-15991-1 (pbk)
ISBN: 978-1-003-25310-5 (ebk)

DOI: 10.1201/9781003253105

Typeset in NewBaskerville
by Apex CoVantage, LLC

For Alfie with many thanks for renewing my interest in football. Lots of love Gwandad.

For my children, Nick, family, friends, students and the hot, messy Elephant in Room 1.

We would like to thank Chris Rispin, Fiona Haggett and Carrie de Silva for all their help and support. We wish we could write as quickly as you three.

Contents

Contents

1 Introduction

Contents

Preface for Volume 2 of *Residential Property Appraisal*

Residential Property Appraisal was originally published in 2001 and this book is the second edition. For many years, Chris Rispin and I resisted updating because of the increasing rate of change in the country generally and the residential property in particular. Much of this change was fuelled by technological developments associated with the internet that would rapidly result in any paper-based publication soon becoming out of date.

However, 21 years later, we still think the core principles of valuation and surveying are as applicable today and are likely to remain so for some time. For example, people still want to live near good schools, away from noisy roads and glue factories, while (to our knowledge) the mechanisms of building movement, the growth of *Serpula lacrymans* and the stability of lath and plaster ceilings are still the same today as they were at the beginning of the 20th century. Consequently, we have tried to identify a series of benchmarks that highlight these core principles, giving you a framework around which you can take account of current and future changes. Hopefully, this will provide you with a firm basis for assessing any residential property and how to report it to your client.

To enable us to do this, Chris and I felt it important ask a number of specialists to join the writing team and we are pleased to welcome:

- Fiona Haggett to add depth and breadth to the valuation content;
- For his knowledge of construction and building pathology, Larry Russen; and
- To demonstrate the importance of the law, Carrie de Silva.

Their short biographies are included on pp. i-ii.

DOI: 10.1201/9781003253105-1

As the writing team began to discuss the revisions required, it became clear that the book would have to be physically bigger if it was to cover valuation, building pathology and surveying themes properly. In partnership with the publishers, we decided one large book would be too unwieldy, so it would be best to split the publication into two volumes: Volume 1 would cover valuation and law, and Volume 2 would tackle building pathology and other aspects of residential surveying. Volume 1 worked well and was produced in a timely fashion, but as the content for Volume 2 was further developed, it became clear it had grown into a weighty tome by itself. For example, the word count for the chapter on building services stood at 54,000 words alone, even after a dramatic edit. As a consequence, a decision was taken to add another volume to what had become a series. This now includes:

- **Residential Property Appraisal. Volume 1: Valuation and Law.** This volume focuses on the valuation of residential property and the law supporting it.
- **Residential Property Appraisal. Volume 2: Inspections, Defects and Reports.** As well as construction and building pathology, this volume also includes environmental matters, safety hazards and report writing.
- **Residential Property Appraisal. Volume 3: Assessing Building Services.** This includes the usual range of building services commonly used in domestic buildings, including low-carbon and micro-generation systems.

We think the three volumes will provide a complete coverage of the dark art of residential property appraisal and so hopefully last for another 21 years. See you again in 2043!

Before you proceed to the main chapters of the book, I would like to pick up on another topic mentioned by Chris in his preface to Volume 1. I would also like to dedicate this book to those chartered surveyors who were central to many RICS initiatives in the residential sector but have sadly died, including David Dalby, Barry Hall and Graham Ellis. Not only did these work tirelessly on our behalf, but they were really nice people too. They were helpful, very open and taught me so much about the residential property sector.

Sadly, the list does not end there. 2021 saw the very sad death of Professor Malcolm Hollis. Most of you will know of Malcolm by reputation and many more of you will have at least one of his excellent books on your bookshelf. His contribution was immense: he set up a highly successful practice in London, contributed to RICS education initiatives for many years and wrote some of the best-selling surveying books ever published. I saw Malcolm speak many times during my own career and learnt so much from him. Although I will never match his knowledge and skills, I hope you will notice his influence in the forthcoming pages.

Phil Parnham, February 2022.

1.1 Context of the second edition and Volume 2

In terms of describing the context for Volume 2 of this series, we would echo the sentiments expressed by our colleagues in Chapter 1 of Volume 1. Not only does this approach save us from having to think up something original (a task we always find challenging), but importantly, it acknowledges what Chris, Fiona and Carrie so eloquently described – the pace of change in the residential survey and valuation sector has exceeded our collective imaginations. In this chapter, we want to highlight the most significant of these changes and how they have and will continue to influence the residential property sector.

1.1.1 Continuing government intervention in the property market

A small number of surveyors reading these words will recall the Home Information Pack (HIP) legislation of the early part of the 2000s. This government initiative aimed to make home buying easier. However, faced with increasing resistance from many parts of the residential sector itself, the Coalition Government finally abolished HIPs in 2010. Since then, although subsequent governments have adopted non-interventionist approaches, little progress has been made and the home buying and selling processes are not too different from what they were in 2001.

The following examples outline some of the more recent government interventions and how they have affected the residential market:

- In the late noughties, *Fallopia Japonica* hit the headlines and caused chaos in the residential market. Otherwise known as Japanese knotweed, a number of high-street lenders refused to approve mortgages on properties affected by this plant. This resulted in several colourful (and often inaccurate) media campaigns in support of the small number of homeowners who were adversely affected. Although professional and financial organisations had some success in 'calming' the market through new guidance and lending policies, because of the continuing controversy, the government's Science and Technology Committee reviewed the matter in an attempt to urge the sector to bring in more effective policies (Science and Technology Committee 2019). This resulted in the publication of a new RICS Guidance Note (RICS 2022).
- There has been an increase in what might be described as 'arm's-length' control of building work and technical standards including the well-established government-approved Competent Person Schemes. Other more recent initiatives place the onus of showing compliance with regulations and standards on the property owners themselves. For example, 'General Binding Rules' for small sewerage discharges (Environment Agency 2021) and 'Ready to Burn' regulations for solid fuel heating systems are examples in which owners are directly responsible for ensuring appropriate standards are maintained (HETAS 2021). This places a greater emphasis on the residential surveyor liaising with the vendor.
- The fire at Grenfell Tower in London on 14 June 2017 killed 72 people. This tragic incident and the public campaign that quickly followed resulted in the government setting up the Grenfell Tower Inquiry 'to examine the circumstances leading up to and surrounding the fire at Grenfell Tower on the night of 14 June 2017' (Grenfell Tower Inquiry 2022). Although the inquiry is still ongoing, the interim findings have already had a profound effect on the residential market, especially for medium- and high-rise properties. At the time of writing, although the inquiry is far from finishing its work, most commentators agree that significant changes in building control legislation will continue into the future (Guardian 2021).

These three examples show how the relationship between legislators and the housing market is both subtle and complex, making the role of the residential practitioner even more challenging. We can no longer rely on that worn-out old phrase of 'legal adviser to check appropriate approvals are in place' when in reality the assessment of building standards is a more nuanced process.

1.1.2 Global crises

Three years ago, we could have used the singular form of this noun, but as we put the finishing touches to this volume, the increasing numbers of global challenges have caused us to include the plural form of 'crisis'. These global threats include:

- The COVID-19 pandemic;
- The climate crisis;
- The volatility of the energy supply market; and
- Armed conflict in Eastern Europe.

Some of these challenges may not directly affect the property market, but others will. Take two examples:

- **Climate crisis and 'net zero'.** The threat of climate change has resulted in governments of many countries announcing a variety of target dates by which they will decarbonise their economies. In the UK, the current plan is to reach 'net zero' by 2050, although there is some debate on whether current trajectories will achieve that aim (BBC 2021). In terms of residential property, the practical changes are likely to include:
  - Stopping natural gas, liquid petroleum gas, oil and some forms of biomass fuels from being used to heat our homes;
  - Setting legal minimum energy efficiency standards for properties that are rented. Current policy discussion papers suggest that all homes will have to achieve EPC band of C by 2030 at the latest; and
  - Introducing voluntary targets for lenders to improve the average EPC rating of the properties in their lending portfolio to at least band C also by 2030. This may lead to mortgage offers being linked to EPC ratings.
- **The volatility of the energy supply market.** At the time of writing, domestic energy bills are likely to double in April 2022, as the cap on energy prices is lifted by Ofgem (Ofgem 2022). This increase is likely to continue throughout 2022 and into 2023. Although this is partly linked to the global climate crisis, other factors have also played a role:
  - Armed conflict in Ukraine;
  - The cold winter across Europe in 2020–2021 reduced the amount of stored gas in Europe;
  - Low wind levels during the summer of 2021 reduced the contribution from renewable energy sources; and
  - Increased demand for gas in Asia (especially China) has pushed prices up.

The resulting increase in energy prices has been dramatic. Many millions of households are facing between a 50 to 100% increase in their fuel bills.

Although we can only guess, the likely impact of these two energy-related crises is a change in the public's attitude towards energy efficiency. As fuel bills soar and lower EPC ratings begin to affect saleability and possibly even value, home buyers will want better information about the energy efficiency of the homes they are planning to buy. Many will look to their surveyor to provide this.

1.1.3 Technological changes

In Volume 1, our colleagues identified the increasing pace of technological advancement over the last 20 years. This included 'big data' playing a larger role in the industry as well as large software suppliers becoming more influential players in the market.

Technological advances have also affected how individual residential surveyors do their job on a micro level. For example, if you were to shadow one during a typical working day, you may witness some of the following:

- Setting up the relationship with their client and managing each instruction through a cloud-based software system;
- Carrying out desktop research into properties using free to use websites ranging from satellite-based imagery through to searchable local authority databases;
- Recording on site information using tablet computers, digital cameras and smartphones capable of storing hundreds of high-quality images and videos;
- Inspecting normally inaccessible parts of the roof, chimney and gutters with easy-to-use drones;
- Assembling the report using speech recognition software, a large database of standard phrases and the ability to include a large number of images; and
- 'Meeting' the client using an online video conferencing system after the report has been delivered.

This scenario is a world away from how we worked in 2001, when the vast amount of the process was paper-based and far fewer photos were taken.

1.1.4 Changing professional standards

In terms of professional guidance, life in 2001 was a little simpler. The Royal Institution of Chartered Surveyors (RICS) was the main standard setter for the residential sector and it published prescriptive guidance that members of the Institution had to follow. Although surveyors would often debate the extent of the inspection for a Home Buyers Report or the content of a building survey report, both products were well known, supported by prescriptive guidance documents and largely well understood (by surveyors at least).

Today, professional standards are less strictly drawn. For example, in 2019, RICS published the Home Survey Standard (RICS 2019) to help their members cope with the unprecedented '. . . scale of social, economic, political and technological change . . .' and allow RICS members to adapt and innovate by providing '. . . a clear, flexible framework within which RICS members and RICS regulated firms can develop their own services the public can recognise and trust, that are consistent with the high standards expected by RICS' (RICS 2019, p. 5). Although a standard RICS format product is still available, a range of other survey report products has been produced by a variety of providers, from corporate organisations to 'sole trader' surveyors.

Accounting for this variety of professional practice in the writing of this second edition has been particularly challenging.

1.1.5 Increasing consumer demands

Over the last 20 years, the ever-increasing consumer standards and expectations have forced residential surveyors to become much more 'customer-facing'. Free-to-use online review platforms allow any consumer with a buying or service experience to review any company, and these reviews can impact a practitioner's reputation. In addition, the increase in consumer expectations has also resulted in a more litigious industry where 'no win, no fee' legal practices compete for business whenever a new scandal develops. Recent examples include cavity wall insulation, Japanese knotweed and spray foam in lofts.

But some things never change

Despite all of these dramatic and unpredictable economic, social and technological developments, one thing has not changed since the first edition of this book – the process of assessing the physical condition of a residential property. This is still based on the physical inspection of a property by an appropriately qualified and experienced professional who collects sufficient information to enable the property's condition to be assessed. This assessment forms the basis of the report that is designed to satisfy the client's specific requirements. In this respect, the approach and content of this second edition matches very closely to that of the first edition, albeit much updated and extended.

1.2 Objectives of Volume 2

After reading this volume, you should be able to:

* Work closely with a client while setting up an appropriate service. This typically includes terms of engagement, knowledge of locality and nature of the property, and helping the clients choose the appropriate service level;
* Carry out the service including researching the locality and property, choosing the suitable equipment, carrying out an appropriate inspection and recording the inspection outcomes;
* Use the information assembled during the previous stages, assess the condition of the principal elements of a range of domestic residential property using suitable diagnostic processes and methodologies;
* Identify and assess risks and hazards in residential property;
* Report the outcomes of the research, inspection and assessment of a property to the clients;
* Assemble, organise and close project files in accordance to current recommended practice once the service is complete.

Additionally, throughout this volume, we will highlight the changing nature of the residential property appraisal process and identify some techniques and mechanisms that may help you adapt to this changing environment.

1.3 Definitions

Clearer definitions of the principal terms employed in this volume may be useful:

Residential property is any property that is used as, or is suitable for use as, a residence. This book is restricted to single domestic dwellings owned by an individual(s), rather than a corporate entity.

Customer or client, in this book, these two terms are used to describe the end user of any survey or inspection report and will usually be a private individual.

'Residential surveyor', 'residential practitioner', 'practitioner' or 'surveyor' These generic terms which has been used to describe chartered surveyors or other suitably qualified practitioners.

1.4 Who this book is for

Like Volume 1, this part of the series has been written to suit a broad range of residential surveyors whose primary interest is with the physical condition of residential property. The book assumes that the reader has:

- Already satisfied the academic requirements of their chosen professional institution, or is close to doing so. This would have included a course of study that introduced participants to how dwellings are designed and constructed and the principal agents responsible for the deterioration of the building fabric; and
- Had some professional experience of assessing or at least inspecting a range of different properties.

In terms of qualifications and level of experience, this book should be suitable for:

- Student residential surveyors in the later stages of their academic course or on the sandwich placement or year-out stage;
- Building surveyors with experience in the commercial building sector who are looking for an introduction to the assessment of residential properties;
- Surveyors who are working towards their professional assessment and need to refer to written guidance and technical information on a regular basis;
- Those more experienced surveyors who may be changing their professional emphasis towards residential survey;
- Qualified and experienced surveyors who need to carry around a source of reference so they can refer to standard guidance when novel or unusual situations are encountered.

1.5 The philosophy of the book

The guidance contained in this publication aims to be challenging to the reader in two ways:

- To outline processes and techniques that may potentially take the surveyor beyond the parameters of level two property assessments. This will enable surveyors to provide level two services more effectively and better cope with any changes to standard practice in the future; and

- To engage with the surveying process and positively advise clients about the suitability of the condition of the home that is new to them.

This book is not a guide to any particular standard form of survey promoted by any particular corporate body, lender or professional institution. For something more specific, the reader should refer to the publisher of the particular product.

References

BBC (2021). 'What is net zero and how are the UK and other countries doing?' Available at www.bbc.co.uk/news/science-environment-58874518. Accessed 23 February 2022.

Environment Agency (2021). 'General binding rules'. Available at www.gov.uk/guidance/general-binding-rules-small-sewage-discharge-to-a-surface-water. Accessed 23 February 2022.

Grenfell Tower Inquiry (2022). 'The Grenfell tower inquiry'. Available at www.grenfelltower inquiry.org.uk/. Accessed 23 February 2022.

Guardian (2021). 'Fire risk "cover-up" one of "greatest scandals of our time", Grenfell inquiry hears'. Available at www.theguardian.com/uk-news/2021/dec/06/fire-risk-cover-up-one-of-greatest-scandals-of-our-time-grenfell-inquiry-hears. Accessed 23 February 2022.

HETAS (2021). 'Ready to burn. A look at the legislation in more detail'. Available at www.hetas.co.uk/ready-to-burn-a-look-at-the-regulations-in-more-detail/. Accessed 23 February 2022.

Ofgem (2022). 'Check if the energy price cap affects you'. Available at www.ofgem.gov.uk/information-consumers/energy-advice-households/check-if-energy-price-cap-affects-you. Accessed 23 February 2022.

RICS (2019). *The Home Survey Standards*. RICS, London.

RICS (2022). *Japanese Knotweed and Residential Property*. RICS Guidance Note, 1st edition. RICS, London.

Science and Technology Committee (2019). Available at https://committees.parliament.uk/committee/135/science-and-technology-committee/news/100943/uk-approach-to-japanese-knotweed-overly-cautious/. Accessed 18 February 2022.

2 Setting up the service

Contents

2.1 Introduction

This chapter follows the structure of a typical instruction and concentrates on those matters associated with setting up the service before a contract is agreed. It will look at the following issues:

- Terms of engagement;
- Conflicts of interest;

DOI: 10.1201/9781003253105-2

- Client liaison;
- Qualifications and experience of the practitioner;
- Knowledge of locality; and
- Levels of service.

These will be considered in turn.

2.2 Terms of engagement

The terms of engagement is a very important document that goes to the heart of the contractual relationship between the practitioner and their client. Before the Home Survey Standard (HSS) was published (RICS 2019), RICS's standard guidance for the different reports was very detailed. Consequently, the majority of the terms of engagement were contained in the standard documentation. To complete the terms, all the practitioner had to do was to send the client a covering letter that included the specific conditions of the contract, such as the client's name and address, any special instructions, conflicts of interest, the fee and so on. Please see Volume 1 of *Residential Property Appraisal* for more information on the general principles of contract law (Rispin et al. 2021).

The HSS is not as prescriptive as previous guidance and the new standard formats of the RICS's Home Survey levels one, two and three contain few of the necessary contract terms. Consequently, the onus has been placed on residential practitioners to formulate their own terms of engagement properly. To support their members and help them meet the appropriate standards, RICS published what is called a 'Surveyor's toolkit'. This consists of a number of general guidance documents that have no formal status but provide help on how the HSS can be met. The one most relevant to this section is titled 'Terms of engagement' (RICS no date).

We have used the document from the Surveyor's Toolkit and the relevant section from the HSS to sketch out the typical content of the terms of engagement for a Home Survey service. As stated in the 'Terms of engagement' document from the Surveyor's Toolkit, this part of the book is for '. . . guidance only and is not a full client engagement letter, or intended to be suitable for every client or every service provided. It should not be relied upon as a precedent or as legal advice' (RICS no date). In other words, you have to take full responsibility for assembling your own terms of engagement document because it will be very particular to you, your client and the subject property.

2.2.1 *Terms of engagement in the HSS*

The HSS states that the client **must** receive an up-to-date document that describes the terms of engagement (TofE) matched to the specific instruction. It points out that at levels one and two on less complex properties, the TofE are likely to be standard documents amended to take account of the property type, intended future use and any specific client requirements.

However, for level two services on older and/or complex properties, historic buildings and those in a neglected condition and all level three services, residential practitioners will have to give careful consideration as to whether any variations to the usual TofE are required because of the special nature of the property and the particular requirements of the client. For example, a thatched property may require some specific information to be gathered or the client may have requested a more extensive inspection of a roof space. In each case, the TofE will have to be adjusted to suit.

Where these are varied, the adjustments must be described and explained by the practitioner during the pre-inspection discussions with the client. This illustrates what we think is one of the main priorities of the HSS – to encourage more direct contact between the practitioner and the client both before the contract is signed and after the report has been delivered (see Section 2.3).

Once the terms of engagement have been issued, ideally these must be signed and returned by the client before the inspection is carried out. Because many clients want a quick response, this is not always possible. In these circumstances, the terms of engagement must be agreed upon '. . . before the delivery of the service, ideally before any professional advice on the property is given'. What this usually means is that the report should not be sent until the client has accepted the TofE, signed and returned it.

The HSS goes on to identify two further issues:

- Practitioners supplying professional services to consumer clients must be aware of applicable regulations including the Consumer Contracts (Information, Cancellation and Additional Charges) Regulations 2013; and
- The terms of engagement must point out that the service does not include an asbestos inspection and it falls outside *The Control of Asbestos Regulations* 2012. However, the report should still emphasise the suspected presence of asbestos-containing materials if the inspection identifies that possibility (see Chapter 13 Assessing Safety Hazards for occupants).

A minimum requirement for the TofE has been included in Appendix D of the HSS (RICS 2019) and these have been listed in the following section. In addition, we have added what we thought were the most useful parts of the commentary from the RICS's TofE document in the Surveyor's Toolkit.

2.2.2   *Typical content of the terms of engagement*

Regardless of the level of the service, the terms of engagement must address the following matters:

- Client's name, address and appropriate contact details;
- Practitioner's name (where known at the time of instruction) and appropriate contact details;
- Subject property's address and postcode;
- Nature and type of service required. This is the section where you must benchmark the service you are offering against those described in the current edition of the Home Survey Standard RICS professional statement. In addition, this should include any additional particular features requested by the client;
- Nature and the intended future use of the property. This is very important. For example, if the client is planning to completely refurbish the property or intends to let the property to tenants, then the nature and content of the report will

vary accordingly. This is another example where the report must be 'property specific';

- Details of any special instructions and/or additional services (linked to the previous point);
- Likely inspection date (if known) and the anticipated date the report will be published.
- Style and delivery format of the report;
- Agreed fee and the fees for any additional work (including VAT);
- Details of any referral fees, inducements and potential conflicts of interest.
- Payment arrangements, payment period;
- Cancellation rights;
- Forewarning of any restrictions due to health and safety implications that may arise on the day of inspection. For example, thick layers of insulation in the roof space, stored occupants' possessions and so on;
- Evidence that the client has confirmed acceptance of the terms of engagement. Ideally, this can be achieved by the practitioner sending a digital or paper copy of the terms of engagement to the client and for these to be signed and returned to the practitioner before the inspection goes ahead;
- Confirm that as an RICS regulated firm their files may be subject to monitoring and will need to be provided to RICS upon request.
- Confirm that any fees taken in advance are not client money and not subject to the RICS client money protection scheme; and
- The RICS regulated firm operates a complaints-handling procedure, details of which are available upon request.

You may also want to consider a few other points raised in the Surveyor's Toolkit document:

> **Limitation of liability:** In addition to the preceding points, you may want to consider discussing with the client any limitation of liability. Preferably, this should be done by speaking to the client directly and explaining the impact of any limit of liability. In addition, this should be confirmed in a letter or by email afterwards. Examples of these clauses can be found in the current edition of the '*Risk, liability and insurance in valuation work*' RICS guidance note (RICS 2021), along with suggestions on how best to limit liability to the client. At the time of writing, there has been considerable debate about 'limitation of liability'. Although it is widely used in other professional sectors (for example, legal and accountancy), its use is still new to the residential sector and many are still unsure about what the courts would find acceptable. For example, a few years ago, practitioners attempted to limit liability to a multiple of the fee charged to the client. Many commentators felt this sometimes-low limit would be considered 'unreasonable'. In more recent times, practitioners have been setting the liability at the same level as their maximum professional indemnity insurance limit. We cannot comment either way because it is outside our area of expertise, but to our knowledge, it has not yet been tested in the courts. This is another illustration of why it is not possible to issue standard terms of engagement documents across the sector and why it is important to get appropriate legal advice on what provisions suit your particular service.

Conflicts of interest and referral fees: RICS and many other professional institutions have been promoting initiatives designed to encourage property professionals to be more transparent and open with their clients. Although referral fees and other inducements may seem part of acceptable business practice to us, our clients may consider them as questionable 'backhanders' that put our independence and objectivity at risk. Here are just two examples:

- Staff at the local estate agency refer potential buyers to a local residential practitioner. For this service, the practitioner pays the estate agents a fixed fee for each referral that results in an instruction;
- A firm of chartered surveyors run a business that offers both estate agency and residential survey services. They regularly accept instructions to act for the sellers of a property as their estate agent, while the residential survey section of the business often advises potential buyers about the very same properties.

The Home Survey Standard states that RICS members must declare any potential conflicts and how these will be managed (RICS 2019, p. 30). More detailed information on this issue is discussed in the current edition of RICS's 'Rules of Conduct and Conflicts of interest', a RICS professional statement published in 2017 (RICS 2017).

Where a practitioner does pay or receive a referral fee to or from a third party, the HSS reminds RICS members they must acknowledge this practice. This can typically be achieved by providing clients or prospective clients with a written statement (which should be included in the terms of engagement) stating one of the following:

- The Practitioner does not pay a referral fee or equivalent to any party who may have recommended them; or
- That a payment has been or may be made, either individually or part of a third-party commercial relationship.

These are just two typical examples and the Conflicts of Interest practice statement includes more typical situations together with advice on how these can be managed. What is clear is that these declarations need to be up front and explained to the client before contracts are signed.

2.3 Client liaison

One of the main objectives of the Home Survey Standards was to '. . . set out a series of concise mandatory requirements. These establish 'benchmarks' around which RICS members and RICS regulated firms can design and deliver services that meet their clients' needs in a changing environment' (RICS 2019, p. 5). In our view, putting the client's needs at the centre of the service will help deliver that.

Under the section of the HSS titled 'client liaison', RICS members must ensure that clients:

- Understand the differences between the levels of service, including the extent and limitations of each option;

- Are advised of the range of options the RICS member can offer, together with the key features and benefits of each;
- Are aware of the fee that will be charged for the service;
- Agree the terms of engagement;
- Agree report format and method of delivery; and
- Explain the intended future use of the property (for example, buy to let).

In our experience, most practitioners are well aware of the importance of working closely with the client and have this level of liaison built into their processes. Not only will this help meet the client's needs, it will also help reduce misunderstandings and the likelihood of complaints and possibly even claims.

Despite these advantages, a minority of practitioners and organisations have very little contact with their clients either before or after they carry out the service. Instead, they rely on standard responses to 'online' forms. In some cases, these can automatically calculate a fee, send standard terms of engagement and even propose a date for the property inspection and report delivery. How can this be done if no one has asked the client what he or she actually wanted from the service? This does not mean that before any contract is signed, you must have an in-depth face-to-face interview with the client. Instead, you should have an appropriate method of establishing your clients' needs at the earliest opportunity. For example, this can be done in a number of different ways:

- By well-trained admin or support staff who are familiar with the Home Surveys process and the service offered by the practitioner;
- Direct contact through a range of communication methods including telephone, email, social media and virtual meeting programs over the internet.

All of these approaches can be supported by standard explanatory documents provided by RICS or created by the practitioners themselves.

Much of the information listed by the HSS can be delivered to the client in a straightforward manner apart from making sure they understand the differences between the levels of service as making them aware of the extent and limitations of each option can be a challenge. Consequently, we have outlined a possible approach in Section 2.4 below.

2.4 Choosing between the different service levels

The Home Survey Standards (HSS) identify a number of different factors that influence the choice between different service levels and we have described these as follows (RICS 2019, p. 8).

2.4.1 Qualifications and experience

The HSS makes it clear that providing a high level of service will depend on the practitioner being qualified, experienced and able to deliver the agreed service (RICS 2019, p. 8). This is much more than having a professional qualification. It also includes:

- Having knowledge of the tasks to be undertaken and the risks involved;
- Possessing the experience and ability to carry out the duties in relation to the appropriate level of service; and

- Identifying their own limitations and taking appropriate action where their knowledge and experience is found to be inadequate.

The professional profile of any practitioner will develop over time as individuals enhance their competence through a mix of initial training, on-the-job learning, instruction, assessment and formal qualification. Active engagement with continuing professional development is especially important in these rapidly changing times.

2.4.2 *Knowledge of locality and nature of property*

To be able to offer clients a competent service, the HSS goes on to say that practitioners must be familiar with the '. . . nature and complexity of the subject property type, the region in which it is situated and relevance to the subject instruction' (RICS 2019, p. 8). The components of this familiarity are taken to include:

- Common and uncommon housing styles, materials and construction techniques. This is particularly important where services are offered for older and historic buildings where understanding the interaction of different building materials and techniques is essential;
- Current advice and guidance relating to asbestos and other common deleterious materials;
- An awareness of the main principles of modern methods of construction;
- Environmental issues including publicly available information;
- The location of listed buildings and conservation areas/historic centres, the implications of these designations especially in relation to legislation that affects repair and improvement work;
- A basic understanding of the type of tenure for the subject property;
- Relevant requirements specified by local and regional government organisations and structures; and
- Awareness of the social and industrial heritage relevant to the instruction.

Hopefully, most readers of these volumes are well on their way to achieving this level of knowledge and the information on these many pages will help fill any gaps. None-theless, this does not mean you will automatically be able to offer all levels of service on any property. For example, carrying out a level two service on a semi-detached interwar house will require a very different level of knowledge than that required to offer the same service on a dwelling built in the 1990s.

The HSS describes this difference in the following way:

> Although an RICS member with this knowledge may be able to provide all levels of service, those who provide level two services on older and/or complex prop-erties, historic buildings and those in a neglected condition and all level three services will require a broader and deeper technical knowledge. Where appropri-ate, the RICS member must decline the instruction if the subject property type is beyond his or her knowledge and skill level.

> (RICS 2019, p. 8)

In our opinion, distinguishing between more straightforward dwellings and those that are 'older and/or complex . . ., historic . . . and those in a neglected condition' is the correct distinction. This can help you judge whether you have '. . . the experience and ability to carry out their duties in relation to the appropriate level of service'.

The first step in helping you make this judgement is understanding what 'older and/or complex properties, historic buildings and those in a neglected condition' actually means. Here is our interpretation:

'Older' and **'historic'** buildings: We have grouped these two terms together because they are closely associated with each other. Although the HSS mentions the date of 1850 (RICS 2019, p. 23), we don't think this should be a rigid point in time. For example, a traditionally built interwar semi-detached property may be approaching 100 years old and could share many characteristics of a much older property. Please note, in our view 'historic' does not necessarily mean the property is formally listed under the planning legislation.

'Complex property': We agree with the HSS that complex properties are those that have been extended and altered over the years. For example, we inspected a larger semi-detached interwar property that had side and rear extensions, including an open balcony. A number of load-bearing walls had been removed to form a separate 'grandparent's' flat on the ground floor. To properly assess this property, a practitioner will have to have enhanced knowledge of alteration work of this type. If not, it is likely that a large number of referrals for further investigations would be the result.

'Neglected condition': We think this is more than poor condition. Imagine a property that has lacked proper care for some time. Typically, ongoing defects may have resulted in serious problems that make parts of the property potentially unsafe and often uninhabitable. A level two service and/or a practitioner who is not familiar with the process of repairing a dilapidated property is unlikely to provide a satisfactory service to the client.

2.4.3 Practitioner knowledge

To offer services on these older, historic, complex and/or neglected properties, a practitioner '. . . will require a broader and deeper technical knowledge' (RICS 2019, p. 5). However, it does not offer an explanation of what this should include. A clearer explanation was provided by the forerunner of the HSS. The RICS's Surveys of Residential Property Guidance Note (SORP) offered the following definition:

> . . . practitioners who provide . . . services on complex properties, historic buildings and those in a poorer condition will require a broader and deeper technical knowledge. A structured, relevant and recent lifelong learning strategy can achieve this. This knowledge can be enhanced by other professional activities, such as:
>
> - Project managing further investigative work on behalf of clients;
> - Defect analysis and advising on appropriate repair methods;
> - Organising building/repair work; and
> - Dealing with consent applications such as building regulations, planning permission, listed building consent, party walls and other neighbourly matters.
>
> (RICS 2013)

In our view, this does not mean that only Chartered Building Surveyors can offer this type of service. On the contrary, any residential practitioner, regardless of their RICS designation, can do this work as long as they have this deeper and broader level of technical knowledge. Consider this illustration. Imagine a practitioner who has a valuation focus to their education, training and employment where the majority of their professional experience has been providing valuation advice. Although their knowledge of condition will be substantial, it could be considered as secondary to their valuation skill set. Additionally, if most of a practitioner's experience is at the 'front end' of the sale and purchase of property, they often don't have the opportunity of seeing how condition-related problems are actually resolved. Conversely, if a practitioner provides a broader range of condition-related services such as defect analysis or organising repair work on behalf of residential clients, then they are likely to have that broader and deeper view. This is because there is no better way of improving your condition assessment skills and knowledge than taking a building to pieces and then putting it back together again.

We realise that these illustrations are very much black and white in a world where there are many different shades of grey. Despite this, we think the days when a practitioner thought they could offer any level of service on any property just because they have a professional qualification are gone.

2.5 Helping the client choose the appropriate level of service

This can be a complex process. Here are some typical vignettes that show how confusing this choice can be:

- The clients who think that procuring a 'survey' is like going to a carwash where the level of service largely depends on how much they want to pay. For example, imagine a potential client who phoned up and says, 'I'm buying this lovely 19th century cottage that needs a bit of doing up and I would like you to carry out a level one service for me';
- The practitioner who says they carry out one type of inspection and it is only the report that varies; and
- A level two service on an older, complex property that is over 50 pages long and contains advice that would otherwise be considered suitable in a level three report.

Each one of these illustrations is based on real examples we have encountered. The following process may help you make this more objective.

2.5.1 Levels of service

The different levels of service are defined in Appendix A of the HSS (RICS 2019, p. 23). Of this description, we think the most important paragraphs are those that outline the type of property each level of service will suit:

Level one: '. . . a survey level one report does not include advice on repairs or ongoing maintenance and this, combined with the less extensive inspection, usually means it is better suited to conventionally built, modern dwellings in satisfactory condition. It will not suit older or complex properties, or those in a neglected condition'.

Level two: '. . . this level of service suits a broader range of conventionally built properties, although the age and type will depend on the knowledge and experience of the RICS member. This level of service is unlikely to suit:

- Complex buildings, for example those that have been extensively extended and altered;
- Unique or older historic properties – although survey level two services may be appropriate for some older buildings, the decision will depend on the RICS member's proven competence and knowledge and the nature of the building itself. For example, a survey level two report on homes with traditional timber frames or those built much before 1850 is likely to be inconclusive and be of little use to the client; or
- Properties in neglected condition.

In such cases, a survey level two service will often result in numerous referrals for further investigations, an outcome that many clients find disappointing'.

Level three: '. . . this level of service will suit any domestic residential property in any condition depending on the competence and experience of the RICS member'.

These definitions, combined with the client's particular requirements, provide five different criteria that can help define a suitable choice of service:

- The age of the property;
- Its heritage status;
- The complexity of the property;
- The property's condition; and
- The clients' particular requirements.

The output from this analysis can then be placed against the knowledge and experience of the practitioner who plans to offer the service. Like many of the professional judgements described in this publication, we think the level of service recommendation can be supported by placing each criterion on a continuum that ranges from level one service at one end to level three at the other (see Figure 2.1).

Before we go through the process in detail, it is important to be aware of a number of general principles:

- Each criterion has been split into three sections. Level 1 is scored between one and three, level two four and six, with level three between seven and nine;
- For each of the criterion, we have included a number of descriptors to help you place the property in the right 'ballpark';
- This is not a mathematical process that gives a 'right' answer. Instead, the resulting score places the property in one of three indicative service-level bandings. At this point, you should review the total score to see if it 'feels' right and then set it against your own knowledge and experience profile. It is at this point you must make a professional judgement about what level of service to offer or, alternatively, whether it's appropriate to decline the instruction.

Whatever the outcome, the level of service recommendation you give your client is a more nuanced process than it used to be and will be influenced by the interplay of a wider range of factors.

2.5.2 *The criteria*

Each criterion is weighted and these are described as follows.

Age of the property

This continuum for this criterion does not follow a linear scale. Instead, we have decided on three date bandings:

* Score of 1 to 3 – Built between the years 2000 to present day;
* Score of 4 to 6 – Built between the years 1920 to 1999; and
* Score of 7 to 9 – Built before 1919.

We debated these bandings for some time. Rather than waste more printer ink explaining the choice, here are the headlines:

* **2000 to the present day:** this was our version of 'modern'. Also, the year 2000 saw the increasing use of modern methods of construction (see chapter 12.10 for more detail);
* **1920 to 1999:** Because of the social and economic impact of dramatic worldwide events we thought the end of World War I was the most appropriate watershed;
* **Before 1919:** properties of this age are now over 100 years old and require a greater level of consideration as the risk of defects and deficiencies increases with age.

We were not deflected by the inclusion of the 1850 date in the Home Survey Standard (RICS 2019b). If a practitioner is very familiar with older properties in their area and has well-developed technical skills and knowledge, then it may be acceptable for them to provide a level two service on a 150-year-old property. However, that is a later judgement and shouldn't skew the decision at this stage.

The complexity of the property

This is about how much the property has been altered since it was first built.

* **Score of 1 to 3**: There have been few alterations to the property apart from some straightforward 'like for like' repairs/renewals;
* **Score of 4 to 6:** The property has undergone a small number of additions/alterations/repairs such as a simple extension, new conservatory, replacement windows, and boilers and so on;
* **Score of 7 to 9**: The property has had a number of extensions and alterations many of which would require building regulation approval. A number of elements have been repaired/replaced, such as a recovered roof, repointed walls, replacement windows and so on.

Condition

This criterion is about condition and will have to be based on information received from the client, agent's online details and from the vendor. Additional insights can also be gained from Google Street View and other virtual sources.

- **Score of 1 to 3**: From information received, the property appears to be in a satisfactory condition for its age and type;
- **Score of 4 to 6**: From information received, the property is in need of some significant repair and updating although not unusual for its age and type;
- **Score of 7 to 9**: From information received, the property is in need of considerable repair and updating and this may be unusual for its age and type. Although this may be due to age and neglect, this score may also apply to younger buildings that have suffered traumatic events such as flooding, small fires and so on.

The level of information requested by the client

This is about any particular client's requirements that may extend the 'normal' scope of the service offered. In our view, assessing this criterion is a challenge to the defensive culture that has developed in the property sector over the last few decades. For example, one of the main objectives of the Home Survey Standards is to encourage closer liaison with the client, so any additional requirements beyond the normal scope of service should not be seen negatively (see Chapter 2.2.2).

- **Score of 1 to 3:** The client had no special or particular requirements and is happy to accept the standard scope of service;
- **Score of 4 to 6:** Although the client was satisfied with the standard scope of service, they did want the practitioner to look at a small number of specific matters mainly associated with the condition of the property;
- **Score of 7 to 9:** Although the client was satisfied with the standard scope of service as a starting point, they did want some additional advice on how the property could be further developed. Examples include a possible loft extension and increasing the level of energy efficiency of the property.

The heritage status of the property

The Home Survey Standards mentions 'historic buildings' a number of times. Consequently, the 'Heritage' continuum uses the property's heritage status as a measure. This includes the following scores:

- **Score of 1 to 3:** the property is relative modern, it is not in a conservation area, national park or area of outstanding natural beauty;
- **Score of 4 to 6:** although the property is not itself protected, it is in a conservation area, an AONB or a national park;
- **Score of 7 to 9:** the property is in a conservation area with an Article 4 designation, or is listed (grade I, II* and II).

Total scores

Based on a simple mathematical calculation, the relationship between total scores and the level of service is as follows:

- 5 to 15 points: level one service;
- 16 to 30 points: level two service;
- 31 to 45 points: level three service.

Although this is a simplistic approach, you may also want to account for any 'outliers'. In other words, a score against a criterion that is very different to the rest. For example, in Case Study A, although the overall score might suggest a 'low' level two service, the 'complexity' criterion scores very highly. For this reason, alone, you may consider a level three more appropriate.

2.5.3 *The knowledge and experience of the practitioner*

Once you have scored each criterion, calculated the total score and made a provisional decision on what service level would be appropriate, you should set this against your own skill and knowledge level. We have provided a number of descriptors that may help you.

- **Green (level one).** The practitioner will have a clear valuation focus to their qualifications and experience. They carry out mainly valuation work and any condition assessment is ancillary to the valuation instructions. Their experience is dominated by relatively modern properties in a reasonable condition;
- **Amber (level two):** Although the practitioner carries out valuation work, they have considerable experience of providing level two services on a range of properties built mainly after 1920;
- **Red (level three):** At this level, the practitioner may offer a wide range of services with a clear focus on condition-related work such as defect diagnosis and organising repairs. Alternatively, they have extensive experience of offering level two services on older and more complex property as well as technical focus to their CPD.

This is our view about skills and knowledge. However, residential practitioners come in all shapes and sizes and the assessment of your competencies will be your own.

2.6 Completing the continuum and recommending the level of service

Figure 2.1 shows a continuum for each of the criteria. The column on the left describes the criteria. The row across the top shows the scale from one to nine split into three sections that correspond to green, amber and red.

When considering each criterion, make an assessment and mark the position in the box on the graduated scale. Once complete, add up the total and place this in the 'Total score' box. Review the profile and the total score, reflect on the results and

Property address:									
Criteria	1	2	3	4	5	6	7	8	9
Age of the property.									
Complexity of the property									
Condition of the property									
Level of service requested by the client									
Heritage status of the property									
Total score									
Provisional recommendation for service level.									

Figure 2.1 Typical assessment matrix for deciding the level of service.

make a provisional recommendation for the service level. Compare this to your own assessment of your ability to offer this level of service on the particular property. A 'comments' box has also been included so the factors that led to your decision can be recorded and kept on file.

Although many practitioners may find this process overly bureaucratic, we think this approach has a number of advantages:

- It helps you objectively establish a suitable level of service for the proposed instruction;
- The outcome will help you assemble your reasoning for your recommendation and so help explain to your client the differences between the levels of service; and
- This step-by-step approach will help you decide whether you have the appropriate knowledge, skills and experience to provide the agreed service.

An instruction that meets these criteria is also likely to satisfy the client's requirements, keep them happy and minimise the possibility of complaints and claims.

2.6.1 Recommending levels of service – worked examples

In an effort to make this process less abstract, we applied it to three different properties. Assume the following information has been gleaned from the client, the selling agent's particulars and a quick check on Google Street View.

Case study A

Case study A. This is a larger-than-normal two-storey detached house built in 2006. It is positioned on a large and steeply sloping site in a conservation area.

It has a flat roof, extensive glazing and air conditioning throughout. The property was 'architect designed'. The client had no special instruction but wanted to know whether the dwelling 'was properly built'.

Property address: Case study A									
Criteria	1	2	3	4	5	6	7	8	9
Age of the property.	X								
Complexity of the property								X	
Condition of the property		X							
Level of service requested by the client		X							
Heritage status of the property					X				
Total score	18								
Provisional recommendation.	The property is a one-off building to a non-standard design. Although the building has no heritage value, the site is within a conservation area. The building is highly serviced and is likely to include some innovative materials, techniques and sophisticated services. A service level three is appropriate provided the practitioner is knowledgeable about newer building techniques.								

Figure 2.2

Case study B

Case study B. This is a mid-terrace traditionally built property built in 1890 with solid walls, a part cellar and an original rear, single storey 'off-shot' kitchen. There is a 'room in the roof' bedroom on the second floor that is original to the property; the roof covering has been replaced in the not-too-distant past. The clients are a young couple new to the property market. They are already stretching their budgets and so '. . . want to make sure that we won't have to spend a lot of money on the property for a few years'.

Property address: Case study B									
Criteria	1	2	3	4	5	6	7	8	9
Age of the property.								X	
Complexity of the property					X				
Condition of the property					X				
Level of service requested by the client							X		
Heritage status of the property	X								
Total score	26								
Summary comments and provisional recommendation.	This property is typical of hundreds in this part of the city. The type and nature of the construction is well known and although a number of repairs and replacements have been carried out, the property remains as originally built. The age and reassurance required by client on level of repairs places this is on the border of a level two/three service. On balance, a service level two will be recommended as long as the practitioner is experienced at providing services on this age and type of property and the client is made aware the identifiable risk of potential or hidden defects in areas not inspected increases with property age.								

Figure 2.3

Case study C

Case study C. This traditionally built detached house was constructed in 1990. From all appearances, it is in reasonable condition for its age and type. There are no extensions or conservatories and the property appears to be unaltered since it was originally built. The client made no requests for additional information.

Property address: Case study C									
Criteria	1	2	3	4	5	6	7	8	9
Age of the property.				X					
Complexity of the property		X							
Condition of the property			X						
Level of service requested by the client	X								
Heritage status of the property	X								
Total score	11								
Summary comments and provisional recommendation.	Although this was outside the 'level one' age band, the property appeared to have 'Level one' characteristics on all other criteria. However, a score of three was allocated against the 'condition' criteria because the house was over 30 years old making some level of defects more likely. In our view this is a marginal decision where level one or two could be offered depending on the professional judgement of the practitioner.								

Figure 2.4

Hopefully these three case studies illustrate a number of issues:

- Selecting the appropriate level of service is a complex process;
- It is more than a numbers game. Even though a total score is obtained the impact of 'outliers' may have a disproportionate effect on your final recommendation.

Whatever the shortcomings of this process, we hope it helps you look at the process of recommending an appropriate service level in a more objective and analytical way.

References

RICS (2013). *Royal Institution of Chartered Surveyors. Surveys of Residential Property, RICS Guidance Note*, 3rd edition. RICS, London.

RICS (2017). *Royal Institution of Chartered Surveyors. Rules of Conduct and Conflicts of Interest, RICS Professional Statement*. RICS, London.

RICS (2019a). *Royal Institution of Chartered Surveyors. Home Survey Standard, RICS Professional Statement*. RICS, London.

RICS (2019b). *Royal Institution of Chartered Surveyors. Home Survey Standard, RICS Professional Statement*, 1st edition. RICS, London, p. 23.

RICS (2021). *Royal Institution of Chartered Surveyors, Risk, Liability and Insurance in Valuation Work. RICS Guidance Note*. RICS, London.

RICS (no date). Available at www.rics.org/globalassets/rics-website/media/qualify/home-survey-standard – toe-guidance-for-members.pdf. Accessed 27 August 2021.

Rispin, C., Haggett, F., de Silva, C., Parnham, P. and Russen, L. (2021). *Residential Property Appraisals: Volume 1*. Routledge, Abingdon.

3 Carrying out the service

Contents

DOI: 10.1201/9781003253105-3

3.1 Introduction

In a similar way to Chapter 2, this part of the book follows a similar structure to that used in Section 3 of the Home Survey Standards called 'Carrying out the service' (RICS 2019, p. 12). Although some aspects haven't changed much since 2001, others have. The use of the internet and new technologies have made dramatic differences to the way residential practitioners collect information both before they leave the office and during the inspection.

This particular section of the HSS begins with a long list of matters that RICS members must do. Rather than going through these one by one, we have highlighted matters we feel are most important.

3.2 Locality

3.2.1 Scope

Volume 1 of this publication described, in broad terms, what should be included in a desktop study for a valuation (Volume 1, Chapter 8.2.1). Although there are differences between a valuation and a level one and two condition service, much is the same. These differences are defined in the Home Survey Standard (HSS) that states:

> RICS members must be familiar with the type of property to be inspected and the area in which it is situated.
>
> (RICS 2019, p. 12)

It goes on to explain that the depth and breadth of the research will depend on a range of factors, including:

- **The practitioner's knowledge and experience**. More experienced practitioners who have worked in an area for some time often have a greater knowledge;
- **The locality**. Well-established urban neighbourhoods may require a greater level of investigation because of the complexity of urban development dating back centuries; and
- **The client's specific requirements**. Some clients may have specific requirements, such as access to public transport and other facilities, sensitivity to noise or pollution and so on.

In terms of levels of service, the HSS states:

> At levels one and two, the amount of research is likely to be similar. Research for level two services on older and/or complex properties, historic buildings and

those in a neglected condition, and all level three services is likely to be more extensive, and also if the client has requested additional services.

(RICS 2019, p. 12)

Further details are included in Appendix C to the HSS, which includes typical types of information about the general environment, neighbourhood and subject property (RICS 2019, p. 30). Before we look at some of these matters in detail, we would make the following observations:

- The HSS introduces the use of online sources for the first time and states these should be *freely available*. Although using the internet is not mandatory, there can be few practitioners who do not use its facilities as it has become an accepted part of our professional lives;
- Although the amount of online information has grown considerably, the nature, quality and accuracy of the data varies between suppliers. Therefore, you should treat this information with great care;
- The range of freely available online resources will change over time. Recently, we have seen 'indispensable' websites disappear without a trace, while other 'all singing and dancing' apps have been developed but offer little value. Therefore, we have specifically NOT included actual websites, as these will soon be out of date.

As with all aspects of the internet, you must use your own professional judgement and critically evaluate what information is offered.

3.2.2 Specific sources of information

Looking back at the first edition of this book, the list of information we recommended practitioners should check during their desktop research was very small – only four items. The situation is rather different now. Building on these original topics, we added additional sources you should be checking for both a level one and level two service.

- **Flooding:** This should include surface (pluvial), river (fluvial), reservoirs and sea;
- **Radon:** These are regularly updated, so we recommend you check for every instruction even if you know the neighbourhood well;
- **Noise from transportation networks:** Although publicly accessible web sites are available, the quality of the data may vary;
- **Typical geological and soil conditions:** Although these can give you an insight into the general nature of sub-soil and underlying geology, the free websites don't usually give much property specific information;
- **Landfill sites and relevant former industrial activities:** Historical maps can give an insight into past use;
- **Former mining activities:** At the time of writing, all underground and most of the open-cast coal extraction in the UK had ceased. However, old shallow and surface workings and former mine shafts are still causing problems in former mining areas;
- **Future/proposed infrastructure schemes and proposals.** Usually based general internet searches for a property rather than any specific property search;

- **Planning issues:** For example, conservation areas, areas of outstanding natural beauty and Article 4 direction, listed building status;
- **General information** about the site including exposure to wind and rain, risk of frost attack, and unique local features and characteristics that may affect the subject property.

3.2.3 Other hints and tips

We would also add the following advice:

Look at the various websites that give access to historical and current survey maps for the area. For several urban areas, maps from the mid-1800s have been published commercially. At the time of writing, you can still view these on screen without having to pay for a subscription. If you do this, make sure you do not use any images in your reports, as this would be a breach of copyright. However, some parts of the country will have very few if any historical images, so be flexible.

View the property details on the various online estate agents' websites. Although you should take this information with a pinch of salt, useful background information can be gleaned, such as a summary of the EPC, a general description of the property and in some cases, a reasonable number of images – sometimes these can be used to identify 'trials of suspicion' to follow during your inspection.

Although we have avoided mentioning specific websites and other online resources, we will make an exception for Google Maps. Although other web mapping platforms and apps are available, Google Maps has proved to be a very useful tool over the last few years. For example:

- Google Street View. In most circumstances, you can 'virtually' inspect the outside parts of the property that can be seen from the street. This can help you decide whether it is a property you have the competence to inspect; help identify those issues you need to research before to get to the property; and in some cases, begin to spot some of the problems;
- By using the 'Time' symbol in the top left-hand corner of most Street View views, you can get access to a range of images built up since about 2008/9. This can give you all sorts of insights and the following examples are just a few of the issues we have identified:

  - As described in Chapter 12 on modern methods of construction, a series of images over a two-year period revealed the nature of the construction system used to build a small development of several houses. The natural stone outer skin of these properties concealed the true nature of the structure;
  - A large tree in the front garden of an interwar semi-detached house had been removed and the area paved over; and
  - On many different properties of all ages and types, we were able to identify the approximate dates of various repairs including new windows, roof covering, new boiler flues and so on.

We've even seen a stand of Japanese knotweed grow in size during a seven-year period.

- Google satellite. Depending on the quality of the satellite images for the particular property, the 2D and 3D views can help you embellish the information gained from Street View. Google satellite images can also help you identify local features that you might otherwise see, such as railway lines, waste ground and streams, and all features that should put you on notice of Japanese knotweed.

Although it can produce varying results, general internet searches using the address of particular properties can produce some interesting insights. Here are a few examples we have come across:

- Reports of a 'cannabis farm' in the next-door property on the website of the local newspaper;
- A local campaign about traffic noise on the same road as the subject property;
- A disgruntled neighbour who had set up a blog complaining about the local water company doing nothing about the overflowing drains in the neighbourhood.

Obviously, you have to take a measured view on these matters, but they can give you some background information about which you otherwise may not have known.

3.2.4 *Using the information from desktop research*

Over the last twenty years, the internet has changed from what some commentators called the 'world-wide wait' to an essential tool that supports the work of residential practitioners. However, it is of little value if you discover all this information and then do not use it to inform and enrich your report. In our view, good desktop research sets the agenda for the inspection. It helps forewarn you of likely problems and gives you a range of matters to follow up on when you are at the property.

Here is how we approach the matter:

- To capture the information from your desktop research, consider using the 'print screen' function, pasting the images into a suitable software package and saving alongside your site notes in the property file;
- In the office, start noting down any desktop issues on the site notes you will be using for the inspection. These can act as a series of reminders or action points. Typical examples might be *'check over the back fence for Knotweed on the railway embankment'* or *'look at neighbour's loft extension from the back garden for party wall issues'* and so on;
- It is one thing to retrieve data, but you must also consider how the information may affect the property. For example, if the dwelling is in a high-radon area, you may not be so concerned if it has well-ventilated suspended timber floors, but you will be a little more worried if it has solid concrete floors and replacement glazing without trickle vents.

Caution: The internet can also be distracting. To avoid wasting time chasing meaningless details, we suggest you have a clear list of useful websites on which you focus.

3.2.5 *Information from the property owner/agent*

The HSS acknowledges the usefulness of the information you can get from the person at the property (RICS 2019, p. 13). Hopefully it will be the owner and seller (who will be one and the same person), but it could be a tenant, an agent or the property could be empty.

In Volume 1 of this publication (Chapter 8.3.4), we outline effective ways of developing a good relationship with the vendor, as this can help get the information we need. For your convenience, we have re-presented them here and added additional points relevant to level one and two condition reports:

- In most cases the occupier will be the vendor and your first role is to put them at ease and explain what it is you intend to do and how long you intend to be there. Not all vendors understand the complexities of the house sale and purchase process. Have some form of identification ready and make sure your shoes are clean (very important, although some practitioners use plastic overshoes);
- Check for any issues that may affect your inspection, such as dogs, children, sick residents, people who are shielding or someone asleep from a night shift;
- It is important to explain how you wish to undertake the inspection and what tools you may need to use. Ask if it is OK to take photographs, but reassure them that no residents will appear in the photographs (very important if there are children in the house);
- Ask the vendor to briefly show you around the property. This can give you the opportunity to ask gentle but probing questions. Once this walk-around has finished, politely inform the person you will be carrying out the rest of the inspection on your own, but you will be happy to discuss any issues at the end;
- At this stage, you may also want to ask the owner to open any access hatches and clear any cupboards that you want to inspect.

Other matters you may wish to ask:

- General details about the property, for example its type, size, age, and so on. These are useful facts that can help confirm the office-based preparations;
- Whether any work has been carried out to the property, such as extensions, alterations, repairs, and so on. If yes, ask them to sort out any documentation they may have. This could include guarantees and warranties, planning and building regulation permissions;
- If they have any evidence of any service agreements, lease details and so on.

It will also be appropriate to ask whether, to the owner's knowledge, any building insurance claims have been made and whether the property (or neighbouring properties) has been flooded or affected by Japanese knotweed. Other matters should include ownership of boundaries, existence of any neighbour disputes, rights of way and so on.

Some agents may provide a range of information about the transaction and the property at an early stage in the conveyancing process. A typical example would be a set of comparables and other information details the agents consider 'useful'. The HSS reminds us that we should carefully evaluate this information before incorporating it into the report.

You should record all information offered by the vendor in your site notes, as these will help you write your report.

If the house is empty, it might be appropriate to take a few photographs of the positioning of any expensive-looking ornaments on a date-stamped digital camera.

3.3 Equipment

In Chapter 8.3.2 of Volume 1, our colleagues described the type of equipment that should be available to the residential practitioner on a valuation. They make the point that not having the proper equipment on a survey is no defence against an action of negligence.

The Home Survey Standards (HSS) also makes it clear that RICS members should have access to suitable equipment required to complete the service (RICS 2019, p. 13). Consequently, we have combined the listing from Volume 1, the requirements of the HSS and other previous publications of ours to produce a listing of equipment that should be available to practitioners who carry out level one and two services.

The list of surveying equipment will vary depending on personal preferences, the organisation you work for and the sort of work to be tackled. The list has been split between what is most likely to be required for the inspection itself and for health and safety items.

3.3.1 Inspection equipment

A typical list could include:

- **Measuring tape:** A laser measure, 5–7.5 metre retractable steel tape is suitable for most jobs but occasionally a 20–30 metre fabric tape will be needed, especially to measure longer distances in the grounds;
- **Electronic moisture meter** (with spare batteries): This should be calibrated every time it is used;
- **Spirit level:** A small hand-held level (sometimes called a boat level) around 250 mm long is useful for checking the levels of window sills and door heads but a longer one (minimum 900 mm long) will be needed for floors and checking the verticality of walls;
- **Ladder:** There are a number of types available including slot together sections; telescopic and folding to name but a few. The choice will often depend on the type you can most easily operate. Whatever your type, it must be long enough to give you access to a viewing point 3 m high. To conform to health and safety requirements, it will have to be 3.75 to 3.9 m long;
- **Powerful inspection torch** (and spare batteries);
- **Clipboard and paper:** For notes and sketches. Although many people use electronic devices to record their site observations, a clipboard, paper and pen can be a great back-up if the batteries run out or you drop your tablet from the loft hatch;
- **Suitable camera** (with flash): Most modern phones can provide good quality digital photographs, but some may prefer a stand-alone camera;
- **A plumb bob and line**: A quick and effective way of checking the verticality of walls although this can be difficult to use when you are on your own;

- **Binoculars** (at least × 8 to 10 magnification);
- **Compass**: So you can judge which side of the property is exposed to the prevailing weather.

3.3.2 Equipment for simple 'opening up'

This will depend on the level of service and your specific terms of engagement but will typically include:

- A robust claw hammer;
- A large flat-head screwdriver;
- A 'wrecking' or crowbar (450 mm long);
- Bolster or cold chisel;
- Two large and two small drainage inspection keys;
- A bradawl or other suitable probe.

3.3.3 Health and safety equipment

This will typically include:

- Personal attack alarm and spare battery;
- Mobile phone;
- First aid kit;
- Safety helmet;
- Face mask with disposable filters for loft inspections, etc. These must be a suitable specification for the dust and fibres that can be expected;
- Safety goggles;
- Pair of protective gloves for lifting inspection chamber covers, etc. Many organisations are now recommending disposable gloves because once the gloves have been used, if they are put back into the survey tool kit, they can contaminate other tools;
- Disposable rubber gloves for dirty or unhealthy locations;
- Plastic overshoes;
- Appropriate steel top capped wellingtons or other suitable safety shoes for inspecting overgrown gardens, dilapidated buildings and building sites.

Although general precautions are no longer required for Covid 19 and its many variants, the pandemic showed how flexible practitioners must be when faced with similar public health crises in the future. Additionally, a significant number of people are still shielding because of their particular vulnerabilities and may require visitors to wear appropriate personal protective equipment.

3.3.4 Other equipment

Get a group of residential practitioners together and ask them what surveying equipment they use and they will talk for hours. As technologies develop at an ever-increasing pace, what once was a luxury often becomes central to the practitioner's kit. Additionally, because the HSS provides a set of broad benchmarks, many practitioners have used its flexibility to provide additional and niche services.

Although the following list will almost certainly be out of date by the time this volume is published, here are some of the newer items we have seen appearing in people's kit bags:

- Thermal imaging camera. Some practitioners use these to assess heat loss and also investigate moisture-related problems;
- Laser glazing measuring devices that can assess the characteristics of glazing units;
- Borescopes/endoscopes. Miniature digital versions that can be plugged into smartphones are now available;
- 'Selfie sticks' and camera poles. This can help you view normally inaccessible places such as underfloor areas, dormer flat roofs and chimney tops;
- Drones – a few months before we wrote this volume drones were impractical and wieldy 'toys for the mainly boys' that brought little benefit. Technological developments now make them easier to operate and practitioners are using them to enhance their service;
- Digital angle measuring app on smartphones. However, we have always found trigonometry to be more reliable;
- Metal detectors for identifying the location of wall ties, pipes and wires and so on.

We think these developments are potentially of great benefit to both clients and practitioners alike. However, we would make two points:

- Whatever is used, you need to have adequate training and competence to use the equipment. In other words, you need to know what they do and how to use them; and
- You may need to adjust your terms of engagement to reflect the change in scope of your service.

It makes us wonder what sort of technologies practitioners will be using in 20 years' time when we write the third edition of this book.

3.4 The inspection

3.4.1 Introduction

As with most aspects of the Home Survey Standards, the description of inspection routines is less prescriptive and focuses on a number of 'critical aspects' in an attempt to set a clear framework for the scope of inspections at the various levels of service while retaining flexibility for those who want to bring different products to the market.

3.4.2 Scope of inspection

According to the HSS (RICS 2019, p. 14), the extent of an inspection will depend on a range of circumstances, including the nature of the property, the level of service offered, specific client requirements and health and safety considerations.

Survey level	Description
General	The RICS member will carry out an inspection of roof space that is not more than three metres above floor level, using a ladder if it is safe and reasonable to do so. Energy efficiency initiatives have resulted in thick layers of thermal insulation in many roof spaces. Usually it is not safe to move across this material as it conceals joist positions, water and drainage pipes, wiring and other fittings. This may restrict the extent of the inspection and the scope of the report. Consequently, this matter should be discussed with the client at the earliest stage.
Service level one	The RICS member will not remove secured access panels and/or lift insulation material, stored goods or other contents. The RICS member will visually inspect the parts of the roof structure and other features that can be seen from the access hatch.
Service level two	In addition to that described for level one, the RICS member will enter the roof space and visually inspect the roof structure with attention paid to those parts vulnerable to deterioration and damage.
Service level three	The RICS member will enter the roof space and visually inspect the roof structure, with attention paid to those parts vulnerable to deterioration and damage. Although thermal insulation is not moved, small corners should be lifted so its thickness and type, and the nature of the underlying ceiling can be identified (if the RICS member considers it safe to do so). Where permission has been granted and it is safe, a small number of lightweight possessions should be repositioned so a more thorough inspection can take place.

Figure 3.1 Inspection of roof spaces from the Home Survey Standard 2019.

Survey level	Description
General	In all cases, the RICS member only opens windows where: • permission has been given and • any keys/locks are available and are easy to operate without force or damage. The presence of occupier possessions and heavy curtains will often restrict level one and two inspections. For level three, a small number of possessions/curtains will be repositioned. Where inspections are restricted, the RICS member **must** inform the client.
Survey level one	Include one on each elevation
Survey level two	Include one on each elevation and one of each different type of window where there is a variety.
Survey level three	Attempt to open all windows where possible.

Figure 3.2 Inspection of windows from the Home Survey Standard 2019.

We have not reproduced the full list of 'critical aspects' from the HSS but have chosen two examples to give you an insight (see figures 3.1 and 3.2).

We would make the following comments about these 'critical aspects':

- They are based on similar criteria first written for the RICS Guidance Note titled 'Surveys of Residential Property' back in 2013 (RICS 2013). They are not meant to give a comprehensive description of the different levels of inspection. Instead, they are there to provide a clear set of benchmarks around which RICS members can build their own product;
- Despite this flexibility, many practitioners have chosen to stay close to the scope of inspection for the former Condition and HomeBuyer Reports;
- These different levels of inspection are incremental. For example, take the windows element. At level one, it is recommended that just one window on each elevation is opened. At level two, this increases to one on each elevation and one of each different type of window where there is a variety, while at level three, all windows are attempted. These aim to help the clients discriminate between the different options and choose the level of service they find most suitable. However, many practitioners find these distinctions artificial. We have come across a number who say 'I carry out the same inspection at all levels; only the depth of the report varies'. In many ways, that is acceptable as long as the client clearly understands the scope of the service.

For more discussion on this point and how you can help your client choose between survey levels, please see Chapter 2, Section 5.

3.4.3 Inspection procedure

The following section is based on the procedure described in Volume 1 and updated to suit condition reports. It is important to point out that this is a generic approach to inspecting a building regardless of the level of service offered. Inspections and levels of service are discussed in more detail in Section 3.4.2.

The method described is just one way to complete an inspection. If you asked the writing team how they approach an inspection, you will get five different responses (especially as one of the authors is a lawyer). Whatever your approach, it must be methodical and systematic. You should adopt the same routine on every inspection, as this can help create an impression of competence if challenged in court.

3.4.4 Carry out an appropriate health and safety risk assessment

This should have already begun when you received the instruction and travelled to the property. This becomes more important as you approach the dwelling. This is discussed in more detail in Section 3.4.6 of this chapter.

3.4.5 The inspection

Work around the dwelling **internally**:

- Inspect the loft space first (if there is one). Whether you go into the loft depends on your terms of engagement and if it is safe to do so (see Chapter 2.2 for more information);

- Inspect the rooms on the uppermost floor by working around that entire floor in a clockwise direction (or anticlockwise if you prefer);
- Finish on the landing and inspect the stairs down to the next floor;
- Follow the same process on each floor down to the lowest one;
- Inspect any cellar, basement and/or sub floor voids if accessible.

In each room, inspect the various elements in the following sequence:

- Ceilings;
- Walls (including skirtings);
- Floor;
- Windows and doors;
- Heating;
- Electricity;
- Plumbing;
- Other amenities (for example, toilets, basins, sinks, etc.);
- Fittings and fixtures (for example cupboards, fitted wardrobes, fireplaces, etc.);
- Any unusual or special features.

External inspections. Going outside towards the end of the survey avoids the problem of tramping muddy boots around the house. However, many practitioners prefer to do the outside first because it can give an indication of what to look for on the inside. The rule is 'there are no rules' – it is down to what makes sense to you.

The outside inspection should include:

- The main elevations including all secondary elements (doors, windows, and so on);
- Observable roof surfaces including chimney stacks, valley gutters and so on. Flat roofs may be inspected from the windows of upper rooms. If they can't be seen from the main house, consider using your ladder if it is safe to do so. Rainwater goods should be assessed at this point:
- Drainage inspections including above ground soil and waste pipes, gullies, inspection chambers, cesspits and septic tanks (if it is safe to do so);
- Any significant garden features such as retaining walls, paved areas and hardstandings and so on;
- All outbuildings;
- Boundaries, fences and gates;
- Vegetation in the garden. This should focus on significant matters such as invasive plants, large trees and shrubs and so on;
- Any special features such as rights of way, facilities shared with other property owners and so on.

Once the inspection has been completed, always inform the owner that you are leaving.

In our site notes, we include comprehensive lists of features to be inspected. These act as a memory aid and also a checklist that can show the inspection has been properly carried out (see Section 3.5 of this chapter).

3.4.6 *Property inspections and health and safety matters*

Since the early 1970s, health and safety legislation has become increasingly central to our everyday activities. Although residential practitioners do not use dangerous machinery or go down deep holes, we do use ladders every day and regularly visit building sites. We often meet aggressive dogs, not to mention their occasionally belligerent owners.

Fragmentation of our sector is another threat to health and safety. As larger surveying organisations have downsized, more of us have become self-employed 'consultants'. Although this provides a more responsive and flexible workforce, it weakens sector-wide policies and procedures, such as pension provision, health and safety, quality assurance, and so on. Self-employed surveyors worried about continuity of work can regularly accept too many instructions in an attempt to build up reserves to help them get through leaner times ahead. In this situation, quality will suffer, and more importantly, hardworking and tired surveyors will often take silly risks.

All these issues have been brought into sharper focus by the COVID-19 pandemic. At the time of writing, the restrictions are beginning to be 'eased' and the world (literally) is trying to work out what the 'new normal' will look like. This goes for the residential sector too. RICS have recently published new COVID-19 guidance for their members and gives important updates on procedures to be followed before, during and after property inspections (RICS 2021). Because this is such a fluid situation, we have decided not to reflect this guidance in this book because in a few years' time, the global pandemic will either be:

* A distant memory; or
* A central and continuing feature of our personal and professional lives.

In either case, anything we put in this book will be of little value. Despite this, we will offer a few short reflections:

* A significant proportion of practitioners continued to inspect properties throughout the pandemic. In our view, this has resulted in an increased awareness of personal health and safety issues that has embedded itself into the sector as a whole;
* The greater use of the developing technologies during the pandemic (especially 'virtual' communication methods) has becoming a 'norm' rather than an exception;
* A greater proportion of practitioners are now experienced home workers and it is unclear what will happen to office working in the future.

Whatever the future brings, these influences have already had an impact on how we work. Hopefully, this will be reflected in this book.

3.4.7 *Health and safety legislation*

This section reviews the legal framework and some of the most important issues that affect practitioners in their daily work. This is important for newly qualified surveyors and experienced practitioners alike. Our experience suggests that older practitioners developed their inspection methodologies many years ago and although satisfactory, they might not have accounted for current philosophies and regulatory changes (although COVID-19 may have helped them update their knowledge).

The Health and Safety at Work Act 1974 is the most important single piece of legislation and sets out the general duties of both employers and employees and the obligations they owe to the public. The Health and Safety Executive (HSE) is the government agency that administers this legislation, and its website includes additional information on the Act and other types of guidance (www.hse.gov.uk).

This is a complex piece of legislation, supplemented by a plethora of secondary Acts, regulations and codes of practice. It requires practitioners to identify health and safety risks and take measures to minimise those risks.

The Act invokes a principle of 'so far as reasonably practicable', which means that not all risks have to be eliminated. Instead, it requires measures that are 'proportionate' to the risk faced (for example, 'sensible and reasonable', taking into account a series of factors including cost).

For example, if an accident is very likely to occur, then an employer/employee must take action to prevent it happening. A brick falling on an operative's head on a building site would be very likely and therefore, all operatives need to wear protective headgear. Conversely, encountering irritating dust particles during general domestic inspections, although possible, is not very likely. Therefore, surveyors do not have to wear a protective facemask all the time but only if they go into areas that may be contaminated. Of course, COVID-19 precautions may continue to change this approach.

It is the application of 'common sense' rather than a prescriptive set of inflexible rules that underpins health and safety policies.

Although it may appear the main legislation applies only to employers and employees, the self-employed should adopt a similar approach to safety – not least to ensure that they do not break the terms of their public liability insurance. Moreover, the self-employed have a clear duty to protect their customers and members of the public whom they meet. They also have a 'moral' duty to protect themselves, especially if they have dependents who rely on their income.

It is beyond the scope of this book to offer detailed advice on establishing an appropriate health and safety policy. But what we can do is signpost useful publications and highlight some important health and safety issues around which you can build your own policies.

There are some basic steps that everyone should follow and the Health and Safety Executive's publication *'Risk assessment – A brief guide to controlling risks in the workplace'* is a good starting point (HSE 2014). This sets out a risk assessment process that provides the main method of identifying hazards in the workplace. Although the publication is primarily aimed at employers, the HSE remind self-employed people that if there is a likelihood of someone else being harmed or injured (for example, members of the public, clients, contractors and so on) as a consequence of your work activity then health and safety laws will apply.

3.4.8 Assessing the risks

This assessment can be as simple as a '. . . careful examination of what could cause harm to people in the workplace'. Once the assessment is complete, you should weigh up whether existing precautions are sufficient or take further action, if necessary. There are five steps to this process:

- Identify the hazards;
- Decide who might be harmed;
- Evaluate the risks;
- Record your significant findings; and
- Regularly review your risk assessment.

For more information, see *HSE 2014* and other up-to-date guidance on the HSE's website.

3.4.9 *Identifying the hazards*

The previous section established the context for health and safety planning. This section is designed to help you identify the hazards in your working environment.

The office environment

Although you will spend much of your time inspecting properties or travelling between them, you will also be office-based. Normal health and safety requirements will apply:

- Adequate ventilation and air changes (especially post COVID-19);
- Comfortable working temperatures – although there are no mandatory requirements, the government recommend the temperature should be at least 16°C for office workers (www.gov.uk/workplace-temperatures accessed 28 august 2021);
- The provision of adequate light;
- Facilities for waste disposal and maintaining cleanliness;
- The avoidance of cramped conditions;
- Regulations about workstation design and seating;
- Maintenance of equipment and services;
- Safety of circulation space;
- The provision of toilet and washing facilities, drinking water, rest and changing facilities; and
- Maintenance of first aid materials and a file for recording all accidents that occur.

Much of your office-based time will be spent using display screen equipment (DSE) or what used to be known as visual display units (VDUs). The HSE advises that you should use a well-designed workstation with a chair that you can adjust to suit the table height and to encourage good posture. Laptops are particularly problematic because manufacturers do not always design them ergonomically, so you should always use a desktop computer if one is available, or use a laptop with a separate screen and keyboard. You should also take regular breaks when using the computer. A number of surveyors use DSEs to record site notes and upload the finished report to a databank through a mobile phone from the front seat of a car. This can result in poor posture and back problems. For more information, the HSE leaflet Working with display screen equipment (DSE). A brief guide, INDG36 (revision 4) is useful (HSE 2013).

Travel and general vehicle safety

Most practitioners spend a considerable amount of time in the car travelling between the office and inspections. You should ensure that the car is well maintained and roadworthy. In addition to the formal servicing, read the car maintenance manual and carry out routine checks.

It is also important to carry an emergency car kit in case of a breakdown, as well as a map or navigation app to reduce the need to stop and ask for directions. You should plan your journeys so that you do not get tired. Take regular breaks on long journeys and follow the Highway Code. Accident statistics suggest that younger males are at greater risk during this part of the inspection, so kill your speed. Just a reminder: it is against the law to drive while under the influence of drugs or alcohol.

The HSE's publication *'Driving at work: Managing work-related road safety'* INDG 382 rev1 provides very detailed guidance on how this risk can be assessed and managed (HSE 2014).

Home working

As we mentioned previously, a growing proportion of surveyors now work from home, even those that are employed by a large organisation. In these circumstances, home workers are covered by health and safety legislation, and employers still have the same range of duties as if their staff were working in the head office, including:

- Regular risk assessments of the home worker's circumstances – this should not be a generalised, standard procedure but one that takes account of the employee's particular circumstances;
- Consideration of handling of heavy goods or equipment – surveyors will have to load and unload heavy sectional ladders and other equipment;
- Provision of suitable equipment for the task in hand, and responsibility for its safe use and maintenance;
- Periodic testing of electrical equipment – although the employer is not responsible for the condition of the electrical system in the employee's home;
- The safety of any hazardous materials provided for the work (for example, toner for printers and copiers); and
- Consideration of the hazards of display screen equipment.

If you are a self-employed practitioner, you will be responsible for your own health and safety, but these pointers will help you build your own policies.

For more information, see the home working section of the HSE's website (2021).

3.4.10 Risks during the inspections

This part of the chapter moves from the general to the specific and describes the health and safety issues associated with the inspection process from start to finish. You may also find the following publications useful:

- RICS's guidance note '*Surveying safely: health and safety principles for property professionals*' (RICS 2018).
- *Protecting lone workers. How to manage the risks of working alone.* HSE INDG73(rev4) (HSE 2020).

Pre-inspection health and safety preparation

Before leaving your workplace, make sure somebody knows where you are going, what your plans are, what time you expect to finish the inspection and return to work or arrive home. When the inspection is over, don't forget to check in with someone in the office to confirm that you have finished, even if you then go straight home. A call from the safety of your armchair may stop your colleagues worrying about your whereabouts. If you do not work in an office, make a suitable arrangement with another associate, family member or friend.

Always carry a fully charged mobile phone and your personal attack alarm. Go to the inspection in daylight wherever possible and organise your day so that you are travelling in darkness rather than inspecting buildings in a fading light. During winter months, this may mean completing the inspection by mid-afternoon.

On arrival at the property

You should undertake a risk assessment prior to starting an inspection. This does not need to be a formal process; you just need to take time to look around so you can make an initial assessment.

We recommend you have a section in your site notes titled 'Health and safety hazards' where you can note down some of the more obvious potential risks. Once the risks are recorded, the level of risk assessed should be assessed (high/medium/low) along with how you plan to minimise the problems. This gives evidence that you have considered health and safety issues and may be useful if an accident does happen. Examples include:

- A roof in poor condition with a number of slates slipped and on the verge of falling off. This may be particularly dangerous during windy weather and may restrict your external inspections;
- An empty property where moisture may have been penetrating for some time. This can result in timber defects to the floors, making the building dangerous to inspect;
- Large dogs or other threatening animals loose in the garden.

If you decide that the identified risks will limit your inspection in any way, it is important that you explain this in your site inspection notes and in your report.

Problems posed by the building

You must judge each property on its own merits, but the development of an inspection routine will help you recognise and minimise risks. Safety awareness has to be a top priority when entering any type of property, and experience will result in checks becoming automatic. If safety is not an essential consideration from the earliest stages, then this lack of awareness could lead to increased risk.

3.4.11 Health and safety in occupied properties

Health and safety rules and regulations are usually associated with working environments in which the dangers are more obvious. Inspecting residential properties is usually considered low risk. However, working alone in other people's property does have its own particular challenges that occasionally result in serious injury and even death. Organisations such as the Suzy Lamplugh Trust (www.suzylamplugh.org) and the HSE (www.hse.gov.uk) have produced excellent guidance for people whose work involves visiting people in their homes.

Based on this information, we have outlined what we feel is a framework for a safe working environment for residential practitioners. If you follow these suggestions, they will become a part of your normal inspection process and help you stay safe.

Arriving

Before arriving at the property, think about its location and if there are any associated dangers (for example, a high-rise block, a country lane or a one-way street). Consider whether you have to go alone. This may be difficult when resources are scarce and workloads are heavy, but two people will be much safer than one. If this is difficult to arrange, two-person inspections could be limited to areas where the risk is the greatest.

When you arrive at the property, park in a position where it will be easy to leave quickly. Do not park in a tight spot where it might be difficult to get out of in a hurry. Always lock your car and put any valuables out of sight in the boot.

When meeting the occupants, state who you are, why you are there and show the person your ID or business card. Check who you are talking to and make sure it is the same person with whom you arranged access. If not, consider carefully whether you should go in. In some cases, the property may be occupied by a tenant who may know very little about the building or even that it is about to go on sale. In these circumstances, an accompanied inspection with the property owner will always be better.

If the only person in the property during the inspection is below the age of 16, or is someone whom you judge to be vulnerable in some way, you should postpone your inspection until a responsible adult can be present.

During the inspection

When carrying out the inspection, do not let the occupier lock the doors behind you. Try to familiarise yourself with the layout of the property so you can get out quickly if necessary and do not leave possessions in several different rooms, as this may delay your departure if you need to leave in a hurry.

Do not take documents into the property that you do not want the occupant to see, and be careful when using a personal recorder to record your site notes. You may offend the occupiers by even the most objective assessments of their home.

Different people live in different ways, so try not to react to the condition of the property, even if it is very untidy, smelly or just plain dirty.

Make every effort not to damage the property. We know this is stating the obvious, but it is inevitable that you will occasionally enter a property with dirty shoes or knock an ornament off a shelf with your clipboard. Here are a few tips:

- If it is wet outside, or you have dirty shoes, either use non-slip plastic overshoes or take along a spare pair of lightweight shoes such as trainers or plimsolls to change into;
- If there are ornaments stopping you using your measuring tape, ask the occupier to move them. Remember, using a laser measuring device reduces the chances of a mishap and makes the process quicker;
- Do not 'force' things to open – you might not be able to shut them again. Ask the occupier to do it instead.

If you do damage anything, it is essential that you inform the owner or occupier immediately, because in any case you will have public liability insurance that will cover major incidents. Where the damage is less serious, you need to settle the matter quickly and effectively to make sure you maintain a good relationship with the seller/occupiers.

If you accidentally enter bedrooms, bathrooms or WCs where occupants are sleeping or changing, apologise and leave the room quickly.

If you notice valuables around the property, you should consider asking the seller to remove them. This highlights the importance of walking around the property with the occupier so these problems can be spotted and resolved early on. If the property is unoccupied, you should make a record in your site notes of the time and circumstances when you first noticed the valuables. Taking a time and date stamped photograph of the items will also be beneficial.

If any of the occupants appears to be drunk, drugged, aggressive or overly attentive, then give your apologies and leave immediately. Most commentators say that if you feel uncomfortable about a person's attitude or behaviour, make an excuse and leave straightaway. Do not worry what people may think of you – it is important to maintain your own safety. A number of organisations provide their employees with personal attack alarms, and although these can be effective, the Suzy Lamplugh Trust takes a more balanced view:

- In an emergency, will you be able to find the alarm and use it quickly and effectively?
- Most alarms will emit a very loud noise but will passers-by take any notice? Some can sound like a car alarm and can be ignored;
- A loud shout or shriek can be more effective.

For more advice, see the Suzy Lamplugh Trust (www.suzylamplugh.org/personal-safety-out-and-about accessed August 2021).

If you slip, fall, injure or cut yourself, consider postponing the inspection until you have recovered. The smallest of cuts or bumps can get worse if they are not properly treated. Also, carefully evaluate any offer of first aid assistance from the occupier. Although they might offer this with the best of intentions, unless they are properly qualified, they could make matters worse.

3.4.12 Specific inspection issues

The previous sections covered general aspects that will help you operate within a framework that protects your health and safety. This section provides more detailed advice on how to stay safe while inspecting two of the most hazardous environments: lofts and derelict property.

Safety during loft inspections

Inspecting a roof space can present the greatest and most regular risk during a practitioner's day. Here are a few tips:

- Never stand on chairs, tables, radiators or other furniture. If you need to inspect at a higher level, you should always use a ladder;
- Only use fitted loft ladders if you are satisfied that they are safe – this is very important, as DIY-installed loft ladders can be very unstable. Take care when lowering the hatch mechanism, because the ladders can often slide down unexpectedly. Make sure you fully extend and lock the ladder into position, and test stability by putting your weight on a lower rung and shaking it. If you feel that the loft ladder is not safe, it is unlikely that you will be able to position your own ladder to get into the loft. This may well limit your inspection and you should state this clearly in your report.

Most surveyors use aluminium sectional or telescopic ladders that extend to between 3 m and 4 m in length (see section on equipment in Section 3.3.1). Whatever the type, the ladders should conform to the appropriate standards, be complete and you know and follow the manufacturer's instructions. The HSE have published a guide called 'Safe use of ladders and stepladders. A brief guide' (HSE INDG455 2014). Here are a few from that document:

- Each time the ladder is used, you must check it is in satisfactory condition and has not been damaged, especially if other people may have been using it;
- All the sections should be properly slotted together, all the bolts and nuts hand tightened and telescopic sections extended and engaged;
- Safely position the ladder using a 1:4 ratio. For example, set the ladder at 25% of its height away from the wall;
- It is often difficult to assemble a sectional ladder on a small landing – three sections can fail to reach the hatch, but there is not enough height and space to fit four sections together with the hatch still in place. You will face the dilemma of how to open the hatch itself. Here are two options:
    - Prop the ladder against an adjacent wall, climb the ladder so you can remove the loft access hatch and set it to one side within the loft space. Caution: some loft hatches can be very heavy, partly painted or securely fixed shut. You will have to evaluate the risk and if you think it is unsafe, record the circumstances in your risk assessment section of the site notes. Climb down and add an additional section to the ladder so it extends well into the loft space. To be safe, there should be at least three rungs (preferably 1 m) extending above the top of the ceiling joists;
    - Alternatively, fully extend the ladder and use it to push the loft access hatch upwards just enough to allow the ladder to extend into the roof space. Once

the ladder is securely in position, you can climb up and simply lay the hatch to one side. The drawback to this method is that unless you are able to balance the hatch on the ends of the ladder, you might easily dislodge it, and it may fall towards you.

The choice will depend on which you think is the safest.

Other considerations are as follows:

- Place the ladder on a firm and non-slip surface as carpet mats and laminate flooring are particularly dangerous. Using a proprietary rubber mat beneath the ladder feet can improve safety on uneven, soft and slippery surfaces;
- Make sure you wear suitable footwear that gives a firm grip on the rungs.
- When on a ladder, never over-reach in an effort to inspect a specific feature. Get down and reset the ladder;
- Always keep 'three points of contact' with the ladder. For example, both feet and one hand or one foot and two hands should always be on the ladder;
- Do not make notes on a ladder: writing on a clipboard or a display screen device means you have only two points of contact. Get down off the ladder and then make your notes;
- If possible, carry any equipment in your pockets and/or in an appropriate shoulder bag or backpack.

If you have to get off the ladder in the loft space, carry out a visual risk assessment before you leave the safety of your ladder. Look for the following:

- Can you see the top of the ceiling joists? If they are covered with insulation or stored items, you should not enter the roof space;
- Is there any boarding over the joists? If yes, will it take your weight and has it been properly fixed?
- Are there signs of dangerous or deleterious materials?
- There can be a variety of animals and insects living in the loft. There are few experiences in life that can match the adrenalin 'rush' of disturbing roosting pigeons in a dark roof space. Although this sounds amusing, it can easily cause a serious fall. Other signs to look out for are flies, wasps, squirrels and even rats and mice. The droppings can also present a health risk;
- You should be alert to the presence of bats. Although they pose no serious health risk (do not worry, they will not bite your neck), they are protected by the Conservation of Habitats and Species Regulations 2010 and Schedule 5 of the Wildlife and Countryside Act 1981. If you do discover bats, end your inspection. Carefully and quietly withdraw from the loft and inform the owner of their presence and make sure your client knows as they can restrict the use of the roof space;
- Because lofts are often dirty you should wear disposable gloves, overalls and an appropriate face mask.

Unoccupied and derelict properties

A percentage of the properties you inspect will be empty and this category includes buy-to-let properties that are not tenanted; owner-occupied homes where the occupants have moved out before they have sold the house or where they have vacated to allow you to do the survey; and homes where the owners have died.

If the property is empty, you need to clarify how you are going to gain access to the property. Getting the answers to the following questions may help:

- Where are the keys and how long can you have them for?
- Are there any security alarms and who keeps the alarm codes?
- Are the services activated and if not, will anyone be able to turn them on?
- Do you have permission to open windows, access hatches, and so on?
- Who will answer all the pre-inspection enquiries?

If someone is meeting you at the property to open it up, make sure you ask them for proof of identity. This may appear a little formal, but it is important to make sure they are who they say they are so that any potential risks are minimised.

Being alone in an empty dwelling increases the level of risk. In addition to the general precautions identified earlier in this chapter, be aware of risks specific to empty properties:

- When you enter the empty building, carry out a cursory inspection of all rooms while loudly announcing your presence. The purpose is to make sure you really are alone in the property. Make sure you look in all spaces, including basements, stores and large cupboards;
- If you are on your own, lock the external doors when you are inside and keep the keys with you. Some commentators do not agree with this approach because it might hinder access if you have an accident. However, we would prefer this to a stranger casually wandering into the house when our head was in a loft;
- If you discover an unauthorised person(s), briefly and calmly explain who you are, why you are there and that you are leaving right away. If the agent has told you that it is empty, then that is how you should find it. You should immediately phone the person responsible for the property and ask them to contact you when it is safe to inspect;
- The same approach is appropriate if you discover signs of unauthorised occupation and/or the property is unsecured. Typical examples would include a broken window or half-open door and evidence that someone has been staying there (for example sleeping bags, open food tins, and so on). In this case, leave the property immediately and notify the person responsible.

Properties left empty for long periods of time can become neglected, allowing defects to develop quickly. Stagnant conditions can allow dormant wood rot to weaken timber floors and roof leaks will cause ceilings to collapse. Vandalism will often result in broken glass and sharp edges throughout the dwelling and drug-using squatters may leave contaminated syringes. Take the following precautions:

- Dress in robust and durable clothing including footwear that is suited to the conditions;
- Ensure you have a powerful torch with you, as many rooms may have their windows boarded over and will be in virtual darkness;

- Look for dangling wires, bare cables and a smell of gas, as electric and gas services in empty properties can pose a danger especially if damaged through vandalism; and
- Be prepared during the winter, as empty properties can be cold.

The final point might be stating the obvious, but standing in sub-zero temperatures can be more than uncomfortable; it can be bad for your health. A pair of 'fingerless mittens' that allow you to use your pen and help prevent your hands becoming cold are invaluable. If you are 'follicularly challenged', a woollen hat is just the job. A flask of tea or coffee and sandwiches can help maintain concentration when the temperature drops.

3.5 Site notes

3.5.1 Introduction

A comprehensive record of your investigations is vital. Commonly known as 'site notes', this document will include information from different phases of the service, including setting up and carrying out the service through to discussing the findings with your client after they have read the report.

These notes can be in hard copy format or recorded digitally. A comprehensive set of site notes will enable you to respond to questions after the service has been delivered. In addition, they are your evidence in the event of a complaint or challenge.

Depending on personal style and preference, practitioners will either make written notes and sketches to record their impressions of a property, use a form of computer tablet, or record their verbal thoughts into a personal recorder.

This ability to reflect on a recent inspection is an important part of the process and to be able to look at the dwelling as a whole. Seeing how the different elements interact and draw reasoned conclusions is what we are paid to do.

3.5.2 Site notes and the HSS

This approach is supported by the Home Survey Standard, which stipulates:

> The RICS member or RICS regulated firm must keep a record of the inspection, including:
>
> - The construction, condition and circumstances of inspection (including any limitations);
> - The checks made to the fabric and structure and what was found; and
> - Appropriate dimensions and diagrams, sketch plans, and any images captured during the inspection.
>
> (RICS 2019, p. 14)

As level two services on older and/or complex properties, historic buildings and those in a neglected condition and all level three services will include more detailed and technical assessments of the building, the amount of recorded information will be greater.

In keeping with the rest of the Standard, this allows flexibility to meet the benchmarks in a number of ways. Over the years, we have encountered a number of different approaches:

- Handwritten notes on standard forms with varying amounts of sketches and drawings on pre-printed sheets;
- Recorded observations that are later transcribed to a typed copy or downloaded as a digital audio file as permanent record;
- Highly structured inspection forms that use coding systems and abbreviations and are completed during the inspection by pen and paper.

Increasingly, hand-held tablets and even smartphones are being used. Other general principles include:

- Clear handwritten notes should be taken on site with enough sub-headings to indicate to which parts of the building the notes refer. Sketches can add to the descriptive value of the record;
- The notes should be legible and set out in a logical order. Ideally, they should be in pencil in wet weather and re-written in the dry, if necessary;
- If a personal recorder is used then the tapes should be transcribed as soon as possible or the digital audio file saved. These should be used when compiling the report;
- Photographs should be taken to illustrate the notes;
- Any additional information from either the pre- or post-inspection phases should be included in the record of inspection. This includes the details of conversations with your client and any agents you consult.

The site notes should be filed along with the rest of the client information for as long as practicably possible. They can be in either hard copy format or scanned as a digital record, but in all cases, they should be secure from being hacked or destroyed by fire or flood.

3.5.3 Site note packages

Over the last 20 years, we have used these principles to design a series of site note packages. These were originally designed as paper forms, but a number of practitioners have integrated them into a variety of software packages so they can be used on tablet computers.

When designing this package, we followed a number of general principles:

- We tried to avoid 'tick boxes'. This is a common feature with many valuations and 'level one' products, but this approach is inflexible when non-standard features are encountered. Wherever possible, we prefer a simple text box to allow for additional notes and sketches;
- An inspection checklist has been included for each element and this has two important functions:

Project ref:

Section E – outside the property
E1 Chimney stack

Checklist:	Describe construction and condition:
• Description, number and location • Flue terminals • Chimney pots • Flaunching • Condition of stacks (leaning, cracked, sulphate attack, etc.) • Pointing, render, other finishes • Aerials and satellite dishes • Flashings and soakers at the junction with the roof covering (but not including the roof covering) • Any party wall issues arising from chimney condition • Listed building/conservation areas issues • **Restrictions on inspection**	

E1   Reflection and rationale for report (including CR)

Links to sections I and J:

E2 Pitched roof coverings

Checklist:	Describe construction and condition:
• type of covering and general condition including: *the slope/pitch of the roof; fixing of tiles, slates; temp. repairs* • roofing felt/sarking boarding – *presence, type, condition* • ridges and hips tiles • verge and eaves details • open valleys • valley gutters ('butterfly' roofs), parapets gutters, their linings and outlets • lead flashings to up-stands, dormers, etc. (*but not the flashings to the chimney*) • roof lights/roof windows • dormers (*usually including the flat or pitched over the feature*) • any party wall issues arising from the roof condition • listed bldg/conservation issues • **Restrictions on inspection**	

E2   Reflection and rationale for report (including CR)

Links to sections I and J:

Figure 3.3 Sample site notes page for chimney stacks and pitched roof coverings.

- To act as a memory aid to make sure you look at everything during the physical inspection;
- To show you have considered these items during the inspection. The individual item can be ticked or line drawn through the whole lot;

- The checklist, should only include only those features included in the inspection for that particular level of service;
- Along the bottom of each element box, we have included what we call a *'Reflection and rationale for report'* box. This can be used to summarise the pros and cons of each element in preparation for writing your report. Although this will mean different things to different people, we use it to note down the condition rating and

Project ref:

Section F – Inside the property
F1 Roof structure and roof space

Checklist:	Describe construction and condition:
Pitched roofs • design and condition of structure • sample moisture readings • alteration of the roof structure (full attic conversions can be included under F9 Other) • roof strengthening • lateral restraint, spreading of roof structure • the use and misuse of the roof space • wood rot and wood boring beetle • party walls – lack of fire breaks, gaps between dwellings • flooring in loft • possible asbestos • access into loft • ventilation in roof space • level of insulation, • vapour gaps in ceiling • birds, bats, vermin, insects, etc. **Flat roof structure** • access/restrictions to flat roof structure; • sample moisture meter readings • ventilation to flat roof void • wood rot and wood boring insect	

F1 Reflection and rationale for report (including CR)

Links to sections I and J:
Other notes

Figure 3.4 Sample site notes page for roof structure and roof space.

the bare bones of the text we will be using in the report. It also provides evidence of reflection and helps us stand back and take a broader view of the property.

Rather than include all the pages of our standard site notes package, we have included just a few typical pages in Figures 3.3 to 3.5.

Although our package may seem long and complicated, we think it helps us meet the HSS requirements, namely that the inspection records:

- The construction, condition and circumstances of inspection (including any limitations);
- The checks made to the fabric and structure and what was found; and

Section G – Services

G1 Electrics	
Checklist: • Description and general condition; • Evidence of installation, alteration or maintenance certification • **Main issues** • Is there a mains supply?; • on-peak/off peak supply; • location of the meter and consumer unit/fuse board; • separate from gas meter? • Is there a RCD/MCB?; • metal consumer unit or fire protected enclosure? • Type and condition of visible wiring; • condition of a sample of the range of light fittings and switch gear; • type and condition of fixed electrical appliances including heaters, storage radiators, electric showers, instant water heaters, etc. (see also G5 water heating); • bath and shower rooms – nature and suitability of electrical fittings; • external installations – garages, outbuildings, external sockets, garden lighting, water feature pumps, etc. RCD protection? • Automatic gates and evidence of servicing (see G7 for blocks of flats) • **Restrictions on inspection**	**Describe construction and condition:** *The following general advice will be printed in every report:* *Safety warning: Periodic inspection and testing of electrical installations is important to protect your home from damage and to ensure the safety of the occupants. Guidance published by the Institution of Electrical Engineers recommends that inspections and testing are undertaken at least every 10 years and on change of occupancy. All electrical installation work undertaken after 1 January 2005 should be identified by an Electrical Installation Certificate.*
G1 Reflection and rationale for report (including CR)	
Links to section I and J Other notes	

Figure 3.5 Sample site notes page for electrics.

- Appropriate dimensions and diagrams, sketch plans, and any images captured during the inspection.

However, you must 'own' your approach to recording information on the inspection. This should match the current sector-wide standard but also reflects your own style of working. Only then will you be able to develop that consistent and thorough inspection methodology.

References

HSE (2013). *Health and Safety Executive. Working with Display Screen Equipment (DSE). A Brief Guide.* INDG36 (revision 4). HSE, London.

HSE (2014). *Health and Safety Executive. Risk Assessment – A Brief Guide to Controlling Risks in the Workplace.* INDG 163. HSE, London.

HSE (2014). *Health and Safety Executive. 'Driving at Work: Managing Work-Related Road Safety'.* INDG 382 rev1. HSE, London.

HSE (2014). *Health and Safety Executive. 'Safe use of Ladders and Stepladders. A Brief Guide'.* INDG 455. HSE, London.

HSE (2020). *Protecting Lone Workers. How to Manage the Risks of Working Alone.* HSE INDG73(rev4). HSE, London.

HSE (2021). Available at www.hse.gov.uk/toolbox/workers/home.htm. Accessed August 2021.

RICS (2013). *Royal Institution of Chartered Surveyors. Surveys of Residential Property RICS Guidance Note,* 3rd edition. RICS, London.

RICS (2018). *Royal Institution of Chartered Surveyors. Surveying Safely: Health and Safety Principles for Property Professionals. Guidance Note,* 2nd edition. RICS, London.

RICS (2019). *Royal Institution of Chartered Surveyors. Home Survey Standard, RICS Professional Statement,* 1st edition. RICS, London.

RICS (2021). *Royal Institution of Chartered Surveyors. COVID-19 Guide to Surveying Services: Physical Inspections for the Purpose of Residential Valuations and Condition-Based Surveys (England) Version 4.* RICS, London.

4   Diagnosing building defects

Contents

4.1 Introduction

Identifying and analysing defects and deficiencies is usually the main reason why clients come to us in the first place, and it is also why many of us got involved in the property sector.

Before you get too involved in the following chapters, it is important to think about the process of diagnosing building defects before you focus on the details. This is because diagnosing the cause of building defects is more than simply matching the visual signs of a problem to a possible cause. It is about a broader understanding of how buildings are designed and built and how they perform.

In this respect, we have been influenced by the work and writing of David Watt and his colleagues at Hutton and Rostron building surveyors. In his book titled '*Building Pathology: principles and practice*', Watt defined the methodical and often forensic practice of building pathology as:

The holistic approach to understanding buildings. Such an approach requires a detailed knowledge of:

- How buildings are designed and constructed;
- How they are used and changed;
- The various mechanisms by which their material and environmental conditions can be affected.

It requires a wider recognition of the ways in which buildings and people respond to each other.

(David S Watt, 2nd edition, 2007)

DOI: 10.1201/9781003253105-4

4.2 The diagnostic process

Using this broader approach, we have identified a number of stages in the diagnostic process:

Step One: Analyse the signs and symptoms

Not only will this involve the investigation on site (inspecting the building, making notes, sketches, measurements, photographs and so on), it also includes the collection of other useful information through desktop research before you even leave the office. This might typically include 'as-built' drawings of the building; O.S. maps; mining activity; former uses of the site; details of any previous work carried out; guarantees; telephone discussions with the clients, occupiers and so on (see Chapter 3 for more details).

You might find two definitions useful:

*   **Signs** – The information recorded by the practitioner. These are the physical matters and visual clues collected during the inspection;
*   **Symptoms** – This is the information gathered from people. Typically, this may include the owner, occupiers and neighbours. It is focused on their experiences and knowledge of the building. For example, typical enquiries might include: How long has it been leaking? When did you first notice the cracking? How long do you have the heating on every day?

This information begins to probe that broader question of how '. . . buildings and people respond to each other'.

Step Two: Possible causes

Based on this initial investigation, the next step is to compare the causes commonly associated with the signs and symptoms to see if they match. A close match might enable you to confidently identify the cause of the problem and give the client clear advice. For example, imagine inspecting a property from the outside and you notice a large area of the solid masonry wall has been affected by moisture. This comes from a broken section of the guttering that is still leaking during your inspection. Internally, you notice a large area of disrupted and stained plaster that corresponds very closely to the location of the problems outside.

In this case, the link between the internal and external signs is very close and, all other things being equal, you may feel able to identify the cause as a leaking gutter.

Sadly, it is rare that the c*ause and effect* of a defect are so closely matched. Consider a slight change in circumstances in another imagined building. Although the outside of the south-west-facing cavity wall shows the signs of water staining, the gutter is not obviously damaged, although you inspected on a sunny day in the middle of a dry spell. The mortar joints were in a deteriorated condition and you noticed a pattern of holes in the mortar joints suggesting cavity wall insulation had been injected. Internally, the damage to the plaster is not so pronounced and consists of a series of small stains at regular spacings. In this situation, the possible causes could include poor pointing, high level of exposure to the prevailing weather, mortar 'snots' on the cavity ties or poorly installed cavity wall insulation. Or a combination of all of these.

Based on a point-in-time inspection, it is not possible to make such a clear diagnosis. Therefore, further investigations would be required, as more information is required to establish the true cause to any level of confidence.

Step Three: Further investigations

In cases where stages one and two did not produce enough information on which a decision could be based, further investigations will be required. The extent and nature of these will depend on the circumstances but typically include:

- Opening up works ranging from discrete 'keyhole' investigations using bore-scopes and endoscopes through to removal of large sections of the construction;
- Environmental monitoring including radon, humidity levels, whole house air tightness/leakage assessment, extract fan testing and so on;
- Use of non-invasive technologies such as capacitance moisture meters, infra-red cameras and metal detectors;
- Remote inspection equipment such as pole cameras, drones, CCTV inspection of drains and flues;
- Laboratory-based testing to establish a range of information including moisture content of drilled samples; chemical composition, asbestos, soil samples and bacterial analysis.

Once the results of these tests are available, the additional information can be combined with the outputs of steps one and two. This may allow a probable cause to be identified. We use the word 'probable' because nothing is ever black and white in the diagnostic process and on occasions, practitioners have to make a decision when confidence is sufficiently high.

Step Four: Rectification work

If steps one to three identified a probable cause, then a scheme of repair can then be formulated with a high level of confidence. Depending on the size and nature of the work, the scope of the repair is defined, the repair work specified and quotations from appropriately qualified contractors obtained. If acceptable, the repair work is carried out under appropriate supervision.

Step Five: Monitoring of effectiveness

Although the diagnostic process may have identified a cause, the effectiveness of the repair needs to be monitored (although this is rarely done in practice). Ideally, periodic inspections should be carried out to make sure the problem has not returned. If it does, the diagnostic process will have to start again. A typical example is with a condensation problem. This is notoriously difficult to diagnose because the problem is dynamic and related not only to the nature of the building but also to how that building is used. Therefore, the property should be re-inspected during the following winter to see if the problems have returned.

Figure 4.1 shows how these stages in the diagnostic process apply to pre-purchase condition reports. Because the vast majority of these commissions are non-invasive,

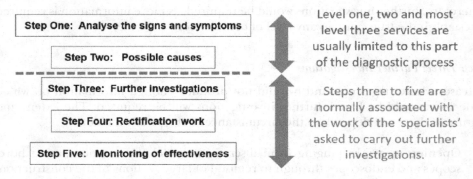

Figure 4.1 The diagnostic process and pre-sale services.

most instructions do not extend beyond Step Two. The only exception is where the terms of engagement include some element of exploratory work such as inspecting cavity walls or removing floor boarding and so on. In reality, even if practitioners offered these additional services, few vendors will allow this level of disturbance.

4.3 Adding value to condition reports

Reviewing this model in relation to pre-purchase condition reports, it is not surprising that many clients are often disappointed with the lack of certainty about the condition of their future purchase. If a condition report is going to call for further investigations, why should they commission one in the first place? This is one of the main themes of this volume: How can we give our clients clear advice without being rash and leaving ourselves vulnerable to future claims? We think this can be achieved using the following approach:

- Help the client choose the most appropriate level of service in the first place. For example, most level two instructions will suit modern and conventionally built property without the need for too many referrals (see Chapter 3 for more advice on how this can be done);
- By giving the client an awareness of the likely scope of the work through a process we call *'safe speculation'*. Although the scope of any repair work will not be known until the results of a further investigation are received, can we give the client an idea of the scope of the likely repair work? See Chapter 14 for more guidance on this;
- Where it is not possible to give an idea of the likely scope of the work, can we give the client information about the level of risk associated with the potential problem?

This is a finely tuned balancing act, but if we get this right, we can provide our clients with the information they need to make a purchase decision.

References

Watt, D.S. (2007). *Building Pathology: Principles and Practice*, 2nd edition. Blackwell Publishing, Oxford.

5 Building movement

Contents

DOI: 10.1201/9781003253105-5

5.1 Introduction

Many surveyors and valuers can become nervous, uncertain and hesitant when they encounter building movement. This often results in recommendations for further investigations, especially by less experienced or less confident surveyors. This is despite the studies by the BRE (1995, p. 2) which have shown that most houses underpinned following the dry summer of 1976 were only exhibiting minimal signs of damage.

The reasons for this overreaction are complex:

- All mortgage valuations and survey reports are 'point in time' inspections. In many cases, it is not possible to give a balanced professional view on just one visit. Indeed, current guidance from BRE (Digests 251 and 475) confirms that the only sure method of arriving at a satisfactory conclusion regarding whether further movement will occur is to monitor the movement for around 12 months (BRE 1995, 2003);
- Many surveyors simply do not have enough experience of assessing damaged buildings;
- Most surveyors feel they do not have the time or resources to carry out a full diagnosis during their standard types of inspection;
- The expectations of the building owner have risen and building damage is no longer tolerated (Evans 1998, p. 18). Owners will often push very hard for underpinning schemes or similar expensive repair works. This can provoke a desire in some surveyors to report in such a way that is sometimes over-cautious.

The role of the surveyor in the process is crucial. Many subsidence claims are triggered by the vendor when a potential purchaser receives an adverse report. Surveyors, therefore, have a clear responsibility to make sure their judgements are measured so that:

- The lender and purchaser are properly advised;
- Unnecessary and wasteful repair works are avoided; and
- The liability of the property professionals involved is protected.

5.2 Incidence of movement

5.2.1 Introduction

The dry summer of 1976 resulted in thousands of buildings developing cracks for the first time. Statistics show that the vast majority of the resulting underpinning claims were concentrated in the southeast and surrounding regions. The BRE (1995, p. 2) put forward a number of reasons for this:

- The main soil type in these areas is clay, which is prone to shrinkage during dry spells;
- The level of rainfall is lower in this part of the country; and
- There were a considerable number of trees in the urban environment.

Despite the greater risks in these localities, building movement can occur in all regions for a variety of reasons. Surveyors must therefore develop an objective procedure that can help them assess the consequences of any building damage.

5.2.2 Definitions

It is helpful to understand the terms used to describe different types of building movement, since the words are sometimes used incorrectly. You will encounter all the following descriptions with our explanation of the terms:

Movement type	Explanation
Heave	Upwards movement of soil, usually clay, caused by swelling due to the water content of the soil increasing. Typically occurs when a tree is removed, since water previously removed by the tree roots remains in the soil, thereby causing the swelling. In one case, over a period lasting 32 years, heave of at least 160 mm was recorded in a clay soil following tree removal (Cheney 1988, p. 13).
Landslip	'Moisture between particles in soil or rock reduces its resistance to sliding, and hence prolonged or intense rainfall is the most common trigger . . . when rock, soil or waste becomes unstable and moves under the influence of gravity' down a slope (Evans 2003, p. 73).
Seasonal	Due to changes in groundwater levels and general levels of water content in subsoils, slight swelling of shrinkable soils can occur in the winter, with shrinkage in the summer months. This can cause a building, particularly if it has shallow foundations, to move up and down with the seasons – rather like a ship in a sea's swell.
Settlement	Movement (i.e., compaction) in the ground as a result of the load (self-weight) applied by a new building, which can result in differential and uniform movement in a structure due to the (sometimes differing) distribution of loading and stresses within the various elements of construction. Pressure (force) is applied by the new structure until the force exerted upwards by the subsoil is equivalent to the force exerted downwards by the structure. Settlement is normally minor, usually occurs in the early life of the building and usually ceases within the first ten years after construction.
Shrinkage	Building materials have differing levels of moisture, especially masonry (bricks, blocks) during the construction stage. Water causes expansion of the materials, particularly in timber. Shrinkage typically occurs when a new building or extension dries out and this process can take a considerable time, sometimes up to 2 years. The resultant reduction in size causes cracks to open up through mortar joints, plaster and joinery.
Subsidence	Movement 'not caused by the weight of the building but by processes occurring in the ground' (Driscoll & Skinner 2007, p. 3). Movement in the ground unrelated to the presence of the building, that is, which occurs in the ground due to something other than the load (self-weight) of the structure. Results in downwards movement of the building foundation due to the loss of support in the ground beneath the building. Associated with volumetric changes in the subsoil and usually occurs as a direct result of some external factor, such as trees, drains, mining. Mostly (but not always) associated with shrinkable, clay, soil, but might be caused by, for example, small particles of soil being washed away by movement of water
Thermal	When building materials warm, they expand. An example of this is expansion of clay bricks on a south-facing wall, which can result in vertical cracks developing and the wall shifting on the DPC, which acts as a slip joint (hence why long walls should have movement joints).

Figure 5.1 Table of descriptions of some different types of building movement.

5.3 Assessing the causes of damage

5.3.1 Introduction

It is important for surveyors to always consider the main structural framework of a property, noting the individual details and considering how they fit into the 'big picture'. The best source of available information is Building Regulation's Approved Document A (BRAD A), which confirms that for a building to be stable under all probable imposed and wind-loading conditions, the following is necessary:

- The three main structural elements of the building: the floors, walls (internal and external) and the roof structure should act together to form a 'three-dimensional box structure'. Imagine each room as a cardboard box and the entire property as a series of those boxes interlinked;
- To prevent excessive distortion under load, each of those boxes should be limited in size – the thicker the walls are in proportion to the size of the box, the stiffer is each box;
- To link the boxes, internal and external walls (the vertical walls of each box) should be adequately connected, for example by 'toothing and bonding' the brick and blocks together where they meet (imagine putting your hands together and overlapping your fingers). This provides vertical stiffening through the resultant 'buttressing' effects;
- The horizontal parts of each box such as the intermediate upper floors and the roof structure, should act as 'horizontal diaphragms' to support the walls against movement and be connected to them; and
- Through this 'interconnectedness', the wind forces are transferred through the horizontal parts of each box and transferred to the 'buttressing elements' of the walls – the corners of each room (box) (where the internal and external walls meet) and corners of the building (OPSI 2013, p. 11).

We will consider this subject in greater depth throughout this chapter.

5.3.2 Extent of inspections

In the ever-changing property world, it is difficult to draw a confident division between where a mortgage valuation inspection ends and a survey begins. The first task of any surveyor is to notice whether the building has been actually damaged. In this context, 'damage' is not only cracking; it must also include distortion and 'serviceability' issues such as sticking doors and windows, disrupted services and lack of weather-tightness.

Once the damage has been noticed, your decision will depend on the type of commission being carried out:

Mortgage valuation – Although the principle of 'following the trail' has been firmly established, the 'trail' in a mortgage valuation may be shorter. With experience, many surveyors will be able to judge quickly when a building is so significantly damaged that further investigations and referrals will be required.

Survey – Because this type of inspection is so much more extensive than an inspection for mortgage valuation purposes, the 'trail' of diagnosis must be followed to a greater length and be better considered. The surveyor has more time to look at the damage to a building and set it in context. A more informed view should help to give balanced advice and so avoid unnecessary referrals to specialists.

Whatever the type of commission, it is essential that all surveyors understand the 'process' of diagnosing the causes of building movement so that:

- Early in their careers, surveyors can follow a step-by-step approach (a 'protocol') giving them greater confidence in their decision making until they develop a more intuitive level of skill;
- When experienced surveyors encounter an unfamiliar or novel form of building damage, they have a clearly defined approach that will help them give appropriate advice, thus ensuring:

    - Their clients get good advice; and
    - Minimising potential liability.

The most popular approaches to this process have been promoted by the BRE (1995, 2003) and the ISE (1994). A combination of these approaches provides a robust, objective diagnostic process for surveyors to follow.

5.4 The diagnostic process

Any structure can be damaged in many different ways and the evidence for the damage will provide the surveyor with information on which an assessment can be based. The approach comprises distinct stages.

5.4.1 Step one – spot the triggers

During any inspection, the surveyor should be continuously watchful for evidence of building damage. For building movement, this is not only cracking and includes:

- Misalignment – sloping sills, lintels, masonry coursing and floors, and so on – all generally known as 'tilt', or 'rigid body rotation' (BRE 2003, p. 5);
- Windows and doors that don't open properly, fractured service pipes;
- Environmental clues such as sloping ground, damage to other properties in the locality – 'knowledge of locality' is fundamental for a surveyor.

Once a 'trigger' is noticed, the surveyor must 'follow the trail' flowing from the defect. This part of the process requires a *present condition survey* following several clearly prescribed stages.

5.4.2 Step two – the present condition survey

The BRE (1995, p. 2) suggests a 'prerequisite for the objective classification of damage in a building is a thorough, well-documented survey', in other words an objective collection of information that helps define the extent of the damage.

Sketch the extent of the damage

A simple, quick, reasonably proportional sketch of each elevation should be prepared, indicating the position and distortion and cracks – many surveyors use a red pen to help their later consideration. These line drawings must show:

- The type of crack – whether it shows tensile, compressive or shear characteristics (see Figure 5.2);
- The width of the crack and whether it tapers along its length;
- Where the cracks are distributed around the building;
- Whether the cracks extend through the width of the wall and appear internally.

Figure 5.3 illustrates part of a typical sketch. Simple floor and site plans should also be included, showing the positions of any drains, trees, slopes, external and internal stop valve positions and so on.

Not all masonry affected by building movement will crack. Most of the housing stock before World War I was constructed using softer sand and lime mortar. Such mortar is more flexible or 'plastic' than a cement and sand mortar. In this situation, it is common to see masonry courses 'flowing' with the movement rather than cracking. This is distortion and is still evidence of movement, so it should be noted.

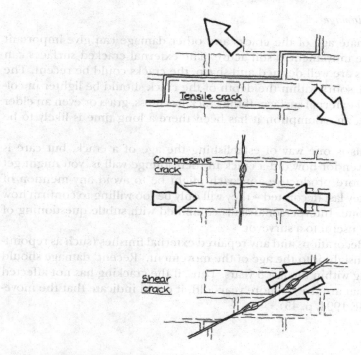

Figure 5.2 Different types of cracks that can occur in masonry walls resulting from tensile, compressive and shear forces.

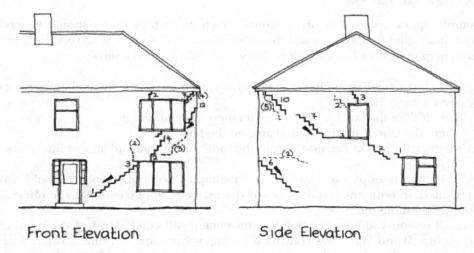

Front Elevation   Side Elevation

Figure 5.3 Typical 'present condition' sketch of a property a surveyor should prepare once damage is noticed. Dotted lines indicate inside cracks. The numbers are crack widths in millimetres.

Determine the age of the damage

Knowing the approximate age of the cracks and other damage can give important clues as to whether the movement is still active. The external cracked surfaces can give a clue. If the edges are well defined and sharp, the cracks could be recent. The exposed faces of the masonry within the depth of the crack should be lighter in colour than the weathered external surface. If the crack has moss, grass or even an elder bush growing out of it, an assumption it has been there a long time is likely to be reasonable.

Asking the occupants is one way of establishing the age of a crack, but care is required. If you ask a vendor how old a crack in their lounge wall is, you might get an evasive answer. A more productive approach might be to avoid any mention of cracks and ask when they last decorated – they will only be too willing to confirm how recently it has been done. Interpersonal skills combined with subtle questioning of the vendor can be very useful to a surveyor.

The age of internal decorations and any repaired external finishes (such as repointing) may also give an insight into the age of the movement. 'Recent' damage should be defined as occurring within the last 5 years. Thus, if the cracking has not affected internal decorations that are five or more years old, it could indicate that the movement has stabilised (ISE 1994, p. 45).

Is the building distorted?

If a building is damaged by ground movement, part of it will rotate or tilt. This can result in walls leaning and floors, lintels and sills sloping away from the horizontal. Measuring the extent of these distortions precisely is a very sophisticated process, but most surveyors will be able to come to a broad view. For a survey, this could be by:

- Using a bricklayer's spirit level (1.0m long) to assess the verticality of walls and slope of floors;
- Placing a smaller spirit level (say 250mm long) on sills, underside of lintels and door heads to see if they are sloping;
- Dropping a simple plumb bob and line out of a convenient window.

This sort of information will be very broad brush. If considered in isolation, it could easily lead to incorrect conclusions, but when matched with the other evidence, it will add depth to the analysis.

Define the serviceability of the building

The functional performance of the various elements of a building will be affected by building movement. Describing the extent of this reduction in performance will assist in the diagnostic process. The features to consider include:

- **Doors and windows sticking** – Door and window openings in any building are weak points since they represent breaks in the structural continuity of the walls and will quickly distort. The frames in the openings will 'rack' or be pushed out of shape. Any opening window sashes and doors will 'bind' against their frames. In extreme cases, they may not open at all. Owners will sometimes regularly plane down surfaces, so doors and windows can still be used. After a time, strange shapes can often result (see Figure 5.4);

Figure 5.4 Distorted window opening in a damaged building. The top sash has been regularly adjusted, so it will close. What will the carpenter do when she gets down to the glass?

- **Cracked window panes** – Significant movement can induce such stresses in window frames that the glass panes will actually crack, particularly in metal frames;
- **Draughts and rainwater penetration** – Where cracks are wide enough on exposed elevations, wind-driven rain can penetrate to the internal surfaces. The gaps that appear around window and door frames can be very vulnerable, especially where the waterproofing sealant has become disrupted;
- **Fractured underground pipes** – Foul or surface water drain pipes (for example, visible in inspection chambers) might be broken, or an incoming water main leaking.

Constructional details

Some types of construction will inherently be more vulnerable to movement by their very nature. Making a note of their characteristics will enable the surveyor to rule out potential causes and thus simplify the diagnostic process. Examples include metal ties in cavity walls and other components with metal reinforcement prone to corrosion, materials prone to expansion and/or contraction such as calcium silicate bricks, and so on. Such issues are so closely associated with clearly defined forms of building damage that prognosis may be quite simple.

Age of the property

The property's age will help determine the construction type and the nature of the evidence recorded. For example:

- A property built with a more flexible, lime, mortar will exhibit evidence that is different to;
- One with a more rigid, stronger but brittle, cement mortar.

The ISE (1994, p. 37) confirms new houses should be treated with extra caution. This is because initial settlement and drying out cracking can be mistaken for subsidence. The definition of 'new' in this case is usually applied to a property less than 5–10 years old. Even though cracking might be more commonly encountered in newer properties, any damage in category 3 and over (see Figure 5.5) should still be referred for further investigation. Lesser damage in 'new' properties should be considered 'normal' unless there is other evidence to the contrary.

5.4.3 Step three – classify the damage

Once all the relevant information has been collected and recorded, the surveyor's analysis can begin. In Digests 251 and 475, the BRE have developed an excellent, objective approach. This gives professionals a common language and criteria that are clearly understood and aids a shared understanding of damage levels in any building. Thus, BRE classifies damage into three main categories (BRE 1995, p. 3):

- **Aesthetic** – This damage affects the decorative finish of the building and has little effect if any on its functional performance;

Category of damage	Description of typical damage (Ease of repair in italic type)	Approximate crack width (mm)
0	Hairline cracks of less than about 0.1 mm are classed as negligible.	Up to 0.1 mm*
1	*Fine cracks that can be easily filled during normal decoration.* Perhaps isolated slight fracturing in the building. Cracks rarely visible in external brickwork.	Up to 1 mm*
2	*Cracks easily filled. Re-decoration probably required. Recurrent cracks can be masked by suitable linings.* Cracks not necessarily visible externally; *some external repointing may be required to ensure weathertightness.* Doors and windows may stick slightly.	Up to 5 mm*
3	*The cracks require some opening up by a mason. Repointing of external brickwork and possibly a small amount of brickwork to be replaced.* Doors and windows sticking. Service pipes may fracture. Weather-tightness often impaired.	5 to 15 mm (or a number up to 3 mm)*
4	*Extensive repair work involving breaking-out and replacing sections of walls, especially over doors and windows.* Window and door frames distorted, floors sloping noticeably. Walls leaning or bulging noticeably some loss of bearing in beams. Service pipes disrupted.	15 to 25 mm* but depends on number of cracks
5	*This requires a major rebuilding job involving partial or complete re-building.* Beams lose bearing, walls lean badly and require shoring. Windows are broken with distortion. Danger of instability.	Usually greater than 25 mm* but depends on number of cracks

Figure 5.5 Table from Digest 251, showing the classification of visible damage to buildings (reproduced courtesy of BRE).

* Crack width is just one factor in assessing the category of damage and should not be used on its own.

- **Serviceability** – Damage is so pronounced the performance of different elements are affected. For example, binding of doors and windows casements; wind and rain entering through cracks in walls and the fracturing of service pipes;
- **Stability** – Parts of the building are so badly damaged that stability might be threatened.

Although this can help a surveyor begin to define the level of damage, the categories remain too broad. To make this approach more useable, the definitions were refined by BRE to produce six different categories of damage (see Figure 5.5). This enables a wide variety of cases to be easily classified. The most important feature of this technique is the written description of the levels of damage, including the '*ease of repair*' and descriptions of the damage itself. This encourages a surveyor to look at factors other than just the cracking, since 'the structural significance of cracks is often exaggerated' (Carillion 2001, p. 133). This helps surveyors avoid relying on subjective personal views and 'gut reactions'.

The way to use this table effectively is to look at the damage to a building and ask the question, 'If the movement has stopped and is unlikely to continue, what work is required to repair the damage?'. Look at the descriptions carefully and classify the damage that can be seen. There are two broad headings:

- In categories 0, 1 and 2, the repair works are mainly cosmetic; involving crack filling, redecoration and repointing;
- Categories 3, 4 and 5 include more significant repairs ranging from the rebuilding of small sections of masonry through to partial or total rebuilding of the whole building.

A few other points should be noted when using this table:

- The surveyor should first focus on the *ease of repair* descriptions;
- Classifications are not based on crack widths, which are given for guidance only – damage to buildings is a much broader concept than just the cracking;
- The assessment must be based on the visible damage at the time of the survey and not how it might progress. The surveyor should not try to speculate subjectively about the cause of the damage at this stage. In other words, you should not jump to conclusions.

Causes of damage

Only when the objective classification of the damage has been completed should the surveyor begin to consider the cause. There are generally two types of movement that cause damage to buildings:

- Movements originating in the building structure above ground; such as bulging walls, thermal expansion and contraction, moisture movements, cavity wall tie failure, chemical changes such as sulphate attack, poor design and workmanship, and so on;
- Movements associated with the ground beneath the building including subsidence caused by shrinkage and swelling of clay, leaking drains, mining activities, settlement of filled ground, and so on.

Relating this back to Figure 5.5, damage associated with movements of the building structure above ground rarely exceeds the damage described in category 2. Movements that originate beneath the building can result in damage in all categories from 0 to 5.
  This distinction is useful and supports two diagnostic principles:

- Unless there is clear evidence the damage will not progress beyond category 2, further investigations are usually unwarranted and all that is required is the restoration of the appearance. The key factor here is judging whether the movement is progressive and this is discussed later.
- If the damage is in category 3, 4 or 5, this is clear evidence that foundation movement is the most probable cause. Because this can be progressive, significant and costly, further investigations are required so the client can be properly advised. Even if foundation movement has ceased, never to move again, it is possible the building could be so significantly weakened that remedial works will be necessary in any case.

5.4.4 Step four – establish the cause(s) and assess whether damage could get worse

To assess whether the recorded damage is progressive, the surveyor should ask two questions:

- Is the damage due to foundation movement?
- If yes, will it get worse?

This difficult stage of the process can be made easier by considering that 'subsidence cracking generally shows a distinctive pattern', and identifying characteristics indicative of foundation movement (Dickinson & Thornton 2004, p. 30). These are summarised as follows and shown in Figure 5.6:

- **Cracks affect internal and external faces of walls in close proximity**. This is especially significant in cavity walls, since 'foundation movement is one of the few processes that can cause cracking in both leaves of a cavity wall at approximately the same location' (Driscoll & Skinner 2007, p. 19). Other causes, such as wall tie failure and expansion problems, usually affect the outer leaf only, except in extreme cases;
- **The damage extends down below the DPC and into the ground**. When support beneath part of a foundation is removed or weakened, the building will virtually 'break' and rotate around the focus of movement, or 'hinge' point. The cracks usually begin from this point, well below DPC level (see Figure 5.6). If the cracking has been caused by superstructure deficiencies, the cracks usually only appear above the DPC because:
  - Brickwork below the DPC is fully restrained by the ground; and
  - The DPC acts as a 'slip joint', allowing the wall above to move independently;

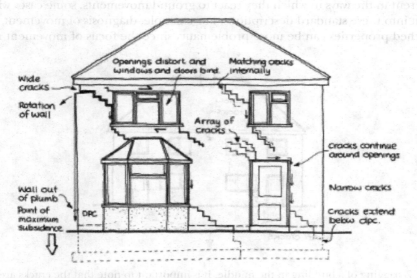

Figure 5.6 Typical cracking patterns caused by subsidence.

- **The cracks are normally tapered, wider at one end than the other**. Because the building tends to rotate, cracks are wider at one end (Driscoll & Skinner 2007, p. 16). For most cases, the cracks will be wider at the top, apart from where the ground 'sags' in the middle of a building causing wider cracks towards the bottom (see Figure 5.7) (BRE 1991, p. 14);
- **Cracks are usually wider and more frequent, closer to the focus of movement**. Most ground problems are relatively isolated, arising from one cause and resulting in damage to one part of the building only (mining subsidence being a notable exception). Other types of superstructure defects can cause damage that is distributed around several elevations;
- **Cracks are normally diagonal in direction**. Other causes of damage cause horizontal cracks (cavity wall tie failure and sulphate attack) or vertical cracks (moisture and thermal movements);
- **Floors and walls tilt and openings distort, causing opening parts of windows and doors to 'bind' and jam in their frames**. This is caused by the rotation of the entire building or parts thereof;
- **Cracks tend to travel around, or perhaps more accurately, 'through' openings**. The ISE (1994, p. 43) state that openings attract cracks because they find it easier to travel around the openings than through the stronger, adjacent masonry panels. This is because all openings or joints in walls are weak points – hence why more wall ties are installed around openings and joints as compared with other parts of a cavity wall, reinforcing the wall at those points (Driscoll & Skinner 2007, p. 16). Door and window frames will twist out of alignment within their openings and any mastic sealant will be disrupted;
- **The focus of maximum damage can be represented by a single large crack or a series of smaller, closely spaced ones or an 'array' of cracks**. This depends on the stress levels in the masonry structure and the way different materials react to it.

Surveyors should consider these indicators as typical symptoms. Because buildings are so different in the ways in which they react to ground movements, some cases will not easily fit into these standard descriptions. For example, diagnosis of movement issues in attached properties can be more problematic, since the focus of movement might

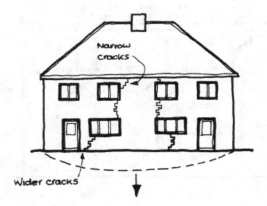

Figure 5.7 Sagging of a building in the middle. It is important to note that the cracks are wider towards the bottom – this is one of the instances where the adjoining semi-detached property must also be considered.

be on another property. In a long terrace, it may be necessary to inspect a considerable distance away from the subject property, either from the highway or by erecting a ladder to look along the mortar bed joints of the whole terrace for evidence of distortion.

5.4.5 Step five – decide whether the movement requires further investigation

Two courses of action are now appropriate:

For damage categories 3, 4 and 5 – the extent of the distortion and damage is usually sufficient to justify further investigation. This is appropriate even if all indications suggest the movement has ceased, since the client will want reassurance that further movement is unlikely to occur, with the resultant effect on value and/or saleability. The exception is if such investigations have already been undertaken and there is an adequate documentary paper trail confirming the current position, thereby providing certainty;

For categories 0, 1 and 2 – if there are no indications the movement will worsen, the damage is likely to be reported as 'longstanding' and 'of no structural significance', unless there *are* indicators of possible further movement.

The surveyor must therefore consider and compare the signs and symptoms on the property and surrounding area against 'risk' indicators that could suggest the likelihood of continuing foundation movement.

The risk indicators typically include the following:

Evidence of 'recent' movement

The surveyor should try to determine the age of the damage, on the assumption that any crack damage that has broken through paint, emulsion or lining paper applied in the last 5 years should be considered as 'recent'. This process can include:

- Asking the occupier when the room where damage is located was last decorated (bearing in mind most people's sense of time is poor);
- Closely inspecting cracks, for example with a magnifying glass:

  - Recent external cracks will look 'fresh' with crisp, sharp, edges; whereas
  - Mortar of older cracks will appear eroded, worn and dirty;
  - Damage on elevations facing prevailing winds will age quicker;
  - Older internal cracks may have emulsion or paint inside;

- Inspect for evidence of damage that may have been repaired but has since reopened, such as repointing of old cracks.

Nearby trees and shrubs

The BRE suggests around 70% of all subsidence claims involve shrinkable clay soil and trees' (Driscoll & Skinner 2007, p. 18). More detailed consideration is contained in Section 5.6, but the most important issues are:

- Trees within 20 m of the property;
- Hedges or rows of large shrubs within 10 m;

- Subsoil in the area is likely to be shrinkable;
- Drains running close to the building that could have been damaged by tree roots;
- Tree stumps or other evidence trees or shrubs have been removed.

Proximity of sloping ground

There is a natural tendency for subsoils to move from higher to lower levels. If this happens, foundations can be affected together with other external features, such as retaining walls, paths, drives and so on. Evidence of such problems includes:

- Trees inclined down the slope of the land;
- 'Fissures' or large 'cracks' in the ground surface;
- Slumped areas of ground, often corresponding to the fissures described previously;
- Garden and retaining walls and other external constructional features being pushed over or showing other signs of distress.

This indicator is likely to be more active during the winter than the summer because of increased ground moisture levels.

Possibility of leaking drains

The BRE suggests 'around 30% of valid subsidence claims . . . involve leaking drains' (Driscoll & Skinner 2007, p. 20). Drains can cause subsidence in two ways:

- By softening the subsoil (especially clays), allowing the building to 'sink';
- Finer elements in the subsoil can be 'washed out' and/or washed into the drains leaving voids that allow subsidence.

Determining whether drains are leaking is a very time-consuming and expensive business. At the initial stage, the following factors may help to establish the likelihood of problems:

- Drain runs close to the affected building – look for routes of inspection chambers, rodding eyes, gullies and so on. If there are no drains within 5 m, it is unlikely to be a cause.
- Whether drains are older than 1970:

  - Clay or similar rigid jointed drains are very susceptible to leaks, as even slight ground movements can cause cracking. Generally, pre-1970 drains tend to be of the rigid jointed type, whereas later pipes have flexible neoprene joints and/or are of plastic;
  - 'Pitch fibre' drains (which may contain asbestos fibres), generally used between 1950 to 70, can deform and collapse;
  - Properties older than around 1970 therefore present a greater risk.

- Trees or shrubs planted directly over or close to a drain near the property. Roots can penetrate even small cracks in their search for moisture, causing damage and further leakage;

- Drains beneath vehicular access areas – drains can be damaged by cars or heavier vehicles if they are too shallow or have not been protected;
- Conditions of and within the inspection chamber – lifting inspection chamber covers can give an insight into the condition of the drain system including:

  - Soil or other granular material running down the drain suggesting a leak further up the run;
  - Tree or shrub roots in the chamber;
  - Condition of benching and brickwork;
  - Cracked drain pipes visible in the chamber;
  - All these signs could increase suspicion that the drains are in poor condition.

Susceptible soils

The very nature of some subsoils will mean ground movement is more likely in some areas than others. Most problems are caused by shrinkable clays, especially when trees and large shrubs are close by. An increasing number of houses have been built on former landfill sites that also pose a high level of risk. Site inspections alone cannot usually determine soil types, so good local knowledge is essential, usually based on the surveyor's interrogation of geological sources in their desktop study (see Chapter 2). This emphasises the need for surveyors to include consideration of 'knowledge of locality' in their deliberations.

If the surveyor identifies *any* of the risk indicators, it may be appropriate to recommend further investigation due to the risk of further movement, even for damage categories 1 or 2. The more indicators are present, the higher is the risk of the damage progressing beyond damage category 2 and therefore the more likely it is the surveyor is justified in recommending the movement is allocated a condition rating 3 – further investigation.

5.4.7 Further refinements in analysis of ground movement

This description of a working protocol on how to analyse structural movement in walls is vital to any surveyor's practice, but it is only an introduction. The surveyor should therefore become skilled in using the protocol based on Table 1 from BRE Digest 251 and then refine their practice by studying:

- Table 2 in Digest 251 – diagnosis of movement in floors; and
- BRE Digest 475 – for supplementary and associated advice regarding how to include consideration of 'tilt' in the diagnosis.

5.4.8 Recommending further investigation

If there are risk indicators present, a recommendation for further investigation is appropriate. Once that decision is taken, who should carry out the investigation?

It is suggested that further investigations into subsidence damage should be organised by an 'appropriately' or 'suitably qualified' professional experienced in this type of work regardless of their designation. Such professionals typically include:

- Chartered structural, civil or building engineers (CEng, MIStructE, MCABE or FCABE);

- Chartered building surveyors (FRICS or MRICS);
- Members of the Chartered Institute of Building (CIOB); or
- Any other 'suitably qualified professional' who has sufficient competence and experience to undertake the commission.

It is worthwhile noting that many lending institutions have clearly defined criteria for who should be used. It is usually therefore appropriate to recommend the client checks to ensure the professional is acceptable to the lender before instructing any particular professional.

5.4.9 Summary on reporting on building damage

The preceding sections have outlined an assessment framework, or protocol, against which building movement can be judged. This is summarised as follows:

- If damage observed is in category 3, 4 or 5, movement probably originates in the ground. At this magnitude, the property is so significantly damaged that further investigation is required even if the movement has stopped.
- If the damage is category 2 or less and:
  - The damage appears recent (less than 5 years);
  - The pattern of the damage suggests foundation failure; and
  - There are 'risk' indicators present

  referral for further investigation is probably appropriate.
- If the damage is category 2 or less, it is older than 5 years and there are no 'risk' indicators, referral for further investigation is usually not required. Restoration of appearance is all that may be required;
- If the damage is in categories of 0 or 1 and no risk indicators are present, it is probably of no consequence. No further action is required, and the damage can be dealt with during normal decoration works.

This guidance is very broad and must be applied in context for each case. There are always exceptions to the rules; however, this framework provides a robust basis for good practice. Providing the surveyor records their thought process in their site notes, the protocol will help ensure:

- Surveyors provide good advice to their clients based on a sound, recognized approach; and
- If a report is questioned later during audit, by a client or in court, the surveyor can prove they used an approach recommended for use by any 'reasonably competent professional'.

5.5 Trees and buildings

5.5.1 Introduction

Trees in urban areas can enhance the quality of the environment in many ways. They can also cause significant damage to nearby buildings if they are too close and are not managed properly – we have referred previously to BRE guidance suggesting a significant proportion of subsidence claims are due to the presence of trees. Surveyors should also remember to not only consider trees but other plants, since a 'combination of shrinkable soils and trees, hedgerows and shrubs . . . take moisture from the ground and, in cohesive soils such as clay, this can cause significant volume changes resulting in ground movement' (our emphasis) (NHBC 2020, Chapter 4.2, p. 1). However, 'trees are regularly seen quite close to buildings causing no apparent problem' (Dickinson & Thornton 2004, p. 35).

5.5.2 *Trees and their effects on properties*

Trees can cause several problems, including:

- Moisture extraction by root systems can cause shrinkage in susceptible soils;
- Pressure of a growing mass of roots can cause retaining and other walls to crack and collapse;
- Roots can enter even the smallest cracks in drainage pipes, causing disruption and blockage;
- Leaves can block gutters causing them to overflow – resulting dampness can lead to wet and dry rot in the building;
- Trees can fall on buildings (and people) and damage them;
- Trees close to boundaries can be a cause of neighbour disputes; and
- Solar panels and similar can suffer shading by trees, affecting their performance;
- Even a single tree will require regular maintenance, including pruning and lopping. On large properties or any property with several trees, annual costs can be significant. Some trees can also lead to soiling of patios, pavements, cars, etc., through the deposit of sticky liquid and fruits. Building owners are faced with the additional costs of clearing up leaves, branches, twigs and so on.

5.5.3 *Factors to investigate*

To avoid over-reaction, when assessing the effect of trees on a property, three key issues should be evaluated;

- **Step one** – Nature of the subsoil;
- **Step two** – Type and maturity of the tree;
- **Step three** – Proximity to buildings, other structures and services.

These are now considered in more detail.

Step one – nature of subsoil

Shrinkable, cohesive soils, such as clay, present the most risk. The extraction of moisture by tree roots causes the clay to desiccate (dry) and shrink. If this occurs too close to a building, especially one with shallow foundations, subsidence can result. With non-cohesive soils (sands, gravels, etc.), trees are less likely to have any effect, emphasising the importance of researching local subsoil types.

Step two – type and maturity of tree

Not all trees pose the same problems. Research has shown (BRE 1985) some trees have a higher rate of moisture extraction than others. In these cases, and where the sub-soil has shrinkable characteristics, greater caution must be exercised. NHBC suggests the most dangerous, high-water-demand trees include all elms, eucalyptus, hawthorn, oaks, poplars and willows (NHBC 2021, Section 4.2.4). More detailed information is included in Figure 5.8.

High water demand species	Mature height (m)	Moderate water demand species	Mature height (m)	Low water demand species	Mature height (m)
Broad-leafed trees:					
English elm	24	Acacia (False)	18	Birch	14
Wheatley elm	22	Alder	18	Elder	10
Wych elm	18	Apple	10	Fig	8
Eucalyptus	18	Ash	23	Hazel	8
Hawthorn	10	Bay laurel	10	Holly	12
English oak	20	Beech	20	Honey locust	14
Holm oak	16	Blackthorn	8	Hornbeam	17
Red oak	24	Japanese cherry	9	Laburnum	12
Turkey oak	24	Laurel cherry	8	Magnolia	9
Hybrid black poplar	28	Orchard cherry	12	Mulberry	9
Lombardy poplar	25	Wild cherry	17	Tulip tree	20
White poplar	15	Horse chestnut	20		
Crack willow	24	Sweet chestnut	24		
Weeping willow	16	Lime	22		
White willow	24	Japanese maple	8		
		Norway maple	18		
		Mountain ash	11		
		Pear	12		
		Plane	26		
		Plum	10		
		Sycamore	22		
		Tree of heaven	20		
		Walnut	18		
		Whitebeam	12		

High water demand species	Mature height (m)	Moderate water demand species	Mature height (m)	Low water demand species	Mature height (m)
Coniferous trees:					
Lawson's cypress	18	Cedar	20		
Leyland cypress	20	Douglas fir	20		
Monterey cypress	20	Larch	20		
		Monkey puzzle	18		
		Pine	20		
		Spruce	18		
		Wellingtonia	30		
		Yew	12		

Table 5.8 Water demand of high-water demand tree species, based on NHBC, Chapter 4.2, Table 3, p. 4 (NHBC 2021).

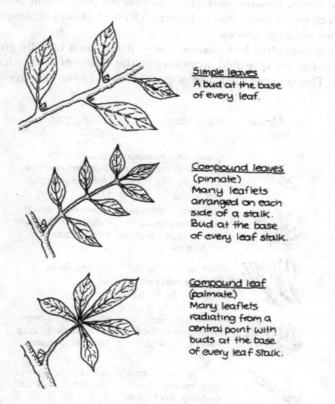

Simple leaves
A bud at the base of every leaf.

Compound leaves
(pinnate)
Many leaflets arranged on each side of a stalk. Bud at the base of every leaf stalk.

Compound leaf
(palmate)
Many leaflets radiating from a central point with buds at the base of every leaf stalk.

Figure 5.9 Sketches of generic leaf shape groups for deciduous trees (reproduced and adapted from *Usborne Guide to Trees of Britain and Europe*).

This list varies among different commentators. Lending institutions may view certain trees with suspicion, so it is important to check their guidance. There are thousands of different tree varieties in the UK and correct identification is the vital first stage in properly advising a client. Like any complex process, it is important to follow a structured approach to tree identification. There are many different sources of information, but Barrett (1981) proposed a method that is still highly relevant and most helpful to the less experienced surveyor. This involves looking at three different characteristics of the tree:

- The leaf;
- Crown and trunk shapes;
- Bark types.

IDENTIFYING LEAVES

Because most trees will be identified by their leaves, this section will look at this aspect only. There are several stages:

- **Categorising into generic leaf group shapes** – all conifers have either needle or scale-like leaves. Broad-leafed trees can have simple or compound leaves. These different types are illustrated in Figures 5.9 and 5.10 and represent the first stage in the identification process;
- **Classifying individual leaf shapes** – once it has been initially grouped, the leaf should be further classified according to the different categories shown in Figure 5.11. This reduces the range of possibilities to fewer manageable varieties.

Leaf Group Shape	Description	Typical tree type
	Single, very narrow leaves attached individually to the twig	• Douglas Fir • Western Hemlock • Norway Spruce • Yew
	Needles in bunches or clusters growing from woody knobs on the twig	• European Larch • Cedar of Lebanon • Atlas Cedar
	Stiff or soft and flexible needles in pairs, 3s or 5s.	• Scots Pine
	Feather-like arrangement of needles in flat rows	• Swamp Cypress • Dawn Redwood • Coast Redwood
	Tiny or large scale-like leaves that overlap and cover twig.	• Leyland Cypress • Western Red Cedar

Figure 5.10 Table of the most common coniferous leaf shapes (reproduced and adapted from *Usborne Guide to Trees of Britain and Europe*).

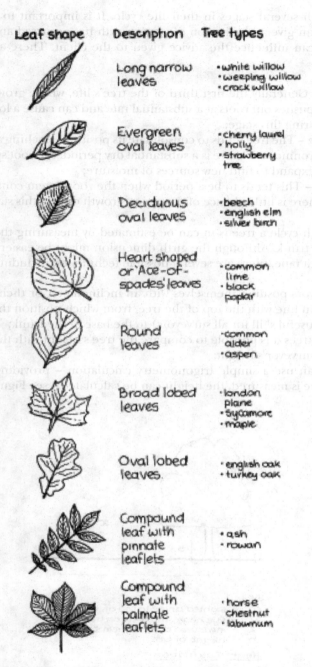

Leaf shape	Description	Tree types
	Long narrow leaves	• white willow • weeping willow • crack willow
	Evergreen oval leaves	• cherry laurel • holly • strawberry tree
	Deciduous oval leaves	• beech • english elm • silver birch
	Heart shaped or 'Ace-of-spades' leaves	• common lime • black poplar
	Round leaves	• common alder • aspen
	Broad lobed leaves	• london plane • sycamore • maple
	Oval lobed leaves	• english oak • turkey oak
	Compound leaf with pinnate leaflets	• ash • rowan
	Compound leaf with palmate leaflets	• horse chestnut • laburnum

Figure 5.11 Table of the most common deciduous leaf shapes (reproduced and adapted from *Usborne Guide to Trees of Britain and Europe*).

MATURITY OF THE TREE

Trees go through several stages in their life cycle. It is important to try to identify this because it can give an indication of what growth has occurred and what can be expected. This can influence the advice given to the client. There are three main cycles:

- **First cycle** – Generally the first third of the tree's life, where growth is the most vigorous. It pushes out roots at a substantial rate and can cause a lot of damage to buildings during this stage;
- **Second cycle** – The tree tends to consolidate its position, reaching an equilibrium with its environment. If there is a substantial dry period, the root system will rejuvenate and expand to find new sources of moisture;
- **Third cycle** – This tends to be a period when the root system contracts until the tree dies. There is little chance of damaging growth during this stage.

Identifying which cycle a tree is in can be estimated by measuring the tree's height and girth of the trunk. Although the girth dimension might be easier for a surveyor to measure with a tape, there are several possible techniques, including:

- Many surveyors position themselves with an inclinometer or their arm raised at a 45° angle in line with the top of the tree, from which position they pace at 1 m intervals (a useful skill for all surveyors) to the base of the trunk;
- For smaller trees it is possible to compare the tree's height with that of the property, or the surveyor's height;
- Surveyors can use a simple trigonometry calculation – providing the distance from the tree is measured, the height can be calculated – see Figure 5.12.

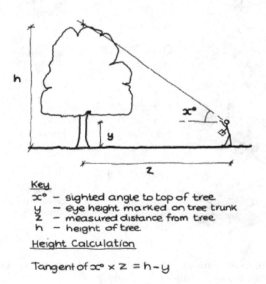

Key
$x°$ – sighted angle to top of tree
y – eye height marked on tree trunk
z – measured distance from tree
h – height of tree

Height Calculation

Tangent of $x° \times z = h - y$

Figure 5.12 Estimating the height of trees by trigonometry.

Tree type	Mature height (m)	Girth (m)
Ash	23	varies
Beech	20	up to **6**
Elm	25	**4–6**
Horse chestnut	20	max 2.8
Lime	21–24	4.5–7.5
Maple	21	varies
Oak	**24**	3–10
Poplar	28	2.5
Sycamore	21–24	1.5
Willow	24	**6**

Note: These sizes will vary according to species and location. The figures have been based on an average taken from a number of sources. This assumes an urban environment where mature height will be restricted.

Figure 5.13 Height and girth of some common trees.

- Another approach is to use a traditional surveying instrument called an Abney level. This measures the angle of inclination and therefore results in a more accurate estimate of the height of the tree (see Figure 5.12). This method is better suited when the size of the garden is limited.

Using the approximate height and girth information in Figure 5.13, a rough estimation of a tree's life cycle can be made. Noting and recording the height of the tree is fundamental in all cases in any event.

Step three – proximity to the building

Once the type and life cycle of the tree have been identified, the safe distance from the building can be established.

The NHBC recommend that the safe distance from buildings and other structures should be a function of their mature height, which is either reduced or increased depending on the water demand characteristics. This is shown in the following table:

Water demand	Zone of influence
High	1.25 x mature height
Moderate	0.75 x mature height
Low	0.5 x mature height

Figure 5.14 Lateral zone of influence of trees related to water demand (NHBC 2020).

ORNAMENTAL TREES AND SHRUBS

Many property owners seeking privacy and seclusion plant a variety of ornamental shrubs and hedges close to their properties. A common variety is the Leyland Cypress or 'leylandii', which, if well managed and clipped, may not present a problem. However, because these trees have a mature height of up to 25 m, uncontrolled growth can cause high levels of water extraction, especially if there are 20–30 of them along a boundary.

5.5.4 Reflecting and reporting on trees

Advising clients on what to do about trees would be easy if the client could simply cut down those trees too close to the building. In many cases, if a tree is removed, moisture will return to the subsoil, causing it to swell. This would then lead to further damage to the building through the heaving up of the ground beneath. This can be more damaging than subsoil shrinkage and can continue for many years. Alternatively, trees can be severely pruned or lopped. However, this is 'a very short-term method of control which needs repeated application' and is therefore costly (Dickinson & Thornton 2004, p. 38).

Consequently, care must be taken over what advice is given. A decision framework is outlined in this section, but this must be set against the clear and preferred guidance given by many lending institutions.

Where the subsoil is known to be a non-shrinkable, non-cohesive type

There is no significant risk from trees unless:

- The branches are either touching or soon likely to touch the building during wind action – pruning and tree management might be needed;
- The tree is in poor condition where parts of it are in danger of collapse;
- Where drain runs pass very close to the base of the tree and/or there is evidence of root growth in the drainage system.

Where the subsoil is known to be a shrinkable cohesive type, or the true nature of the soil is unknown

There are no significant risks from trees unless:

- Any of the three factors previously identified for non-shrinkable or non-cohesive soils apply; and
- The tree is within influencing distance as determined by its type, maturity and water demand.

An example may help:

Circumstances

The surveyor identifies a beech tree where none of the branches are touching or are likely soon to touch the building, the tree is not close to drains and is apparently in satisfactory condition. The soil is known to be shrinkable clay. The tree is situated 15 m from the building and has a height of 16 m. Figure 5.8 indicates that a beech tree has a 'moderate' water demand, with a mature height of 20 m. As a consequence, Figure 5.14 confirms that a beech tree has a zone of influence of its height multiplied by 0.75.

Consideration

The existing zone of influence is likely to be 16 m (the existing height) multiplied by 0.75, which gives 12 m. This is 3 m less than the distance from the building and so is not currently at risk. However, the tree could reach maturity while the client still owns the property, when the root spread will be 20 m x 0.75 = 15 m. Furthermore, whilst 3 m is a reasonable degree of safety, tree roots can become very active in drought conditions and could pose a risk even before the tree reaches maturity.

REPORT

Although this will be a matter of judgement depending on the precise details, in this case you should consider recommending further advice regarding regular management of the tree (including confirmation that this will result in regular expenditure) to help avoid:

- The possibility of damage during drought conditions at the current height; and
- An increased possibility of such damage when the tree reaches maturity and therefore an increased height.

Referrals for further investigation

Where trees pose a danger to buildings or their drainage systems, further investigations are often appropriate. Because of the interrelated nature of these problems, the following suitably qualified persons might be required:

- A structural engineer or building surveyor to deal with the building matters;
- An arboriculturist or tree surgeon to advise on tree management and/or removal;
- A drainage contractor to test and inspect the drainage system.

This level of investigation could be very expensive, so advice should be carefully given.

Liability to other building owners

The surveyor must also ensure they consider the effect trees on adjacent properties might have on the property they are inspecting and vice versa. If damage is caused by trees on neighbouring properties, a building's insurance company may seek to recover the costs of remedial work from the neighbouring owner's insurance company. This can cause a neighbour dispute. Although this is a complex area of law, when advising on trees, the surveyor must also look beyond the boundaries of the property under consideration. For example, are the trees on the public highway or neighbouring gardens far enough away when using the protocol mentioned previously? Are there trees in the property being surveyed too close to the neighbour's house? Such matters should be included in the report.

5.6 Other forms of building movement

The focus of this chapter has so far been on foundation movement, because such a defect can have very damaging consequences. This section will briefly outline some of the other causes of building movement, all of which have causes above ground.

5.6.1 Lateral instability

Section 5.4.3 stated that movements originating in the building superstructure rarely cause damage greater than category 2. Lack of lateral stability to the walls of a property is one of the exceptions to that rule. For new buildings, the Building Regulations require that external walls are securely tied back to the 'horizontal diaphragms' (floors and roof structure) described in Section 5.3.1 (OPSI 2013). This is to ensure that relatively thin walls do not buckle under wind and other loading conditions.

Older buildings were not always provided with this lateral restraint, especially to gable and flank walls. Where joists run from the front to the back of a property, structural integrity is further undermined by the presence of stairwells that represent structural discontinuity. In the worst cases, the wall can begin to 'bow' outwards to form a large 'belly' or bulge. If this goes beyond a certain limit, instability and partial collapse can result. The level of instability depends on:

- The strength of the bond between the bricks, stones, blocks and mortar;
- The level of restraint offered by the rest of the building structure;
- The 'slenderness ratio' of the wall, that is the relationship between the height of the wall and its thickness – the higher and thinner the wall is, the less load it can carry.

The balance between these factors will determine the amount of distortion.

Key inspection indicators

Bulging walls can be difficult to spot; some of the signs include:
Externally:

- All walls can be affected; but it is most common with brick gable and flank walls and any rubble stone wall, so these will be the prime suspects;
- Look *horizontally* along the wall from an external corner, ensuring both corners can be seen at the same time – any distortion should be visible and the amount can be roughly estimated with some practice;
- Consider *vertical* misalignment by placing a level on the wall surface at regular intervals and looking upwards. The longer the level, the better, but this can be misleading, as even a long level extends over only a small area of wall;
- Dropping a plumb line from either the top of a 3-metre ladder or from an adjacent window can give an effective measure of misalignment of the wall – in practice this is difficult in most UK weather;

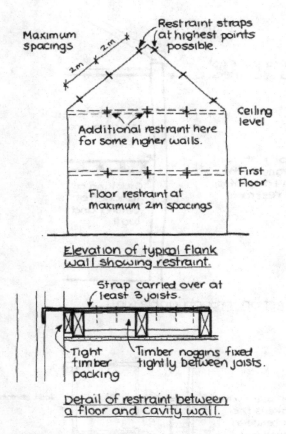

Figure 5.15 Restraint of walls as prescribed in the Building Regulations (2013) for new properties (Crown copyright. Reproduced with the permission of the Controller of Her Majesty's Stationery Office).

- The cracking pattern associated with bulging walls can be varied. There may even be a total absence of cracking, especially with flexible, weaker lime mortars. Try to imagine a large balloon being inflated in the building – the resultant cracking would be like that caused by bulging walls (see Figure 5.17);
- On rubble stone walls, substantial tapping of the wall at regular intervals can reveal a 'hollow' sound where blocks of stone may have become loose and are bulging outwards. This is a particular issue if there are different types of masonry in the same wall or if the stones are uncoursed, since the slenderness ratio of an uncoursed rubble wall is around 25% less than a brick or stone wall laid in regular courses (OPSI 2013, para. 2C7, p. 14);

- Internally:
  - Horizontal cracking at ceiling and external wall junctions;
  - Vertical cracks at internal and external wall junctions and close to corners;

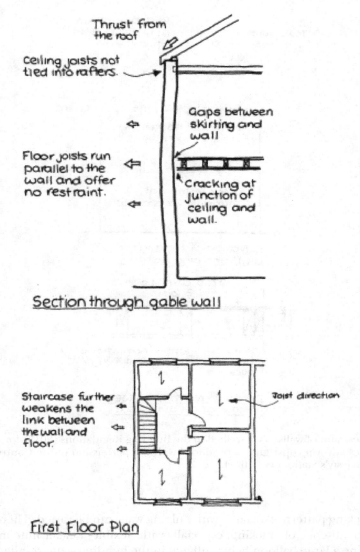

Thrust from
the roof

Ceiling joists not
tied into rafters.

Gaps between
skirting and
wall

Floor joists run
parallel to the
wall and offer
no restraint.

Cracking at
junction of
ceiling and
wall.

Section through gable wall

Staircase further
weakens the
link between
the wall and
floor.

Joist direction

First Floor Plan

Figure 5.16 A typical situation where lateral instability to a gable wall can occur.

- Movement around the stairwell (a weak point in that 'horizontal dia-phragm', the floor), such as a gap between outer wall and stair string (Hox-ley 2016, p. 62);
- Gaps between external walls and adjacent flooring, sometimes masked by timber cover beads (Marshall et al. 2014, p. 233).

The affected wall might also show evidence of previous remedial repair attempts. Restraint plates and associated tie bars secured to floor and roof timbers might be evident (see Figure 5.18). These may indicate an attempt to resolve the defect, but

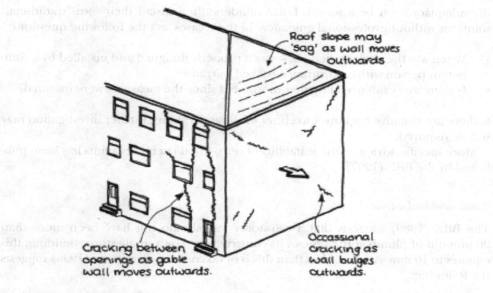

Roof slope may 'sag' as wall moves outwards

Cracking between openings as gable wall moves outwards.

Occassional cracking as wall bulges outwards.

Figure 5.17 Typical cracking patterns associated with lateral instability to gable and flank walls. Now redrawn

Figure 5.18 External restraint plates to these terrace properties clearly indicate the possibility of previous lateral instability.

the adequacy must be assessed. Local builders often install their own 'traditional' solutions without professional guidance. In these cases, ask the following questions:

- When was the work carried out – was it properly designed and installed by a competent person with an insurance-backed warranty?
- Is there any evidence of further movement since the measures were installed?

If there are negative responses to either of these questions, further investigation may still be required.

More specific advice on the suitability of retro-fitted lateral restraints has been published by the BRE (1997).

Assessment and advice

The BRE (1989) suggests that a two-storey wall should not have been more than 20 mm out of plumb at the time of its construction. For a single-storey building, this equates to 10 mm. Anything less than this is of no concern. Parkinson (1996) suggests the following:

- Where the wall is less than one sixth of its width out of plumb (35 mm for a one brick thick wall), no remedial work may be required;
- Where the wall is between one sixth and one half out of plumb (35–100 mm for a one brick wall) then remedial tying back to the main structure will be required;
- Where the wall is more than half of its width out of plumb then stability is threatened and rebuilding is likely to be necessary.

Because the latter two categories involve specialist knowledge, referral to a building surveyor or a structural engineer may be appropriate.

The *'middle-third rule'* is a helpful rule of thumb based on basic engineering principles that surveyors use to assess walls. The rule is based on the relatively and intuitively obvious fact that if a wall is straight and true (in other words, vertical), it is performing satisfactorily. Whereas if the wall is leaning or bulging, it is suffering from some form of failure. The 'middle-third' rule can be used to quickly assess whether the distortion is so significant that remedial work or monitoring may be necessary.

The rule states that if a line (the load path) projected vertically upwards from the mid-point of the bottom of the wall falls within the 'middle-third' of the wall throughout the height, the wall has not distorted or moved to a significant extent. However, if the vertical line falls outside the middle-third at any point, the wall is more likely to require repairs because it has leaned and/or bulged to a significant extent. Therefore, if the wall distorts by more than one-sixth of its width, it is more likely to continue to move and is so likely to require provision of restraint, such as tie bars and pattress plates. Figure 5.19 illustrates the principle.

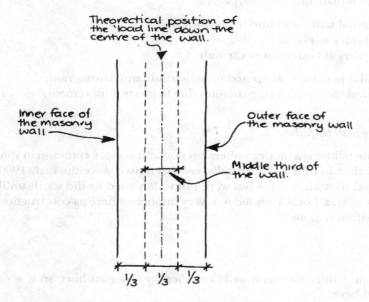

Diagram 5.19 Using the 'middle-third' rule for leaning and bulging walls.

As with most surveying problems, each case must be decided on the particular facts but in a very general sense:

- If the wall has distorted to the extent that the load line is close to the outer boundary of the middle third, it is likely to require repairs and provision of restraint soon or urgently depending on the extent; whereas
- If the load path crosses the outer boundary of the middle third then stability will be threatened as it has gone past the point of no return and rebuilding may be the only option.

Like all approximate 'rules of thumb', this middle third rule is only a guide and the assessment and action will depend on the particular circumstances. For example, walls exposed to high winds and rain will be more vulnerable than those that are sheltered.

Reasons why a wall is more likely to continue lean or bulge include:

- There is no 'lateral restraint' to the wall, such as joists built into the wall (the friction between the bedding mortar and the timber joists helps prevent the wall moving), or restraint straps or tie bars;
- The wall is built of randomly-laid or un-coursed stone or rubble materials, which are more likely to slump and crumble;
- Any part or all of the wall is not properly bonded together – a typical example is where an old wall suffering deterioration such as spalling has been 're-faced' in the past with another half-brick skin that was not mechanically fixed back to the original wall, but instead was simply 'attached' with mortar;

- The wall is inadequately buttressed by:

    - Internal walls built into the outside wall,
    - Chimney stacks, or
    - Corners at both ends of the wall;

- The wall is particularly exposed to high winds and driving rain;
- The foundation is suffering rotation due to ground movement.

5.6.2 Cavity wall tie failure

Cavity wall tie failure is a modern phenomenon and is more common in some parts of the country than in others. Cavity walls have been used since the early 1900s in many northern and western regions but were not widely used in the south until after the Second World War. Local knowledge is very important here as construction methods vary, even within regions.

The cause

Cavity wall tie failure has been well described by other authors, so it is only briefly summarised here:

- Older, thick wrought and cast-iron ties were originally used to tie the leaves of the walls together;
- These were often inadequately protected; many were merely dipped in liquid bitumen before being built into the mortar course;
- In exposed areas long periods of dampness have led to corrosion of the wall tie. Because rusting iron can expand by up to seven times its original size, these ties can literally 'lift' the outer skin of the wall, resulting in a horizontal 'gap' or 'crack';
- In some regions this effect was intensified an acidic environment that accelerated the corrosion process.

Recognition

This problem affects cavity walls. This sounds obvious, but many surveyors have attributed cracking in a solid wall to wall tie failure. It is often unmistakable. Straight horizontal cracking at 450 mm intervals (or every four or five courses) is obvious, especially on rendered cavity walls. Typical signs and symptoms are illustrated in Figure 5.21 and described as follows:

- The horizontal cracking often increases in frequency in the higher areas of the wall. This is because there is less self-weight of the wall at upper storey levels, allowing the pressure of the expanding wall ties to overcome the weight of the wall. A surveyor should ensure to open and close windows to rooms at a high level – this is a good opportunity to look outside along the bed joints for evidence of cracking;
- Look out for recently repointed bed joints. Building owners will often repoint cracks to prevent wind-blown rain from entering their homes;

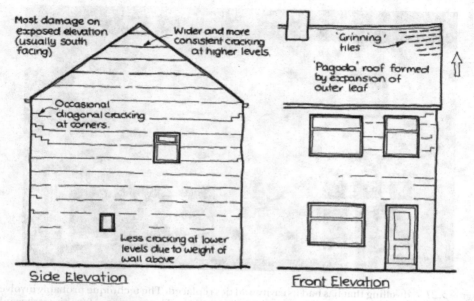

Figure 5.20 Typical signs of cavity wall tie failure.

- Cavity wall tie failure usually occurs on the most exposed elevations (west and south-west facing);
- The cumulative effects can result in the expanding outer skin of the wall lifting the edge of the roof structure. This produces a 'pagoda effect' as the verge tiles rise upwards;
- Where there is a room in the roof, the pressure of the expanding wall can be overcome by the substantial weight of the roof structure and the uppermost floor. This prevents upward movement and forces the external skin to bulge outwards because it has nowhere else to go. This can often be mistaken for lateral instability or bulging caused by sulphate attack;
- Small diagonal cracks can occur around window openings or at the corners of the building where the expansion can be partially restrained;
- The cracking only affects the outer skin. It is rare that it extends to the inner skin;
- Look at similar properties, do they show signs of wall tie failure or evidence of repair?

In the worst cases, the corrosion can be so bad that the ties break or snap. This will leave the outer skin unstable and vulnerable to possible collapse, especially on windy days. Some properties will show evidence of repair. There are two main methods:

- A complete brick is taken out and replaced so that the old wall tie can be completely removed and a new one inserted (Figure 5.21);
- The old wall tie is isolated from the wall by grinding away the portion built into the external skin. A new wall tie is installed through a hole drilled immediately above or below.

Figure 5.21 A dwelling that has had its cavity wall ties replaced. The technique probably involved
the complete removal of the defective wall ties. The guarantee for this work would
have to be checked and should be backed by an appropriate insurance scheme.

Where this has been done, a check should be made to determine whether all elevations have been included and whether the work has been carried out by a reputable contractor with an insurance-backed guarantee.

Assessment and advice

The risk of cavity wall tie failure must be assessed in a balanced way. Here are a few key inspection indicators:

- Pre-war (1939) cavity walls are probably the most at risk;
- The standard of galvanizing (which helps prevent corrosion) to wall ties was significantly increased in 1981, so the risk of corrosion in properties built after that date is reduced;
- In 2004 the Building Regulations required all new wall ties to be of stainless steel, thereby removing the risk of corrosion in most instances;
- In regions 'where it is known aggressive mortars were used in construction, the surveyor must take special care;
- Surveyors should be particularly careful if the property is close to the coast, since sea salts blown onto the outer face of walls can significantly increase the risk of corrosion;
- As time passes, the risk of corrosion of old metal wall ties inevitably increases; but performance is the best indicator. If there is no evidence of cracking, partial repair or repointing on the most exposed elevations, then there is not likely to be a current problem.

If symptoms of failure are noticed, further investigations are required. Although there are specialist contractors and installers in most regions of the country, the surveyor must ensure that a reputable firm is recommended.

5.6.3. Sulphate attack

The cause

This is the term that is applied to the chemical reaction between sulphates and the various constituents of cement. It can lead to the eventual breakdown of the mortar joint. Three ingredients are necessary for sulphate attack to begin:

- Sulphates – these are usually present in the bricks themselves or can be introduced by combustion gases in a brick chimney;
- Tricalcium aluminates (a normal part of cements);
- Water.

In most cases, the sulphates and cements are always present; therefore, the controlling factor is water.

Recognition

Because of the influence of water, sulphate attack can be expected in the following locations:

- Parapet walls, since both sides are potentially exposed to driving rain;
- Exposed brick elevations facing the prevailing wind for the area (usually, but not always, the south-west facing);
- Chimneys and other flues.

Sulphate attack can also affect cellar walls and other retaining structures. Key inspection indicators are:

- Expansion of the mortar, especially the bed joints can cause considerable vertical expansion, giving rise to symptoms of 'pagoda roofs' and/or bulging brickwork similar to cavity wall tie failure. The distinguishing feature is that sulphate attack affects every bed joint, while cavity wall tie failure affects every fourth, fifth or sixth bed joint only.
- Close inspection of the mortar joints reveals:
    - Fine cracks along the length of every mortar joint. This is especially noticeable on rendered walls where the horizontal cracking becomes obvious;
    - Whitish appearance to the mortar joints, especially where they meet the brickwork;
    - In very wet conditions, the mortar may be reduced to a soft powder or a 'mush';
- The edges of the bricks may spall.
- Chimneys and parapet walls may lean away from the direction of the prevailing wind. This is because the damper, more exposed side allows greater rates of expansion than the drier side (see Figure 5.22).

Mortar joints on 'wet' side expand causing the stack to lean over.

Prevailing weather ⇨

Cracks wider this side of the stack.

Figure 5.22 Typical damage to a chimney caused by sulphate attack.

- Because the expansion occurs in all the mortar, vertical mortar joints will be affected too. This can cause the wall to increase in size horizontally, causing the wall to slip on its damp-proof course. Oversailing of the DPC is often noticeable at the corners of the building.

Advice

The type of advice will vary depending on the severity of the attack. One of the key factors will be the amount of disruption to the brickwork.

5.6.4 Expansion of masonry

All building materials will change size after they have been built into a structure. This is due to the influence of thermal or moisture changes. For example, concrete and calcium silicate bricks are formed by a wet process and will shrink after manufacture. Clay bricks are fired in a kiln and so will tend to swell and expand. If this movement is not accommodated in the design, the stresses that build up in the walls can result in cracking. The level of damage is generally at the lower end of the spectrum and is rarely more than an aesthetic issue. This cracking is sometimes confused with other more important causes of movement, so a brief review of some of the main symptoms will be useful:

- Cracking due to expansion is often vertical, consistent in width and can be distinguished from foundation movement, which tends to be wider at one end than the other;

- The cracks are often a straight line (although they can be diagonal) and will often crack the masonry unit as well as the mortar joint;
- Where the wall is expanding, the whole building can 'slip' on the physical DPC. This is because the wall below is restrained by the ground, while the structure above can move more freely;
- Because of this slippage effect, cracking due to expansion rarely extends below the DPC (see Figures 5.23 to 5.24).

Ideally, buildings must be designed and constructed to allow for any expected change in dimensions. Movement joints, which are usually finished externally with a non-hardening sealant, should be present in most walls at centres broadly in accordance with the following rules:

- Clay bricks – not more than every 12 m;
- Calcium silicate bricks – 7.5–9 m at least. Special provision needs to be made to restrain these bricks at openings, returns, corners, etc., as well;
- Concrete masonry – about every 6 m.

It is important to note the type of brick used in a dwelling because this can help in the diagnosis of any cracking.

Figure 5.23 Cracking to a masonry elevation caused by expansion of brickwork. These semi-detached properties are built of calcium silicate bricks that are vulnerable to expansion defects.

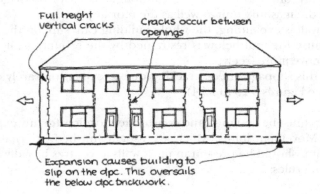

Figure 5.24 Sketch showing typical damage caused by expansion of masonry walls.

5.6.5 Roof spread

Chapter 8 looks at the problems of pitched roofs in more depth. If rafter feet of a traditionally built roof are inadequately restrained, then the rafters can push outwards. This can move the wallplate and the top courses of the wall. Visually, this will cause the top of the wall to lean over and could possibly result in horizontal cracking at a high level (see Figure 8.2).

5.6.6 Embedded timbers

In traditional construction, it was common to build large timbers into the thickness of external walls to provide a form of stiffness to the construction. They were often used as fixing grounds for internal timbers such as skirtings, stair strings, door linings and timber boarding. In some walls, these timbers can be as close as 100 mm to the external surface of the wall. Other examples include gauged brick arches, 'only a half-brick in thickness with a timber lintel' behind the arch, susceptible to water penetration (Lynch 2007, p. 213). In exposed locations, dry and wet rot can result, causing the timber to shrink and distort. The weight of the walling above will cause the timber to compress. This will induce stresses on the external face, where similar movements are prevented. Header bricks and through-stones are often broken, causing the external face to 'bow' outwards.

Visually, this defect will manifest itself as a bowing or bulging in the brickwork and will be difficult to distinguish from that caused by lateral instability. In some cases, the internal face of the wall can remain vertical while the external face bows outwards (see Figure 5.25). The same assessment criteria should be applied to this defect as with lateral instability.

5.6.7 Stone walls

Stone was once a popular building material, but because of its high cost, it became limited to the best quality work. Stone external walls can be more common in some parts of the country, depending on the availability of the material from local quarries. Even where it was more available, it was often used only on front elevations to save

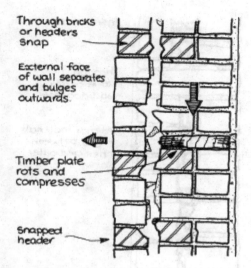

Through bricks
or headers
snap

External face
of wall separates
and bulges
outwards.

Timber plate
rots and
compresses

Snapped
header

Figure 5.25 Damage caused to a solid wall by the failure of an embedded timber.

money. Although many structural principles are the same as for brickwork, there are a few essential differences that require emphasis.

Types of stone walling

Before the typical defects of stone walling are discussed, the different types of construction will be quickly reviewed.

- Types of building stone – these include:
  - Igneous rocks – 'crystallised from molten material underground or as lava flows', the main type is granite (Evans 2004, p. 2);
  - Sedimentary rocks – originally 'mainly deposited as sediment, usually underwater', these include the sandstones and limestones that form the bulk of stone used in construction (Evans 2004, p. 22);
  - Metamorphic rocks – *'formed by the action of . . . heat, pressure or both'*, examples are slates and marbles (Evans 2004, p. 37).

- Classification of walling – stonework is classified by the way the walls are built. These consist of two generic types with variations as described as follows.

RUBBLE WORK

These are stones that have been quarry dressed (cut and shaped) and can range from the cheapest, roughest stonework used on boundary walls to better quality work. The approach is like building brick walls, but because the stones are generally irregular, great skill is needed to ensure the wall is properly constructed. As can be seen in the diagram in Figure 5.26, this type of stone wall is built with the best 'face' of each stone

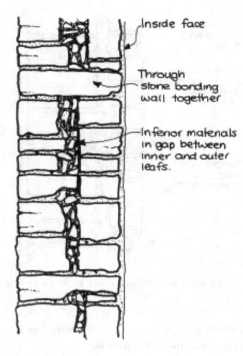

Figure 5.26 Section through typical stone wall.

aligned towards the outside or the inside. Depending on the size of each stone, this can potentially result in voids inside the wall. Stability is usually ensured if enough 'bonders' (stones that reach beyond the middle of the wall) and 'throughs' (stones that extend the full thickness of the wall) are used (see Figure 5.26). In a sense, the bonders act rather like wall ties in a cavity wall, ensuring the outer face is tied back to the inner face. The gap between the internal and external faces is usually filled with inferior materials (for example, rubble, chippings, old mortar and anything else to hand). The category includes:

- **Random rubble** – Including 'uncoursed' and 'built to courses';
- **Squared rubble** – Which has several different types, 'uncoursed', 'built to courses' and 'regular coursed';
- **Miscellaneous** – A variety of different types that are often unique to specific regions including 'polygonal walling', 'flint walling' and 'Lake District masonry'.

ASHLAR

This class includes stones that have been accurately dressed so fine bed and end joints can be formed. The face appearance of ashlar can vary. Sometimes it can resemble Flemish bond in brick walls, arranged in courses that alternate between thick and thin, sometimes set in courses that diminish in thickness from the base upwards. Most ashlar walls are termed 'compound walls' because they can be backed with brickwork or rubble construction to reduce the costs of the construction.

Defects in stonework

Because stone is a natural material, its physical properties may vary even between blocks cut from the same quarry face. Despite this, there are several agents of deterioration that are common to all different types of stone. These include:

- **Incorrect bedding** – All sedimentary rocks were laid down in geological beds and if the stones are not laid with their bedding planes running horizontally, they could be vulnerable to damage. This is especially true where stones have been 'face bedded' or edged bedded. Andrews et al. (1994) suggest an analogy with the pages of a book. If the book is laid flat, it can take a lot of weight without any adverse effects. If the book is placed upright and pressure is applied from above, the pages bend outwards and the book soon collapses. The same principle applies to stone. Water can travel more easily down the vertical sections and salt and frost action can cause the outermost layers to spall away.
- **Salt crystallisation** – Like all other porous building materials, salts in solution can soak into the stone. These salts may come from the atmosphere, other building materials, the stone itself or the subsoil. When the wall dries out, the salts will crystallise and the resulting internal forces can damage the stone.
- **Effects of acid rain** – Rainwater in many urban areas can be slightly acidic because of dissolved carbon and sulphur dioxides. This can have a severe effect on some walls, especially limestones.
- **Expansion of embedded metals** – Metal cramps have been used in ashlar stonework for centuries. When these corrode, they expand and cause the stone to crack and fall away. A common example of this can be seen when park railings expand in their supporting brick or stonework.
- **Frost attack** – Like in brickwork, when saturated stonework freezes the ice crystals can impose significant pressure on the stone, forcing it to spall away. Stones with smaller pore structures tend to be more vulnerable to this effect.
- **Structural instability**:
  - Cracking – stone walls are affected by structural movement in similar ways to equivalent brick walls. One of the main differences is that because the individual stones are usually stronger than their brick equivalents and laid in a more haphazard way, stone walls will tend to crack differently. Cracks will follow mortar joints much more readily and because of the random coursing, the nature of the cracking will be much harder to analyse;
  - Bowing and bulging – stone walls can be affected by lateral instability in the same way as brick walls (see Section 5.7.1). Because of the way stone walls are constructed, deficiencies in the way the internal and external faces are bonded together can lead to a separation of the wall. A typical example is shown in Figure 5.26.

5.6.8 Stability of bays and enclosed porches

Bays and porches are often built on shallower and/or narrower foundations than the rest of the building, using very differential constructional methods. They may have single-skin walls that are poorly rendered. Some may be framed with timber or possibly be of composite construction. Older stone bays may be formed of what are in effect large stone blocks that are highly embellished and very expensive to replace.

Figure 5.27 Face bedded masonry wall. The outer layers of the stone can easily be 'peeled' away by the action of salts and frost. In this photograph, the stones that have deteriorated the most are face bedded.

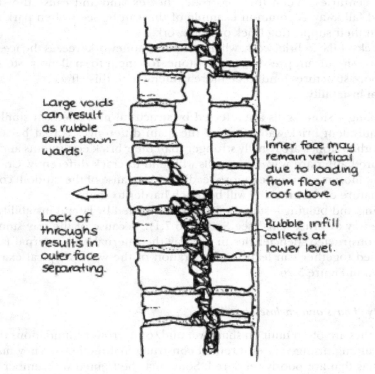

Large voids can result as rubble settles downwards.

Lack of throughs results in outer face separating.

Inner face may remain vertical due to loading from floor or roof above.

Rubble infill collects at lower level.

Figure 5.28 Separation of stone leaves can lead to the bulging of the external face of the wall, leaving the internal side vertical.

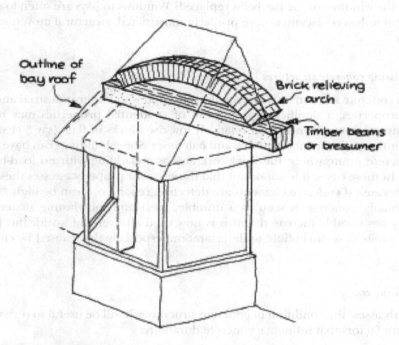

Figure 5.29 Diagram of timber 'bressumer' beams over a typical bay window opening.

The intermediate floor of two-storey bays can be supported by cantilevers of the floor of the main house. Whatever the method, there may be poor structural connection between the bay and the main property.

These deficiencies will often result in differential movement, cracking and further deterioration due to penetrating dampness. One of the key problems is where there is a timber bressumer or beam that spans the bay opening (see Figure 5.29). Water penetration can often lead to timber decay and structural failure.

Key inspection indicators

- Does the bay/porch appear to be constructed to the same standard as the rest of the property? Intermediate columns and piers should be checked for slenderness especially.
- Are there signs of cracking or pulling away at the junction of the bay and main property? This is often caused:
  - Due to the fact the main property, with two or three storeys, exerts a significantly greater load on the soil than the bay and so initially settled downwards to a greater extent; and/or
  - Because the foundation to the bay is usually shallower, it is more prone to seasonal movement, especially on a shrinkable soil.
- Are there signs of water penetration into the bay? Staining to the ceiling is a typical sign.

- Have the windows of the bay been replaced? Windows to bays are often loadbearing and unless replacements are properly constructed, structural movement may occur.

5.6.9 Assessing concrete structures

Although concrete structures are normally associated with larger industrial and commercial properties, a significant proportion of residential properties may include major concrete components. Even relatively low-rise blocks of flats (say 3–4 storeys) can have intermediate concrete floors and balconies. Some buildings can have a complete concrete frame where the brickwork acts as a cladding with no load-bearing function. In these cases, it is important that the surveyor properly assesses these components because if concrete elements are defective, repair costs can be high.

Traditionally, concrete is seen as a durable, inert and long-lasting material. Its popularity has steadily increased and it is now used all over the world. But like all other materials, it is susceptible to deterioration, especially that caused by chemical changes.

Nature of concrete

To properly assess the condition of concrete structures, it will be useful to review a few of the main factors that influence concrete durability:

- The permeability of concrete will affect its durability, low permeability being more durable than one that is highly porous;
- Human factors can affect concrete such as poor design and construction practices, poor quality materials, etc;
- Most concrete is reinforced with steel. This steel is protected from corrosion in two ways (see Figure 5.30):

  - Concrete cover provides a mechanical barrier preventing water and oxygen affecting the steel;
  - Chemical reaction between the concrete and steel creates a highly alkaline environment around the reinforcement, protecting it from corrosion (called passivation).

If there are deficiencies with any of these aspects, then the long-term durability of the concrete could be in doubt.

Corrosion of reinforcement

One of the key factors involved in the breakdown of concrete is the corrosion of the reinforcement. Because steel expands during the corrosion process, it can cause the concrete cover to crack, spall or even delaminate (see Figure 5.31). Corrosion occurs when moisture and oxygen penetrate through the concrete cover to the depth of the reinforcement. Several processes may help to fuel this:

- **Carbonation** – This is where carbon dioxide in the air acts on cement products. When moisture is present, calcium carbonate is formed, allowing a volume

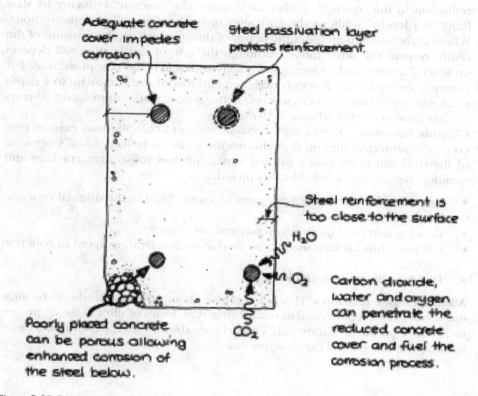

Adequate concrete cover impedes corrosion

Steel passivation layer protects reinforcement.

Steel reinforcement is too close to the surface

H_2O

O_2

CO_2

Carbon dioxide, water and oxygen can penetrate the reduced concrete cover and fuel the corrosion process.

Poorly placed concrete can be porous allowing enhanced corrosion of the steel below.

Figure 5.30 Diagram showing the main influences on the durability of concrete.

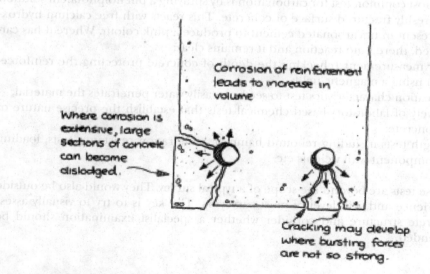

Corrosion of reinforcement leads to increase in volume

Where corrosion is extensive, large sections of concrete can become dislodged.

Cracking may develop where bursting forces are not so strong.

Figure 5.31 The effect of corrosion of reinforcement in concrete sections.

reduction in the concrete. It also 'carbonates' the concrete, reducing its alkalinity to a level that allows the corrosion of steel (often called depassivation). Where carbonation penetrates the depth of the concrete cover, corrosion of the reinforcement will often follow. Although the rate of carbonation will depend on several interrelated factors, good-quality concrete is the best protection. For example, concrete that is poorly designed and placed can carbonate to a depth of 25 mm in less than 10 years, whereas better quality materials may take 50 years to carbonate to a depth of only 5–10 mm.

- **Chloride intrusion** – Where high concentrations of chloride ions exist in concrete, the protective film on the reinforcement can be broken down. Corrosion of the steel can often result, causing expansion and subsequent cracking and spalling. Typical sources of chlorides include:

  - Contamination from the aggregates and water used in the original concrete mix;
  - De-icing salts used on nearby roads and pavements;
  - Calcium chloride that used to be used as an accelerating agent in concrete mixes;
  - Highly polluted environments.

- **Alkali-aggregate reaction** – This is a complex chemical reaction where the alkaline solutions present in cement react with certain forms of silica in the aggregate to produce an alkali silicate gel. The gel can absorb water, expand and cause spalling and cracking of the concrete cover.

TESTING

Where concrete has begun to show signs of deterioration, specialist tests are usually carried out. These can include the following:

- The most common test for carbonation is by spraying a phenolphthalein solution on a freshly fractured surface of concrete. This reacts with free calcium hydroxide present in uncarbonated cement to produce a pink colour. Where it has carbonated, there is no reaction and it remains clear;
- Cover measurement (checking the depth of concrete protecting the reinforcement) using a magnetic field;
- Permeation characteristics test to see how easily water penetrates the material;
- A variety of laboratory-based chemical tests that establish the precise nature of the concrete;
- Strength tests including rebound hammer, penetration resistance tests, loading the component with weights etc.

All of these tests are beyond the scope of a typical survey. They would also be outside the experience and skill level of most surveyors. The key is to try to visually assess the concrete structure and consider whether a specialist examination should be recommended.

REPAIR

The repair of defective concrete structures is a specialist operation that can range from the making good of isolated spalled areas to the reforming of large sections of defective concrete. It can include epoxy resins, polymer latex and polyester compounds. Where carbonation is involved, the whole surface of the concrete may have to be coated with a protective and decorative covering. Depending on the extent of the work, it can be very expensive and have an impact on value. Should a specialist investigation be required, it is essential that the client be referred to a professional person who can investigate the condition of the concrete and specify and organise an appropriate repair scheme.

References

Andrews, C., Young, M. and Tonge, K. (1994). *Stone Cleaning. A Guide for Practitioners.* Historic Scotland, Edinburgh.

Barrett, M. (1981). *Usborne Guide to Trees of Britain and Europe.* Usborne Publishing, London.

BRE (1985). 'The influence of trees on house foundations in clay soils'. *Digest 298.* Building Research Establishment, Garston, Watford.

BRE (1989). 'Simple measurement and monitoring of movement in low rise building, Part 2: Settlement, heave and out-of-plumb'. *Digest 344.* Building Research Establishment, Garston, Watford.

BRE (1991). 'Foundation movement and remedial underpinning in low rise buildings'. *BRE Report BR 184.* Building Research Establishment, Garston, Watford.

BRE (1995). 'Assessment of damage in low rise buildings'. *Digest 251.* Building Research Establishment, Garston, Watford.

BRE (1997). 'Connecting walls and floors: Design and performance'. *Good Building Guide, 29 Part 2.* Building Research Establishment, Garston, Watford.

BRE (2003). 'Tilt of low-rise buildings – with particular reference to progressive foundation movement'. *Digest 475.* Building Research Establishment, Garston, Watford.

Carillion (2001). *Defects in Buildings, Symptoms, Investigation, Diagnosis and Cure,* 3rd edition. The Stationery |Office, PO Box 29, Norwich NR23 1GN.

Cheney, John E. (1988). '25 years' heave of a building constructed on clay, after tree removal'. Available at https://cdn.ca.emap.com/wp-content/uploads/sites/13/1988/07/1988-07_Pages_13-27.pdf.

Dickinson, Peter R. and Thornton, N. (2004). *Cracking and Building Movement.* RICS Business Services Ltd, The Royal Institution of Chartered Surveyors, Surveyor Court, Westwood Business Park, Coventry CV4 8JE.

Driscoll, Richard and Skinner, Hilary (2007). *Subsidence Damage to Domestic Buildings, a Guide to Good Technical Practice.* I BRE Press, Garston, Watford, WD25 9XX.

Evans, D. (1998). 'The root of all evil?' *The Valuer,* pp. 18–19. November/December.

Evans, Gareth W. (2004). *Geology, a Practical Introduction for Surveyors, Estates Gazette, 151.* Wardour Street, London W1F 8BN.

Hoxley, Mike (2016). *Building Condition Surveys.* RIBA Publications, The Old Post Office, St Nicholas Street, Newcastle upon Tyne NE1 1RH.

ISE (1994). *Subsidence of Low Rise Buildings.* Institution of Structural Engineers, London.

Lynch, Gerard (2007). *The History of Gauged Brickwork, Conservation, Repair and Modern Application, Butterworth-Heinemann, an Imprint of Elsevier.* Linacre House, Jordan Hill, Oxford OX2 8DP.

Marshall, Duncan, Worthing, Derek, Heath, Roger and Dann, Nigel (2014). *Understanding Housing Defects*, 4th edition. Routledge, 2 Park Square, Milton Park, Abingdon, Oxon OX14 4RN.

National House Building Council (NHBC) (2020). *NHBC Standards 2020, NHBC, NHBC House, Davy Avenue*. Knowlhill, Milton Keynes, Bucks MK5 8FP.

Office of Public Sector Information (OPSI) (2013). *Building Regulations 2010*. Approved document 'A' Structure 2013. OPSI, Richmond, Surrey.

Parkinson, G., Shaw, G., Beck, J.K. and Knowles, D. (1996). *Appraisal and Repair of Masonry*. Thomas Telford, London.

6 Moisture problems in buildings

Contents

DOI: 10.1201/9781003253105-6

6.1 Introduction

6.1.1 The moisture problem

The UK has a temperate maritime climate. This climate type is characterised by the absence of extreme conditions, with mild winter temperatures and warm summers. The influence of the sea and the weather systems that follow the Atlantic's Gulf Stream make sure that moisture is the dominant element of our weather although climate change may modify these patterns. The western side of the British Isles gets significantly more rain than the east. Scotland and Wales receive more rain than England. As a consequence, the exclusion of moisture is one of the biggest challenges for any building owner.

This is reflected in housing conditions. According to the English Housing Survey in 2017, 897,000 homes in England (4% of the total) had moisture-related defects (MHCLG 2019). Although this was down from 2.6 million homes in 1996 (13% of the total), there has been little change since 2011. North of the border, the Scottish House Condition Survey of 2017 revealed that 9% of Scottish housing was affected by some form of moisture, while in Wales, the Welsh Housing Condition survey found 7% of the Welsh nation's housing was affected by moisture (MHCLG 2019).

6.1.2 The effects of moisture

Due to the UK's climatic conditions, moisture is ubiquitous – it is simply everywhere. During the summer, moisture levels will generally reduce, while in the winter they will increase. Where moisture levels become excessive in buildings, they can have two main effects:

Moisture can damage the health of the occupants. It does this through two main mechanisms (Ormandy & Burridge 1988, p. 272):

Mould growth. Mould is a generic term for a range of fungi that spread by putting out spores. These spores are microscopic particles that are classed as 'allergens' and can cause an allergic reaction in the respiratory tracts of a significant proportion of the population. For example, according to the leading charity Asthma UK, 5.4 million people in the UK (around 1 in 12) are receiving treatment for asthma (Asthma UK 2020).

House dust mites. These thrive in moist environments. These microscopic insects are close relatives of ticks and spiders and live on the dead skin cells that all humans shed. They can be found in carpets, soft toys and soft furnishings and are common in mattresses and bedding. The more moisture in the dwelling, the larger the population of house dust mites. According to the Allergy UK factsheet on house dust mites, the mite itself is not a problem, but proteins in their droppings are and are classed as 'allergens'. Each mite produces about 20 of these waste droppings every day and these continue to cause allergic symptoms even after the mite has died (Allergy UK 2021).

Moisture can damage the building fabric. For example, it can:

- Spoil the decorations;
- Break down plaster finishes;
- Corrode embedded metal components; and
- Cause wet and dry rot in affected timber elements (see Chapter 7).

The financial loss resulting from moisture-related defects can be considerable.

6.1.3 A question of definition

Since the first edition of this book was published, a fierce debate has engulfed the diagnosis and repair of moisture-related problems in buildings. Much of this has been centred on the phenomenon usually called 'rising dampness'. We do not have the space in this publication to do justice to this debate, so if you want more information on the many different perspectives, then you should seek out the discussions on the large number of websites and blogs associated with the issue of 'rising dampness' in buildings.

However, there is one change we do want to make in the second edition and this is the terminology used to describe moisture-related problems in buildings. In the first edition, this chapter was titled 'dampness' and there were sub-sections called 'penetrating dampness', 'rising dampness' and 'traumatic dampness'. Although these types of defects are still as relevant today, it is the term 'dampness' or 'damp' that we consider has been misused, resulting in unnecessary confusion. For example, Howell (2008, p. 49) investigated the origin of the term 'rising damp' and argued it did not appear in technical publications until the late 1950s and only came into common usage in the 1970s.

Consequently, our reason for changing the terminology in this edition is that the term 'dampness' has become one of those casually used, simplistic words that suggests there is a defect that needs to be resolved. By stating that an element of a building is 'damp' (and so the implication is that it is unsatisfactory), there is an implicit assumption that the component has to be 'dry' to be in a satisfactory condition. Simply describing part of a building as being 'damp' provides no useful information to the client.

As described at the start of this chapter, moisture is the dominant feature of our climate, so all parts of a building will contain some moisture to a greater or lesser extent depending on the circumstances. Therefore, the issue is not whether a component is 'damp' or 'dry' but whether it is affected by 'excessive moisture' that could damage the health of the occupants and/or affect the fabric of the building. Identifying 'excessive moisture' in a building is a complex judgement, but it is a skill we want to help you develop. As the first step in this process, we have replaced the words 'damp' and 'dampness' with different phrases involving the term 'moisture'. It is our view that this will help build a more objective approach to a complex problem.

6.2 Measuring moisture – different methods

Any building owner that has suffered from a moisture-related defect will have witnessed a succession of different types of surveyors and contractors who prod and poke the affected parts of their property with a variety of battery operated 'moisture meters'. This is by far the most common method of detecting moisture but is not the only approach. Even when moisture meters are used, the results have to be interpreted with great care. The principal methods of identifying the signs of moisture-related problems are described in the following sections.

6.2.1 *The senses*

This includes:

- **Sight** – This sense is very underrated. Most forms of moisture-related defects have distinct visual characteristics that can help diagnose the cause;
- **Smell** – Although this is a subjective measure, a 'fusty' smell can alert a sensitive surveying nose to a moisture problem in a property;
- **Touch** – This sense can be very misleading. Touching a cold or cool surface can give the impression of moisture when the material might actually be dry, so this is a very unreliable method.

6.2.2 *'Moisture meters'*

There are many different types of handheld instruments on the market that claim to measure moisture levels in building materials. For convenience, we will call these by the generic name of 'moisture meters'. At the time of writing, an internet search listed seven different manufacturers with prices ranging from just over £10 to well over £300. In terms of the residential survey market, one manufacturer continues to dominate the market.

Essentially, there are two types of moisture meter (often combined in the same instrument):

Conductivity meter – This type of meter has two probes or electrodes (usually pins) that are pressed against the surface to be tested. It uses electrical conductance principles to measure the current flowing between the two pins or 'electrodes'. In theory, moisture can increase the current flow. Therefore, greater levels of moisture present will result in higher readings. Older versions had a needle gauge, but newer models usually incorporate digital readouts, various coloured lights and even audible buzzers.

Capacitance meters – These are usually contained behind the flat face of the meter that is pressed against the material being measured. The manufacturers claim that they detect moisture at or near the surface by utilising a capacitive-coupled sensor. This measures the electrical capacitance between two plates, one charged positively and the other negatively. Different materials will have varying effects on the electrical capacitance. For example, air has a very small effect, whereas water is about 80 times more effective than air. Dry brick, on the other hand, is only about four or five times more effective than air. For practical reasons, these two plates are located behind the flat plate and this provides a very small spread of the electrical field. This is sometimes called the 'edge' or 'fringe-field' effect. Oxley and Gobert (1994) state the fringe field effect falls off very rapidly the further it gets from the flat plate sensor, although some manufacturers claim the readings of this type of moisture meter can be effective up to 13 mm deep.

Most modern moisture meters include both capacitance and conductivity meters within the same unit.

A number of manufacturers also produce meters that measure air temperatures and humidity. These aim to assess internal environmental conditions and are beyond the scope of this book.

What a moisture meter actually measures

We try to avoid getting too technical in this introductory book, but in this case, we think it is important to understand what is meant by 'moisture content'. If you understand some of the science involved in the process, then you will be more able to diagnose moisture-related problems.

Timber is a 'hygroscopic' construction material and this means it will absorb moisture from the air. The more moisture in the air, the 'wetter' the timber will be (see also the section on relative humidity in Figure 6.46). Timber can also absorb moisture from adjacent materials, where the two are in contact.

The moisture (water) content in the wood is defined as the weight of the water in the moist material divided by the weight of the same material in a dry state. To obtain a percentage, the result is multiplied by 100.

In other words, the piece of the timber to be measured is first weighed (wet weight). It is then placed in an oven or kiln and heated to just over 100 degrees centigrade for a certain amount of time until it is completely dry and all the moisture has been removed. The piece of timber is weighed again (dry weight). These values are incorporated into the following formula and the result gives the moisture content of the timber in percentage terms in those particular conditions.

% moisture content = ((wet weight – dry weight) ÷ dry weight) x 100

Knowing the moisture content of timber in a building is important because the material can become vulnerable to wood rotting fungi and wood boring insects at moisture levels above 20–24%. Therefore, knowing when the actual moisture content of timber components is approaching these 'at risk' levels is very important during inspections.

In relation to measuring these moisture levels in buildings, the main problem facing residential practitioners is that moisture meters cannot accurately measure the moisture content of all building materials. As a consequence, manufacturers calibrate their meters to measure the moisture content of one material only – timber. This is because timber is a reasonably consistent material and the probes of a conductivity meter can usually be pushed by a few millimetres below the surface into the timber itself. A number of leading commentators agree that the readings give a reasonably accurate measure of the actual moisture content in that timber component in percentage terms. However, Ridout and McCaig (2016) found that conductivity meters are only accurate at lower moisture contents. His research revealed that for moisture contents in excess of 22%, the accuracy of moisture meters 'diminishes considerably'.

The Timber Research and Development Association also confirmed this view. They stated that when used in timber, most of the moisture meter readings will be within plus or minus 2% of the true moisture content within the range of 8 and 25%. Outside of this range, TRADA says the readings should be viewed as indicative only (TRADA 1999). This will be discussed again in the chapter on wood rotting fungi (see Chapter 7).

Once calibrated for measuring the moisture content of timber, it is not possible to reset the meter to suit other materials. The physical characteristics of brick and plaster, for example, are so different to timber, a comparison on the same scale is not possible.

Oxley and Roberts (1987) provide a classic but theoretical explanation to illustrate this issue. Imagine an external solid brick wall that is plastered on the inside and has a timber skirting fixed along its base. In equilibrium conditions where moisture is not a problem, the skirting could have a moisture content of 12%, while the brick might contain about 1% moisture and the plaster about 0.5%. If this construction got wet due to rain penetration, for example, the moisture content of the timber could increase to 22%, the brick between 2 to 5% and the plaster between 1 and 3% depending on their actual physical properties. This shows the accurate measurement of actual moisture levels with an 'on-site' instrument is not possible.

In an effort to overcome this drawback, most manufacturers use a relative scale called 'wood moisture equivalent' (or 'WME' for short). This can allow residential practitioners to make broad assessments of the presence of moisture in a range of materials during their inspections. Although the meters are calibrated for timber, the percentage figures shown on the readout panel are simply matched to colour-coded LED lights that indicate the theoretical moisture levels. Protimeter, one of the leading manufacturers gives the following guide: (GEsensing 2013):

- **Green** – Readings in this band can be considered 'air dry' and no problems. Moisture contents at this level will be of no concern;
- **Amber or yellow** – This suggests that there is excess moisture present and it is coming from sources other than the air. If readings in this category persist, then further investigations may be required;
- **Red** – A serious moisture condition exists. There is a clear source of excess moisture and further investigations are definitely required with the possibility of remedial action being very high.

So what does this mean for the residential practitioner? Here are our recommendations for conductivity moisture meters:

- **For timber components** – Where you are able to press the pins of the meter a few millimetres into the timber, the percentage values shown on the LED are likely to be close to the actual moisture content of the timber. However, where moisture levels are higher (say 24–26% and more), the accuracy of the meter is likely to reduce, so use these values with care;
- **For all other building materials** – Even if you are able to press the pins of the meter below the surface of the material, the figures shown will bear no relation to the actual moisture content. In these cases, record the colour of the LED lights on the read-out panel in your site notes.

In both cases, the information provided by a moisture meter does **NOT** provide you with a diagnosis of the moisture problem. Instead, the readings can provide you with useful data that can be combined with other information to enable you to come to a more balanced judgement. In many circumstances, higher moisture meter readings will be the start of the 'following the trail' process.

Comparing the conductivity and capacitance moisture meters

Before we go on to discuss how moisture meters should be used during an inspection, it is important to outline the differences between the conductivity moisture meter (the one with the pins) and the capacitance moisture meter (the one that usually sits behind the flat plate on the back of most moisture meters).

As previously described, the capacitance moisture meter is a non-invasive instrument. As long as the flat plate is placed carefully against the material to be measured, then it does not damage the surface – a useful feature if the owner of the property is not keen on pin holes being left in their skirtings and wall surface. However, many commentators have asserted that capacitance moisture meters are not as accurate as those using conductivity principles. For example, the advice given on Protimeter's own website about the use of its own capacitance meter has been summarised below:

- Protimeter recommended the capacitance meter could be used in 'Search mode' to quickly find higher than normal moisture levels in a building element and these are often behind wall and floor coverings.
- Once an area of high readings has been discovered, the meter should be switched to the 'pin' mode to more accurately measure the moisture levels as the capacitance meter '. . . does not measure as accurately as a pin measurement'.
- On Protimeter's own instruments, the capacitance meter reads on a scale of between 0 and 1000. This is loosely based on the WME scale, so a reading of 200 in 'search mode' is close to a pin reading of 20%, while a reading of 165 is '*somewhat*' close to 16% WME. As a consequence, Protimeter recommends that only readings obtained using the pin-type meter should be reported (Protimeter 2020).

It is also interesting to note that in Good Repair Guide 33: Assessing moisture content in building materials (Part 2), the BRE do not even mention capacitance moisture meters during their review of the different techniques that can be used to measure content of building materials (BRE 2002a).

In conclusion on this point, an analogy we have always found helpful is that moisture meters could be compared to metal detectors that are used by 'detectorists' to search for buried treasure. Although the metal detectors can help identify the approximate location of a metal object, they cannot tell you whether it is a bottle top left by a picnicker or something more valuable buried back in the Bronze Age. In the same way, moisture meters can help you identify a problematic area, but it cannot tell what is causing the problem. For that, you will have to exercise your professional judgement.

Other restrictions on use of moisture meters

Moisture meters are not only limited by their inability to measure the precise moisture contents of a variety of materials but also by other physical factors:

- The way some materials gain and lose moisture can 'fool' meters. For example, where a material has a pore structure that is very fine, a drop in the humidity levels in the property can result in the surface layers drying out, leaving the heart of the material still saturated. This may result in lower surface moisture meter readings despite a continuing moisture problem. Conversely, if the same material was in a satisfactory state, high internal humidity could result in condensation and only the surface layers would become saturated. This would result in high moisture meter readings while the underlying material may remain relatively dry.

- The physical nature of the contact between the pins of the meter and the material being tested can affect the readings. On hard surfaces, it is essential that the pins are properly seated on the surface under a constant pressure.
- The same applies to capacitance or flat plate moisture meters. The whole of the flat plate must rest evenly against the material being measured. Consequently, this type of meter will not be suitable for heavily textured decorative surfaces.
- As discussed later in this chapter, high levels of 'salts' in the material being measured can affect moisture meter readings because the electrical resistance of the material is lowered by the salt crystals themselves. Although the presence of salts can be a problem in itself, it does not necessarily mean that the material is affected by free moisture.

These limitations clearly demonstrate that the instruments should be seen as tools that help the residential practitioner come to a judgement and not make that judgement on the practitioner's behalf.

6.2.3 Using a moisture meter on a survey

Although moisture meters have their limitations, it is our view that they still remain an essential inspection tool for residential practitioners. For example, the meters can help identify moisture affected areas that normally would not be apparent visually and so can initiate the all-important 'trail of suspicion'.

How you use a moisture meter will depend on the level of service you are providing, but ideally, they should be used at points in a property that are vulnerable to moisture-related defects. Although this must always be left to the practitioner's professional judgement, we have listed those locations where we think a moisture meter should be used:

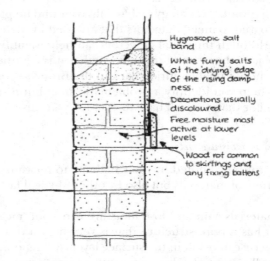

Figure 6.1 Cross section through the base of a typical solid wall that has been affected by moisture and hygroscopic salts.

- **Across the base of the inside face of all outside ground floor walls**: because the risk of moisture-related problems to these walls is high, readings should be taken at 1.0 metre horizontal intervals unless interrupted by furniture or fixings. At each measurement position, readings should be taken at a number of vertical spacings (see Figure 6.2). This is because moisture problems can manifest themselves higher up the wall as well;
- **Around the base of chimney breasts**: although not strictly the outside wall, the sides and face of chimney breasts can be affected by moisture and salts from a number of different sources;
- **Across the inside face of 'exposed' outside walls**: indicative spot measurements should be taken to the inside face of the walls that face the prevailing weather. This is particularly important where the property is exposed (for example, on a hillside, by the coast and so on);
- **Across the base of all other internal and party walls (if applicable) at ground floor**: in most circumstances, the risk of these walls being wet is not so high so only indicative 'spot' measurements should be taken (for the exception to this rule, see Fryer vs Bunney on this matter in RPA Volume 1, section 4);
- **The surface of the ground floors (both timber and solid floors)**: you should take an indicative sample of readings where the floor surfaces are on view or can be

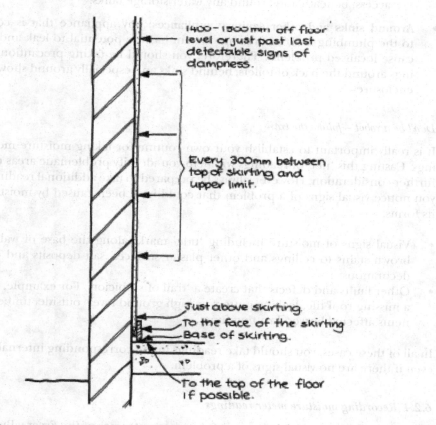

Figure 6.2 Cross section through a typical wall showing the vertical spacings at which moisture readings should be taken.

easily exposed by moving small items of furniture and lifting the floor coverings. This latter issue will depend on the nature and extent of your inspection (see Chapter 3);

- **Below every window sill and across its width**: the junction of the window frame and the wall is vulnerable to leaks so take a number of readings to the plaster below the internal window sill. This will typically include below each end of the sill and at reasonable intervals across its width;
- **The reveals of every window and door opening**: for the same reason as explained previously, regularly spaced readings should be taken from top to bottom of the reveals and across the head where possible;
- **In the roof space**: although this will vary depending on the design of the roof, accessibility (especially health and safety considerations) and the nature of the service, typical locations will include:

    - To the party walls close to the junction with the underside of the roof covering;
    - To any chimney stacks in the same location;
    - To vulnerable timbers especially rafters running close to the party/flank walls; trimming rafters around the chimney stacks and especially any timber back gutters; underneath and around any parapet and/or valley gutters; sample readings to a number of rafters; the trimming timbers around the loft access; beneath and around any water storage tanks.

- **Around sinks and other sanitary appliances**: any appliance that is connected to the plumbing and/or drainage system has the potential to leak and this can cause localised problems. Therefore, you should be taking precautionary readings around the back of toilets, behind sinks and especially around shower trays/enclosures.

Don't be a robot – follow the trail

It is really important to establish your own routine for taking moisture meter readings. Casting this 'net' of readings will help you identify problematic areas that need further consideration. However, always be prepared to take additional readings where you notice visual signs of a problem that could have been caused by moisture in all its forms:

- Visual signs of moisture including 'tide' marks along the base of walls, watery brown stains to ceilings and other plaster surfaces, salt deposits and disrupted decorations;
- Other faults and defects that create a 'trail of suspicion'. For example, if you see a missing roof tile, leaking gutters or high ground levels outside, timber components affected by wood rotting fungi.

In all of these cases, you should take readings in the corresponding internal location, even if there are no visual signs of a problem.

6.2.4 Recording moisture meter readings

According to Murdoch (1995), one of the main outcomes of the Fryer v Bunney case (1982) was that surveyors should '. . . prepare a very simple sketch of the property as it is surveyed, marking roughly where a moisture meter has been applied'. Based on our recommended approach to using a moisture meter, it is clear that a typical inspection will generate a large amount of data.

Many lenders and surveying organisations have their own guidance on this matter, but most usually require the practitioner to record the moisture meter readings in some way and this will include the location and value of the readings. One of the major drawbacks is that there is no common practice across the sector. In an attempt to fill this gap, you may want to consider our approach:

- Mark the location of the readings on some form of plan of the property;
- This plan should show the main furniture positions and/or be backed up by a photograph/video of the room that clearly shows the furniture positions at the time of inspection;
- Because moisture meters are only calibrated for timber, the numerical values obtained are strongly linked to the circumstances at the time of inspection and so the meter readings have no intrinsic value by themselves. For example, environmental conditions in a typical household are dynamic and will vary throughout the day. According to the BRE, water vapour levels early in the morning after the occupants have bathed and left for work/school will be high and can harmlessly dissipate by mid-afternoon (Garrett & Nowak 1991). Consequently, moisture meter readings taken at 9 am can be very different to those taken at 3.30 pm. Additionally, readings of a timber skirting will have different implications to those recorded in the plastered surface above. Consequently, noting down the precise numerical values will be, at best, unhelpful and, at worst, potentially misleading.

A way of resolving this dilemma is to record only the colour shown on the instrument's readout panel rather than the number. This avoids the misleading nature of the 'precise figure' but places the reading in a defined banding that can support your decision-making process.

What do you record?

Once you have decided where to take the readings and what outputs to record, the next issue is how to note this information on your sketch plan. The way of doing this varies greatly. If you were to examine the 'site notes' of 20 experienced practitioners, we are sure you will get 20 different ways of recording moisture meter readings. It is important to develop a system that works for you. Whatever symbols/notation you chose, they should account for the following:

- The position where you took the measurement;
- The relative value of the reading (for example 'R', 'A' or 'G' for red, amber and green);
- At what height on the wall/reveal was the reading taken?
- How does your system differentiate between readings on the floor and ceiling?

Where possible, your sketch plan should include a key to your notation. Although your client will not see this information, an auditor will. If you get a complaint in the future, a clearly annotated sketch will give a favourable impression to any court.

6.2.5 Moisture meter readings and the different service levels

In Chapter 2, we described the different levels of service in RICS's Home Survey Standard (RICS 2019). To help members understand the differences between the inspection routines of each level, RICS included a number of different elements in Appendix B of the document. Although this included 'critical benchmarks' for windows, roof space, floors and underground drainage, the use of a moisture meter was not included and so left to the professional judgement of the practitioner.

In an attempt to fill this void, we have outlined the use of moisture meters on the different levels of service, including valuation.

Valuation inspections

Valuation inspections aim to discover whether the condition of the property 'materially affects value' (see Volume 1, chapter 8 for more information). Consequently, the inspection should aim to identify only significant moisture-related problems. So the use of moisture meters will not be so extensive when compared to the inspections carried out for condition reports. The following guide may be helpful:

- Use the moisture meter at regular intervals to the inside face of all outside walls where the furniture and contents allow. Take readings to the floor (if possible), skirtings and to the wall just above the skirting level;
- Take readings around chimney breasts and other high-risk areas such as kitchens, showers rooms, bathrooms and so on;
- Visually check other parts of the property for signs of moisture problems especially in vulnerable areas. This would typically include below flat roofs and valley gutters, ceilings below bathrooms, showers and WCs, and window reveals and heads in exposed locations. Where visual signs are present and accessible (for example, water stains, salts, mould and so on), check the area with the moisture meter;
- If a 'head and shoulders' inspection of the loft is carried out, take moisture meter readings of adjacent timber sections without getting off the ladder. For example, trimming joists, ceiling joist, reachable trusses/rafters and so on;
- Record high moisture meter reading (red zone) on a simple plan of the property.

This is an approach we think is suitable for a valuation. If you carry out this work for a lender, panel manager or a larger surveying firm, you should follow their processes.

Inspections for level one services

According to RICS, level one services are '. . . better suited to conventionally built, modern dwellings in satisfactory condition' and although services at this level do focus on condition they are designed for clients '. . . seeking a professional and objective report on the condition of the property at an economic price' (RICS 2019). As a result, level one services are less comprehensive than those at levels two and three. Level one services are also designed for properties that are less likely to be affected by moisture-related defects, so the risk is lower (apart from the exceptions). A typical age range might be less than 20 years old.

Therefore, we think the use of the moisture meter should match that already described previously for valuation inspections.

Inspections for level two services

A level two service is an '. . . intermediate level of service and includes a more extensive visual inspection of the building' according to RICS (RICS 2019). The inspection should include a roof space inspection (where it is safe to do so) and the report should '. . . objectively describe the condition of the different elements and provide an assessment of the relative importance of the defects/problems'.

Services at this level are likely to '. . . suit a broader range of conventionally built properties, although the age and type will depend on the knowledge and experience of the RICS member' (RICS 2019). Although this will vary depending on the practitioner, we think this will include properties that go back to the end of WW1 and in some cases before. The incidence of moisture problems to properties of this age and type are likely to be higher and the risks of defects greater.

In this case, we think the moisture meter should be used more extensively than for valuation/level one services. In addition to those areas listed in the valuation inspection, the meter should also be used to take:

- A sample number of readings to all door and window reveals and beneath all window sills on the most exposed sides of the property;
- A sample of moisture meter readings to the inside face of all walls on all floors facing the prevailing weather;
- In the roof space. Although level two services include a roof space inspection, it is generally limited to those parts most at risk of moisture-related problems and will typically include sample readings:

  - To the party walls close to the junction with the underside of the roof covering;
  - To any chimney stacks in the same location;
  - To vulnerable timbers especially rafters running close to the party/flank walls; trimming rafters around the chimney stacks and especially any timber back gutters; underneath and around any parapet and/or valley gutters; sample readings to a number of rafters; the trimming timbers around the loft access; beneath and around any water storage tanks.

In the foregoing descriptions, where the term 'sample' is used, we take that to mean a small number of readings that give an indication of the general condition of the whole of that particular element.

During level two inspections, when high moisture readings are detected, additional readings should be taken on a 'follow the trail' basis to help diagnose the cause of the problem where the defect is relatively straightforward and obvious.

To illustrate this approach, consider two scenarios:

- a solid walled property with a leaking gutter; and
- a solid walled property with a moisture-related problem to the base of a number of the outside walls.

For the gutter leak, assume the brickwork below the gutter is stained and the water has soaked through to the internal face where it has stained and disrupted the decorations.

For the problem at the base of the outside walls, assume there is a wavy 'tide mark' up to a height of 450–600 mm above the internal floor level with white furry salts showing on the edge of the staining. The ground floor is solid and the inside faces of the walls are plastered with a skirting board at the junction with the floor.

Although these are simplistic illustrations, both should trigger the 'follow the trail' process and the moisture meter should be used to assess:

- **The extent of the problem:** is this a localised problem that has little effect or is it a major issue that affects a large part of the element? Moisture meter readings taken at regular spacings (say on a grid of 150 mm) can help establish this;

- **Confirm the cause of the defect (if possible):** It is important to say at this stage that moisture meters do not diagnose the cause of a defect – that is the surveyor's job. But meters **can** help distinguish between defects.

  For example, with the leaking gutter, typical meter readings are likely to change from 'low' (green) to 'high' (red) very rapidly at the boundary of the visual staining. As long as there are no other possible causes in the vicinity (leaking water pipes, for example), these meter readings combined with the visual evidence the gutter is leaking (the staining on the outside), the practitioner can be confident of the cause and give clear advice to the client.

  However, diagnosing the cause of the moisture at the base of the wall may not be so straightforward. Moisture meter readings could gradually change from 'high' (red) at the base of the wall to medium (amber) higher up and then to low (green). In some cases, the band of furry white salts may result in a few higher readings before it drops back to low readings again. The cause of this problem may not be so obvious: it could be high ground levels; a leaking gutter from above; the absence of a link between the damp-proof membrane in the floor with the DPC and even problems with the DPC itself. Many of these causes cannot be identified during a visual inspection and it is at this point the practitioner may have to recommend further investigations by a competent professional.

- **Check whether adjacent timber components are at risk**. As discussed in Chapter 7, timber components in contact with surfaces affected by moisture will be at risk from wood rotting fungi and wood boring insects. Consequently, in both cases, the moisture meter should be used to check the moisture content of any adjacent skirtings, floorboards (if accessible), built-in furniture and so on.

In this way, during level two services, the moisture meter can help support the investigations of the practitioner by determining the extent of the problem, supporting the diagnosis of the defect and assessing the risk of problems in adjacent timber components. However, as RICS state in the Home Survey Standard (RICS 2019), level two services are '. . . *for clients who are seeking a professional opinion at an economic price*'. Consequently, there is a limit to what can be included in a level two report and for older properties in a poor condition, a high number of referrals for further investigations will often be the result. In such cases, level three services may have been required.

Inspections for level three services

As we stated in the introduction, this book is designed to support residential practitioners who offer their clients predominantly valuation services and levels one and two condition reports as defined by RICS (RICS 2019). In our view, the breadth and depth of technical knowledge required for level three services is beyond the scope of any one publication. Despite this, one of the questions we often have to address at conferences and seminars is where level two services end and level three begin. Consequently, we thought it would be important to describe how a moisture meter should be used on a level three service so you can be more confident about your approach to level two.

According to RICS, a level three service is for clients '. . . who are seeking a professional opinion based on a detailed assessment of the property'. It goes on to say that it consists of a '. . . detailed visual inspection of the building . . . and is more extensive than a survey level two'. The HSS explicitly states the level three report should outline the scope of any repair work and where the RICS member feels unable to reach the necessary conclusions with reasonable confidence, they should refer the matter for further investigations. However, '. . . with level three services these referrals should be the exception rather than the rule'. This clearly differentiates between level two and three inspections: level two adopts a sampling approach that provides an insight into the general condition where level three is a detailed assessment. Where high moisture meter readings are recorded, then the practitioner should 'follow the trail' to a greater extent than at level two so the nature and scope of the defect can be identified wherever possible. The scope of the level three inspection should allow for this level of investigation as standard, whereas for the level two inspection, this would be the exception.

6.2.6 Other methods of evaluating moisture problems

Where moisture has been detected and further investigations are required, there are a variety of other techniques and equipment that can diagnose the cause of the problem more thoroughly. These are usually beyond the scope of level one and two services but an awareness of the methods will be useful.

Moisture meter accessories

Many of the manufacturers of moisture meters provide a range of accessories designed to help the surveyor evaluate other signs and symptoms associated with moisture. These include:

Pin/probe extensions – This consists of a pair of moisture pins set in a separate head that can be connected to the moisture meter by a fixed cable. This allows meter readings to be taken in hard-to-reach areas such as small gaps between furniture, underneath sinks and so on.

Deep wall probes – These are long probes that can be connected to the moisture meter. They look like a pair of metal knitting needles with their shanks insulated with PVCu. Two holes can be drilled into the wall and moisture meter readings taken with the deep probes at successive depths, allowing the level of moisture to be evaluated deep inside the wall. This can help distinguish between moisture that affects the surface (for example, condensation) and that which is deep seated (gutter leaks or moisture from the ground). However, as this involves drilling two holes into the wall, it is usually beyond the scope of most inspections (and will upset the owner).

Hammer action probes – These are separate all-metal probes with a weighted and sliding handle that can be used to hammer the robust moisture pins into a wooden surface. Although this will measure dampness in the same way as the pins attached to the meter, the surface is likely to be damaged and so falls outside of most conventional inspections.

Salts analysis kit – Although not strictly a meter accessory, the manufacturer of this kit claims it enables a surveyor to analyse samples of wallpaper and plaster for the presence of salts that may have come from the soil. The kit consists of two chemicals packed in sachets of the correct amount for each test, a bottle of distilled water and two measuring beakers. Once mixed with an appropriate sample, a colour change can indicate the type of salts present and help support the diagnosis. This technique is usually associated with further investigations of moisture problems.

Destructive testing methods

Where a more precise level of diagnosis is required, specialists may use more destructive testing methods. Although these are beyond the scope of most surveys, they may be encountered where the presence of dampness is disputed or during analysis prior to remedial work. These are described in more detail by the Property Care Association (PCA 2020 'The use of moisture meters to diagnose dampness in buildings') and include:

Chemical method – Sometimes called a calcium carbide or 'speedy' method, this approach was originally developed for testing the moisture content of more bulky materials (for example, aggregates and grain). These on-site kits use drilled brick samples. These samples are placed in a pressure vessel along with powdered calcium carbide. Any moisture in the samples reacts with this chemical to produce acetylene gas – the higher the levels of moisture, the more gas is produced. A gauge on the top of the flask is calibrated to give a direct reading of the moisture content of the sample.

Gravimetric method – This is a more precise laboratory-based method fully described by the BRE (1985). Drilled samples are taken away and weighed both before and after drying in an oven. A standard calculation can then determine the true moisture content. Additional tests on the same samples can also determine the hygroscopic salt content of the material, which can help establish the precise cause of the moisture.

What to do with furniture?

During inspections of occupied property, furniture and possessions can restrict the visual inspection of the dwelling and, more importantly, the use of the moisture meter. As moisture levels can be higher behind furniture (for example, beds and wardrobes on outside walls), it is important to formulate your own inspection methodology that takes account of this restriction and clearly communicate this to your client.

Although the Home Survey Standard (RICS 2019) is not prescriptive about what RICS members should do during inspections, it does mention the issue of furniture and possessions in a number of places. For example:

- Under 'Specific inspection details' (p. 10, RICS 2019), the standard states:
  'RICS members should ask the owner/occupier to . . . move furniture and/or possessions where these prevent normal levels of inspection (where practicable)'.
- In Appendix B 'Benchmarking the levels of inspection', the furniture issue is mentioned under a number of headings:

  - Under 'windows', the Standard states that heavy curtains and possessions '. . . will often restrict level one and two inspections' whereas for level three 'a small number of possessions/curtains will be repositioned'.
  - In roof spaces, for level one and two services, stored goods or other contents will not be moved, while during level three inspections a '. . . small number of lightweight possessions should be repositioned so a more thorough inspection can take place'.
  - For 'floors', the HSS clearly states furniture will not be moved during level one and two services. Although it is silent on furniture at level three, it is accepted that the corners of any loose and unfitted floor coverings should be lifted '. . . where practicable'.

Based on this brief review, at level one and two service levels, the Home Survey Standard is clear – furniture and possessions should not be moved by RICS members. Instead, the owner/occupier should be asked to move any items that restrict the inspection and if this is not done, then the client must be advised of the implications. At level three, the HSS states that a small number of lightweight possessions may be repositioned so a more thorough inspection can take place.

Although the HSS is clear, we think it is important to meld this regulatory 'benchmark' with the expectations of the contemporary client. It is our opinion that many clients would be disappointed if their surveyor used the wording of the Home Survey Standard to restrict the scope and extent of the inspection and report on the property on which they are about to spend hundreds of thousands of pounds. For example, imagine a small lounge chair that can be safely and easily pushed to one side, so a moisture meter reading can be taken of the wall behind. As long as the owner/occupier has given their permission, shouldn't this be done on all levels of service? We think it should. However, where a bed or a large wardrobe is positioned against the inside face of the outside wall, then we think it is unreasonable for a practitioner to even attempt to move this even at level three.

The following comments may be helpful when you are deciding on your own approach:

- Before you arrive for the inspection (and again upon arrival), you should ask the owner/occupier to move as much furniture as possible away from the inside face of the outside walls. In addition, during the course of the inspection, if specific furniture/possessions restrict your inspection, you should ask the owner/occupier to move those too. This should be done as soon as the items are noticed;

- Even where the HSS states furniture should not be moved for level one and level two services, we think where the 'trail of suspicion' is strong (for example, the risk of a concealed defect is high) you may want to consider repositioning furniture/ possessions when:

  - You have the owners/occupiers permission;
  - The items are not physically fixed to the wall or floor;
  - They are light enough to be easily moved by one person; and/or
  - The surfaces are substantially clear of possessions and/or where moving a few items is unlikely to cause damage or too much disturbance.

- Because level three services include a '. . . *detailed visual inspection*' of the property, this approach to moving furniture should be seen as the norm rather than the exception.

If you do decide to move furniture/possessions you should ask for permission first. This can normally be included in your discussions with the person in charge of the property on arrival.

Where furniture/possessions could not be moved, this fact should be clearly noted in your site notes and moisture meter readings taken either side of the item. This may enable a speculative judgement of whether a moisture problem is likely to exist behind the furniture.

6.3 Directly penetrating moisture

According to the Property Care Association (PCA 2020), rainwater is one of the three main natural causes of moisture-related problems in buildings: the other two are water from the air within the building and water from the ground. This section will focus on water penetrating through the outside wall.

6.3.1 Solid walls

Whether constructed from brick or stone, we define solid walls as elements that do not contain a formal cavity. In terms of moisture exclusion, they operate on what could be described as a 'sponge' principle. The materials that make up the wall soak up the rain during wet periods and hold on to it until drier periods come along, allowing the moisture to safely evaporate. This relies on the free movement of moisture both in and out of the construction. It also depends on the wall being thick enough to prevent the moisture from reaching the internal surfaces during the wettest of weather (see Figure 6.3).

Problems occur when:

- The mortar pointing between the masonry units is in a poor and/or porous condition;
- The masonry units themselves are porous or have deteriorated through frost action; and
- The building is in a very exposed position allowing a lot of wind-driven rain to hit the wall.

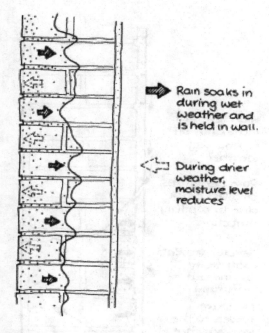

Rain soaks in during wet weather and is held in wall.

During drier weather, moisture level reduces

Figure 6.3 Cross section through a solid wall showing how it resists the passage of moisture.

Any of these deficiencies may allow moisture to travel through the wall and affect the internal surface.

Rendered walls

On many older properties, cement-based renders are often used to improve the performance of solid walls. As long as these are mixed and applied appropriately, they can have a beneficial effect. If they are of a poor quality, they can actually make the problem worse (see Figure 6.4). For example, a cement-rich, strong, impervious render applied to a brick that is not so strong can result in cracking that can channel the water behind. Because a cement-based render can be impervious, the render can prevent the wall from drying out. To be successful, the characteristics of the render must match those of the wall.

Other coatings

Over the last few years, a wide variety of proprietary coatings have been developed and used by homeowners. These include:

- **Bituminous applications** – A traditional solution that used bitumen-based paints and coatings often combined with hessian or other reinforcing fabric (sometimes called 'tunnerising'). They can have short-term beneficial effects. Because

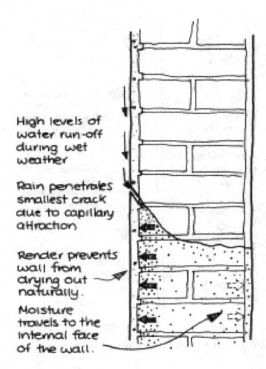

High levels of water run-off during wet weather

Rain penetrates smallest crack due to capillary attraction

Render prevents wall from drying out naturally.

Moisture travels to the internal face of the wall.

Figure 6.4 Sketch showing the effects of a cracked render finish on a solid wall.

bitumen soon breaks down after exposure to sunlight and temperature variations, water can easily get behind the coating and penetrate internally. The coating will also have a permanent visual effect on the building, which may not be pleasing (see Figure 6.5).

• **Thin-coat cementitious paint applications** – Many companies have developed thin-coat render systems. These are usually a combination of paint, cement and sometimes reinforcing fibres sprayed or applied on the surface of the wall. These will change the visual appearance of the property and will often have the same benefits and disadvantages as normal render applications. Where these special coatings have been used, any guarantees, British Standard approvals, Agrement certificates, etc. should be asked for in an effort to assess the quality of the work.

• **Transparent applications** – Difficult to detect visually, manufacturers of these coatings and creams claim the weather resistance of walls can be improved by these silicon- based treatments. The chemicals in the coatings aim to block or line the pores of the masonry in an attempt to prevent capillary action from occurring. Although manufacturers make impressive claims about the lifespan of these products, it is our view that they should be seen as a short-term solution, especially in exposed areas.

PROBLEMS CAUSED BY CONCENTRATED FLOW AND OTHER DEFECTS

Water penetration through solid walls often occurs where water flow has been concentrated by other defects, such as leaking gutters, rainwater pipes and even running

Figure 6.5 A gable elevation of a house that has been treated with a bitumen-based compound in an attempt to improve its moisture-resistant properties.

overflows from toilet cisterns and other appliances. These high levels of moisture can quickly affect internal surfaces.

WINDOW FAULTS

The rainwater run-off from the glazed panels of windows can be very intense, especially on exposed elevations. Most of this will run down over the window sill and either drip clear of the wall or flow down the face, where the throating is inadequate. Wind-driven rain can penetrate around window frames and sills, especially if they are poorly decorated, partly rotten or not properly sealed with mastic to the surrounding construction.

Key inspection indicators

INTERNALLY

Internal symptoms associated with directly penetrating moisture can include:

* Damaged decorations and plaster surfaces especially below and around window openings;
* Well defined and localised areas of high moisture meter readings that clearly match externally observed defects especially after periods of heavy rain. Moisture meter readings will change from low (green) to high (red) very quickly; and
* High levels of concentrated mould growth that closely matches the affected areas.

EXTERNALLY

Where water is penetrating from the outside, then the faults should closely match the location of the signs of damage inside. The exception to this can be where the bricks or stone blocks haven't been fully set in mortar, creating a number of open joints and small 'informal' cavities within the thickness of the wall. In these circumstances, the water may percolate downwards from the point of entry to where it affects the inside face.

The types of faults you should be looking for include:

* The condition of the pointing and the masonry units (for example, the bricks and stones) and any rendered surfaces. Look for any cracks, spalling and porous surfaces;
* Window frames that are in a poor condition with little or no protection at the junctions with the wall construction;
* Defects in the window sills such as cracks, sills out of level, inadequate projection of the sills over the face of the walling, blocked or absent throatings to the underside, and so on;
* Leaking gutters and pipes;
* The orientation of the affected area. West- and south-west-facing elevations are particularly vulnerable in most parts of the UK. For other areas, regional meteorological data should be obtained. In some cases, there may have been changes in the local exposure. For example, a tree might have been removed or an adjacent building demolished, allowing a higher amount of rain to reach the building than before.

SCALE OF REPAIRS TO RENDERED SURFACES

Where an inappropriate render is stopping the wall from drying out, then the owner has two options:

* Remove the existing render and replace with one that is properly matched to the backing brickwork. The BRE in their Good Building Guide 18 (1994)

recommend a number of different cement-based render coatings matched to different backgrounds and the exposure zone of the property. However, the Society for the Protection of Ancient Buildings (SPAB) warn on their website that care should be taken with older buildings (SPAB 2020). The use of modern materials (Portland cements and masonry paints) can prevent the wall from drying out and result in further damage. In such cases, only lime-based renders and paints should be used. Where the property is listed or in conservation area that has an Article 4 designation, there normally is no choice – the local planning authority will require lime-based renders and paints to be used;

- Remove the render completely and leave the original walling exposed. The biggest drawback with this approach is the condition of the concealed masonry beneath. At the very least, some of the brick/stone will have to be repaired and the exposed elevation repointed. However, this should only be done if the dwelling is not in an exposed position. Additionally, this level of intervention may affect the appearance of the wall (see Figure 6.6).

Figure 6.6 This image was taken from Google Street View and dates back to 2008. The original brick walls were covered with a thick render. The adjacent properties were painted.

Figure 6.7 This image was taken of the same property in 2020 after refurbishment and the whole
process captured on Street View. Although we do not know whether this repair was
successful, it shows what an impact render removal can have on the visual quality of
the wall. It is also interesting to note the leaking gutter at the party wall position. This
could be affecting the inside of the wall, creating a trail that needs to be followed.

In both circumstances, it may be appropriate to warn your client they should only use
contractors experienced in this sort of work and as a consequence, this can make the
repairs more costly.

6.3.2 *External wall insulation and moisture problems*

Since the first edition of this book was published, the practice of applying external wall
insulation to older dwellings has become widespread. Concerns over global warming

and the need to save energy have resulted in a number of government initiatives aimed at encouraging improvements to the energy efficiency of residential property. For example, in 2012, the 'Green Deal' was launched. This was a UK government initiative that aimed to give homeowners, landlords and tenants the opportunity to pay for energy-efficient home improvements through the savings on their energy bills. There were 45 different types of improvements available, including solid wall insulation (Which? 2021).

The burden of paying for the work stayed with the property (paid through the electricity bill), regardless of the owner and/or tenant. This meant that new tenants or owners became liable for the payments for the energy efficiency improvements previously installed. The Green Deal was controversial and arguably never achieved its original objectives. In 2015, only around 15,000 Green Deal 'Plans' had been completed, far below the government's original targets. Consequently, the scheme was scrapped in July 2015.

Running alongside the Green Deal was another government initiative called the 'Energy Company Obligation' or 'ECO' for short. This energy efficiency scheme aimed to tackle fuel poverty and reduce carbon emissions for the country as a whole. Under ECO, medium and larger energy suppliers were required to fund various energy efficiency measures in British households, such as loft and wall insulation and heating measures. The scheme began in April 2013, and at the time of writing, the latest policy called 'ECO3' started in December 2018 and is planned to run until 2022. The scheme is focused exclusively on those customers with lower incomes and includes insulating up to 17,000 homes with solid walls.

According to statistics released in January 2018 (Department for Business, Energy and Industrial Strategy 2018) around 2.4 million energy efficiency measures were installed in around 1.8 million properties. This figure included around 157,200 solid wall insulations, which accounted for 7% of all measures. Although this may be a very small proportion of the UK housing stock, anecdotal feedback from practitioners indicates that these insulation systems are regularly encountered during inspections. Consequently, we have included a new section of how external wall insulation can be identified, inspected and assessed.

Moisture problems associated with external wall insulation

In their White Paper titled *'External wall insulation – who are we kidding?'* (PCA 2016), Steve Hodgson, the CEO of the Property Care Association (PCA), expressed his concern about the '. . . lack of quality, accountability and good practice in the external wall insulation (EWI) market'. Although Hodgson considers many EWI products as 'extremely good', he claims they are being applied to the outside of buildings without considering the broader implications of this energy improvement method '. . . with almost no regard to the legacy of problems they are currently creating'.

We share Hodgson's concern and the next section provides a methodology on how external wall insulation systems can be identified, inspected and assessed.

Inspecting and assessing external wall insulation (EWI)

A useful source of information is the Insulated Render and Cladding Association (INCA). According to their website, they are the recognised trade association for the EWI industry (INCA 2021). Like all trade associations, they promote the work of their members, but their website does include a number of useful technical guides that will help you build up your knowledge of these systems.

According to the Energy Saving Trust, there are two types of external wall insulation (EST 2006):

- Wet render systems; and
- Dry cladding systems.

These will be considered in turn.

Wet render systems

This type of system is most commonly used on low-rise residential properties and consists of a number of different components:

- **The insulant** – A variety of materials can provide the thermal insulation including mineral fibre; rigid foam panels (for example, polyisocyanurate or PIR, urethane or phenolic material); expanded and extruded polystyrene and plant-based panels including cork, cellulose, wood-fibre and hemp. To achieve the U value laid down in the Building Regulations (Approved Document L1b), the thickness of the insulating layer can range from 60 mm to 120 mm thick, depending on its thermal properties.
- **The fixings** – These securely fix the system to the wall and usually includes adhesive mortar and/or mechanical anchors such as mushroom-headed nailable plugs.
- **The finish** – A protective layer providing weather protection and a finish. Most installers use a thin coat polymer cement render with glass fibre or polyester mesh with a total thickness of 6–10 mm thick. This usually includes a base-coat render (incorporating the mesh) and a top-coat render, with or without a finish. A range of surface finishes are used, including dry dash and wet dash, mineral renders, brick effect renders and acrylic and real brick slips.

All external wall insulation systems will include other accessories (such as beads, trims and flashings) that offer further protection and connection to other parts of the building, such as windows, doors, roof coverings and other similar features.

Dry cladding systems

Dry cladding systems are usually formed by:

- The insulant, fixed to the wall in a similar way to wet systems;
- A supporting framework or cladding fixing system;
- A ventilated cavity;
- Cladding material and associated fixings.

According to the Energy Saving Trust (EST 2006), dry cladding is seldom used on low-rise dwellings, as the cost can be prohibitive and so are usually associated with high-rise residential blocks. Additionally, following the tragedy at Grenfell Towers in June 2017, fire-related regulations are critical to the performance of these insulating systems and are beyond the scope of this book. Consequently, the remainder of this section will consider the problems associated with wet render systems.

Identifying EWI systems

The first step is to be able to recognise that the property has been insulated in the first place. Here are a few tips on how to recognise an EWI system:

- The outside walls may have a different finish to neighbouring properties;
- There will be an obvious increase in wall thickness usually at DPC level (see Figure 6.8). This could be in the region of 60 mm to 120 mm. Where the wall has been rendered without insulation, this increase in thickness is usually less than 25 mm;
- Preformed metal/PVC flashing sections are usually fitted at the junctions of the EWI system and the eaves and verges of the roof, window and door sills (see Figure 6.9);
- The visible depth of the window and door frames will be reduced where the system has been applied to opening reveals;

Figure 6.8 This shows the base of an EWI system where the insulation is fixed to the outside of the wall. It is then covered with a thin render finish. The thickness clearly sets this apart from a rendered finish alone.

Figure 6.9 This end of the terrace house has been clad with an EWI system. The insulation has been stopped short of the decorative brickwork to the front and the roof verge to the flank wall. A preformed metal flashing has been used to waterproof the top of the insulation. These areas will be vulnerable to rain penetration and cold bridging.

- Although it is difficult to describe a sound in words, if you lightly tap the surface of a EWI system with the handle of a screwdriver or small hammer, it won't exactly sound 'hollow' but neither will it be like tapping dense brickwork.

Inspecting EWI systems

Where you have identified EWI systems, it should be inspected in the same way you would look at any wall surface, but it is important to focus on a number of critical details:

- Look at each elevation from as far back as possible (INCA suggest at least 10 m) using your binoculars. This may seem like stating the obvious, but we have noticed a growing tendency for practitioners to take multiple digital photos of the outside of the property and 'virtually' inspect the images on their hand-held devices or computers when they are writing the report. In our view, there is no substitute for inspecting the wall with your own eyes, as you will be able to see much more than digital images will ever reveal;

- INCA and other organisations suggest wall surfaces should be viewed under 'glancing' light conditions. A term commonly used by paint and plastering manufacturers, 'Glancing light' describes a critical lighting condition which exists when light hits a flat surface at an acute angle and casts shadows that highlight any surface irregularities. Consequently, where possible, you should look 'across' a wall surface at an acute angle. This will help you spot any potential problems (especially on sunny days), so you can make an assessment of its general appearance. It is not possible to do this from a photograph;
- During your visual inspection, focus on the edges and junctions such as:

  - Along the bottom edge (called the 'starter track'), to make sure there are no gaps through which vermin can get into the building. A hand-held inspection mirror will be useful here;
  - Junctions at window reveals, sills and heads to ensure they are sealed correctly and the window sills overhang the face of the cladding by at least 40mm;
  - Junctions at eaves/verge/soffit level, and at verge trim level to ensure joints look satisfactorily made and there is no uneven flow.

Physical deterioration

If the EWI system has been poorly installed and/or located in an exposed position, the render/surfacing may have started to deteriorate. INCA (2015) acknowledge that '. . . *certain finishes (such as brick effect renders) can suffer from drying out cracks or fissures, which although at times unsightly, do not have any adverse effect on the overall performance of the system*'. Although they suggest cracks of less than 0.2 mm are of no concern, you should be alert to all forms of cracking as they might indicate a more worrying underlying cause.

You should also look out for:

- Areas of render coatings that have debonded (come away from its backing);
- Loose and missing brick slips; and
- Cracks that follow insulation panel joints.

These can be potential future problems that allow water to get in behind the insulation. You should be particularly alert to where parts of the system could fall away and pose a safety hazard.

Entrapped moisture

One of the problems with fixing an impervious cladding system on the external face of a building is that moisture within the existing wall can become trapped. This can have a similar effect to rendered walls, which are illustrated in Figure 6.4. There are two main sources of this trapped moisture:

Pre-existing residual moisture

Any solid wall (and especially those in exposed locations) may contain relatively high levels of moisture, depending on prevailing conditions. Fixing an EWI system to the surface will prevent this moisture from drying out. The Insulated Render and Cladding Association (INCA) acknowledge this risk in their publication 'Best Practice Guide: External Wall Insulation' (2015) and advise:

> . . . any large items of furniture should be moved away from external walls to prevent condensation build up behind them whilst the walls dry out. The occupants of the building must also take ownership and be encouraged to engage in air movement.

The guide goes on to state:

> . . . condensation during the initial couple of heating cycles can be addressed through simple changes of the occupants' habits.

This suggests that a newly insulated wall will take time to settle down before the property achieves a new equilibrium (possibly two heating seasons). This poses practitioners with a dilemma: if mould growth is identified, is this a temporary 'adjustment' or signs of an on-going moisture problem? This will be difficult to diagnose on a single inspection, but you might find the following points helpful:

- If the insulation was fitted more than two years ago and the moisture-related problem is still occurring then it is likely to be an indication of a continuing problem;
- If the insulation is less than two years old and the moisture related problem is focused on normal cold bridges (for example, window and door reveals), consider informing the client that it may be a transient problem and the occupants should adhere to a sensible heating and ventilating regime. However (and that is an important word), if the problem persists over the two-year period, the client should ask the original installer to investigate the problem.

Penetrating moisture from building defects

We have seen a number of installations where faults in the existing building have never been properly resolved and could result in water getting behind the insulation and affecting internal surfaces. Examples include:

- Leaking gutters and downpipes;
- Poorly installed cover flashing at eaves and verges;
- Inadequate window sill extensions; and
- Unresolved dampness at the base of the walls (that some would call rising dampness).

In most cases, these faults are likely to result in concentrated areas of moisture and associated mould growth closely associated with the leaking/faulty feature (see Section 6.3.1 for a more detailed description).

The exception is with unresolved rising dampness. If the moisture does come from the ground (by whatever mechanism), then the relatively impervious EWI is likely to result in the moisture being 'pushed' further up the wall (see Section 6.4 for a further explanation of this effect).

Altering the internal moisture balance

In their Best Practice Guide, INCA (2015) point out that before an EWI system is fitted, many properties will not be air tight. For example, gaps and cracks around windows and doors, service pipe entry, gaps in the floor and other junctions can result in high levels of 'adventitious' ventilation. Although unreliable and inherently variable, this accidental background ventilation can contribute to balanced internal environments and so lessen the risk of condensation (see Section 6.5 for further explanation). A properly installed EWI system will seal many of these cracks and gaps, reduce the level of air change and so disrupt the sometimes-delicate internal environmental balance. This could result in increased levels of condensation and mould growth experienced before the improvement. This risk is heightened if windows have been replaced, suspended timber floors covered with laminate and lofts properly insulated.

In these cases, excessive condensation and mould growth return year after year. In these circumstances, you should take a broader view of the property and assess the level of formal ventilation in the property. As a minimum benchmark, you should expect to see:

- Background ventilation in every habitable space (for example, trickle vents in the windows or airbricks); and
- An appropriate extract fans in the 'wet' areas such as kitchens, bathrooms and so on.

Cold bridging

One of the biggest potential problems with EWI systems is the potential for dampness associated with cold or thermal bridging. This occurs when there is a direct connection between the inside and outside through one or more elements of the building that are more thermally conductive than the rest of the building envelope. This leads to a reduction in thermal performance, lower surface temperatures and an increased risk of surface condensation and concentrated mould growth over these specific features (see Figure 6.29 for a further explanation).

In an effort to reduce this problem, INCA have published a set of construction drawings titled 'Thermal bridging details' (March 2017) to highlight common areas of construction where this might occur. INCA cover a large number of possible thermal bridges together with recommended solutions that can help to minimise these effects. Rather than include them all, we have included a number of examples we have encountered during surveys (see Figures 6.9 to 6.10.

Assessment and reporting

If you see any of these problematic features that could result in cold bridge condensation, you should carefully inspect the corresponding surfaces and features on the

Figure 6.10 External wall insulation system has been fitted to a Wimpey 'no-fines' house. The installer chose to stop the insulation at the DPC level and to cut around the fence post and the cable junction box. This creates a number of thermal or 'cold bridges' where condensation and mould growth could develop on the inside.

inside. You should use your moisture meter (see Section 6.2.2 for further advice on moisture meters) and look for evidence of:

- Direct penetration of moisture;
- Surface condensation and mould growth.

If there are no symptoms at the time of inspection, you should warn your client of the potential for future problems. This should be done in a measured way, but the trouble with cold bridge condensation is that once it begins, no amount of additional

Figure 6.11 In this example, the external wall insulation has been cut around the fixing pole for the satellite dish. In addition, the stop-end of the gutter is clearly defective and will allow water to flow down behind the insulant.

ventilation will cure it: the only solution is often expensive alterations to the EWI system itself.

Government advice for EWI systems with brick slips

In December 2017, the Department for Communities and Local Government (DCLG) issued advice to building owners who have EWI systems with a render or brick-slip finish. Although the advice was primarily aimed at buildings that are over 18 m tall or in an exposed location, it is also relevant for low-rise buildings too. Based on advice from the Independent Expert Advisory Panel set up in the aftermath of the Grenfell Tower tragedy, the advice stated that although properly installed EWI systems can perform satisfactorily, inadequate design and/or poor installation can reduce safety factors making the EWI system more vulnerable to defects, including:

- Damage from high winds resulting in parts of the render coating/brick slips falling away posing risks to life safety;
- Poor water tightness of the render that can result in the insulation becoming waterlogged. This will increase the weight of the insulant (especially mineral wool), resulting in the whole system pulling away from the wall. This will often result in horizontal cracking at insulation board joints (DCLG 2017).

Consequently, when inspecting a property with EWI cladding in an exposed position, you should be alert to these issues.

Fire and EWI systems

Since the Grenfell Tower tragedy, fire risks associated with external cladding have become a great concern, especially for taller buildings. Although most EWI systems do not include a ventilated cavity, they still have to conform to the same regulations. In their publication titled 'Fire Protection Requirements for EWI Systems' (INCA 2015a), INCA outlined the main provisions. It is not appropriate to describe these provisions in detail here, as this is beyond the scope of this book. However, some general points can be made:

- For properties up to 2 stories, the system has to be at least euro class C, B or low risk;
- For properties up to 18m (usually 5 stories), combustibility of at least euro class C, B or low risk and non-combustible fire breaks at every floor;
- For properties above 18m, every floor above the 18m level should have a combustibility of no less than euro class A or the whole system conforms to BR 135 classification. There should also be fire breaks at every floor level.

To put this into context, mineral wool usually has a combustibility rating of class A1/A2 and is considered non-combustible. On the other hand, extruded polystyrenes, phenolic foams, polyurethane and polyisocyanate are normally classified as B or C and considered combustible. Consequently, if you know the type of insulant, then you should be more cautious with the non-mineral wool types.

When considering fire risk in buildings, it is important for residential practitioners to make measured and safe judgements.

6.3.3 Cavity walls

Historical context

According to Historic England, 'hollow walls' were first used in early Victorian times and by 1910, most builders had a version of cavity wall construction in their standard pattern books (Historic England 2016). Although the main reason was to provide better protection against the elements (driving rain especially), cavity walls are also an economic use of materials when compared to a solid wall.

In most areas of the country, the prevailing winds come from the south-west. As a consequence, cavity walls were first used in exposed areas in the north, west and in some coastal areas as far back as the 1890s. In the south-east of the country, cavity walls were not commonly used until after WWII, but beware, you can find early 20th century cavity walls along the channel coast. This emphasises the importance of knowing your area.

When cavity walls were first used, in some parts of the country, builders would ventilate the cavity in an attempt to help control moisture levels in the gap. In these cases, rows of brick-sized terracotta air bricks can be seen at the top and bottom of the main walls.

How cavity walls resist moisture penetration

Depending on the exposure of the property, rain will regularly penetrate through the outer skin. This is most common at the vertical joints, where the amount of mortar included between the bricks is normally insufficient. Water will then run down the internal face of the external skin within the cavity to the base where the angled cavity fill allows the water to harmlessly soak into the ground (see Figures 6.12a and 6.12b).

To prevent this water from transferring to the inner skin of the wall, the method and quality of construction are critical. Problem areas include:

* Poorly installed wall ties on which large mortar droppings have lodged (sometimes called 'snots');
* Absent or poorly installed cavity trays and DPCs above and around doors and windows, meter cupboards and other types of opening through the cavity wall;

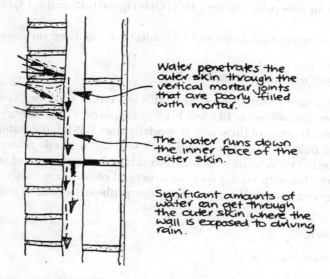

Water penetrates the outer skin through the vertical mortar joints that are poorly filled with mortar.

The water runs down the inner face of the outer skin.

Significant amounts of water can get through the outer skin where the wall is exposed to driving rain.

Figure 6.12a: Cross section through a cavity wall showing the normal route of water in a properly functioning cavity wall (upper level).

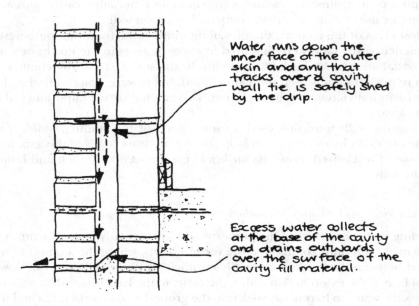

Water runs down the inner face of the outer skin and any that tracks over a cavity wall tie is safely shed by the drip.

Excess water collects at the base of the cavity and drains outwards over the surface of the cavity fill material.

Figure 6.12b: Cross section through the base of a cavity wall showing the normal route of water in a properly functioning cavity wall.

- In more recently built cavity walls, poorly installed cavity wall insulation batts (BRE 1985);
- A build-up of mortar droppings and other debris at the base of the cavity wall that can bridge an otherwise effective DPC (see Figure 6.13 and 6.14).

Issues relating to retrofitted cavity wall insulation are discussed on page 151.

Diagnosis of moisture problems in cavity walls

The first stage is to identify whether the walls are either solid or cavity. For experienced surveyors, this will seem like teaching grandparents to suck eggs, but mistakes have been made. In parts of the south (especially after WWII), the outer skin of some cavity walls was built to resemble a solid brick wall for aesthetic reasons. In parts of the north, external walls were built with two skins of stretcher bond set between 10 and 20 mm apart but without any form of structural connection. In some areas, these are called 'finger cavities' and can cause stability problems as well as moisture defects.

Identifying cavity walls

The main characteristics are:

- **Using brick bond.** The outer skin of a cavity wall is usually constructed using stretcher bond (see Figure 6.15), but in some regions, headers were incorporated in the outside skin of early cavity walls to mimic the more attractive solid wall

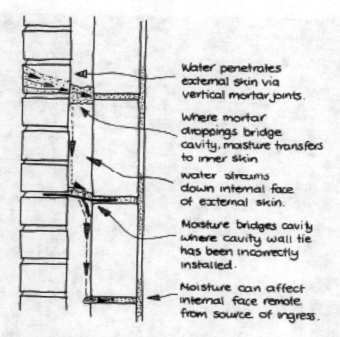

Water penetrates external skin via vertical mortar joints.

Where mortar droppings bridge cavity, moisture transfers to inner skin

water streams down internal face of external skin.

Moisture bridges cavity where cavity wall tie has been incorrectly installed.

Moisture can affect internal face remote from source of ingress.

Figure 6.13 Section through a typical cavity wall showing possible moisture routes.

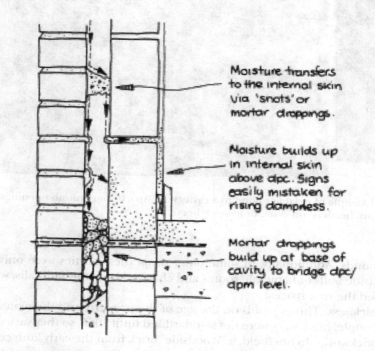

Moisture transfers to the internal skin via 'snots' or mortar droppings.

Moisture builds up in internal skin above dpc. Signs easily mistaken for rising dampness.

Mortar droppings build up at base of cavity to bridge dpc/dpm level.

Figure 6.14 Section through the base of a cavity wall where the build-up of debris and other faults can result in internal dampness problems. This can easily be mistaken for rising dampness.

Figure 6.15 Example of stretcher bond to a cavity wall. Apart from around openings, there are no 'headers' on show, only 'stretchers'.

brick bonds. Additionally, in some older properties, cavities were only used for the outside walls of habitable rooms and changed back to solid walls, where they enclosed the roof spaces;

- **Wall thickness.** This depends on the age of the property and the materials used. For example, brick sizes were not standardised until 1965, so they varied between local brickworks. In Sheffield, a 'Woodside' brick from the early 20th century was typically 240 mm long, 110 mm wide and 80 mm deep – a monster of a brick.

A report on 'hard to fill' cavity walls for the then Department of Energy and Climate Change produced a useful timeline for cavity wall construction (DECC 2012). The most relevant information included the following:

- Although the use of brick and brick cavity walls continued into the early 1960s, clinker/concrete blocks inner skins became increasingly common from the 1950s;
- In new builds, partial and full fill cavity insulation started to be used from the early 1980s;
- Wider cavity widths of 75mm were common from the mid-1990s.

Further increases in thermal requirements in the building regulations resulted in a variety of solutions, including wider cavities (100 mm or more) and thicker 'solar' blocks. Therefore, identifying post 1995 cavity walls by their thickness alone is impractical. The use of prefabricated timber frames and modern methods of construction (such as structurally insulated panels or SIPs) have further complicated matters.

In summary, up to the 1990s, the thickness of most cavity walls is likely to conform to the following (Parnham & Russen 2008):

- Outer brick skin – 102.5mm thick;
- 50mm cavity;
- Inner skin (either brick or a variety of concrete blocks) – around 100mm;
- Plaster finish – assume 15mm;
- **Total thickness = 265 to 280mm** (depending on the materials used).

However, after the early to mid-1990s, the thickness of a wall became a less reliable indicator of its construction.

- **The presence of weep holes**. A cavity tray is a damp-proof course (DPC) that crosses the cavity in order to prevent dampness running down the inner face of the outer skin from getting inside the building and affecting the inner skin. To prevent water from ponding at the base of the cavity tray, a number of vertical mortar joints are left open (weep holes) or fitted with plastic sleeves or 'weep vents'. These can vary between narrow grille types of slots the full height of the vertical joint or small tubes that protrude from the mortar joints (see Figures 6.16, 6.17 and 6.18). These can be very difficult to spot, especially at higher levels.

The presence of weep holes/vents can help confirm a cavity wall; however, cavity trays did not become common until the 1970s, so there are hundreds of thousands of cavity walls without these important damp-proofing features.

Cavity trays and weep holes are not limited to doors and windows; they should also be installed wherever the cavity is 'closed', including along the base of the wall, over gas or electric meter boxes, over sub floor vents and at roof junctions.

Signs and symptoms of moisture problems

In many ways the signs and symptoms of moisture defects in cavity walls will be similar to those of solid walls with the following differences:

- If the fault is associated with wall tie problems (for example, mortar on the wall ties), then the moisture problem will be more isolated although the moisture

Figure 6.16 This shows a weep vent to a DPC tray above a ventilator to a suspended ground floor.

signs (including high meter readings) may be remote from where the water actually gets into the cavity;

- Where the DPC/cavity tray around openings is absent or faulty the defect will be very specific to those features (BRE 1987);
- The build-up of debris in the base of the cavity can give symptoms that resemble 'rising dampness'. Telling the difference between the two is difficult and is discussed in more detail in the section on moisture problems to the base of walls (see Figure 6.13 and Section 6.4).

Moisture problems at the junction of cavity walls and extensions

Where an extension is added to an existing property with a cavity wall, a cavity tray must be inserted into the existing cavity wall and linked with the flashing of the new roof. This is a requirement described in Approved document C of the Building

Figure 6.17 The outer skin of this cavity wall will sit below a coping and it incorporates a DPC tray. The tray is drained by a small weep vent tube. Can you see it? It will be difficult to spot from ground level 3 m below.

Figure 6.18 This shows part of the cavity wall to a single storey extension to a property. The outer skin has been rendered and partially blocked a number of the weep vents. This could result in a build-up of water on the cavity tray and if this has not been properly installed, the water could affect the inside of the property.

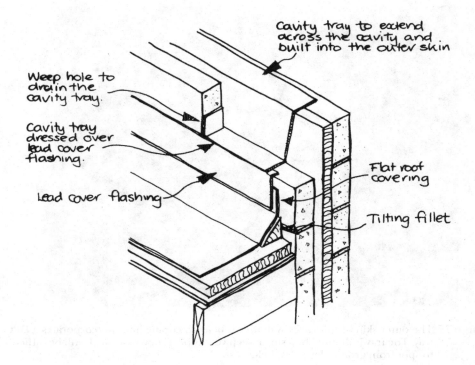

Weep hole to drain the cavity tray.

Cavity tray to extend across the cavity and built into the outer skin

Cavity tray dressed over lead cover flashing.

Lead cover flashing

Flat roof covering

Tilting fillet

Figure 6.19 This is a section through a typical junction of a flat roof extension and a cavity wall. If a cavity tray is not fitted at the same time as the extension, the space below will be vulnerable to leaks especially in stormy weather.

Regulations and included in the technical requirements of most of warranty schemes. Figure 6.19 shows a typical cross section. However, we have come across an increasing number of cases where this important feature has not been included even where the extension may have received building regulation approval.

If a cavity tray is not provided then there will be a risk of water running down the inside face of the outside skin above the new extension and affecting the rooms below especially where the property is exposed to the prevailing weather.

When inspecting a property that has an extension, look out for the following features that could indicate it has been fitted with a cavity tray:

- The wall/render shows signs of disturbance for between one to three courses above the flashing to the extension roof covering;
- There are a number of weep holes/vents above the roof flashing level.

Where there are no signs that a cavity tray has been inserted and the flashing to the roof covering has simply been dressed into an adjacent mortar joint or simply stuck to the wall surface, there will be a risk of moisture penetration. In such a circumstance, you should 'follow the trail' and closely inspect the underside of this feature for any visual signs of water ingress and take a number of moisture meter readings even if that means getting your ladder out.

Where a cavity tray has NOT been fitted, you should warn your client about this potential problem but in a balanced way. The following scenarios may help you formulate an appropriate response:

- Where there **IS** evidence of water penetration (for example, staining on the adjacent plaster surfaces and high moisture meter readings) then urgent action will be required including a new cavity tray and any making good;
- If there is **NO** visual evidence of water penetration and low moisture meter readings then consider the following:

  - Where the property is in an area with relative high exposure zone for driving rain (say zone 3 or 4 as defined in diagram 12 of Approved document C of the building regulations) or in a position that is locally exposed, then although it is not leaking at the moment, the likelihood of this happening in the future is considered significant. To reduce this risk, the client may want to consider installing a new cavity tray in the short to medium term;
  - Where the property is in a lower exposure zone for driving rain (say zone 1 or 2 as defined in diagram 12 of Approved Document C of the building regulations) or in a position that is locally sheltered, the likelihood of this leaking in the future is low. However, if the client wants to be sure they may want to consider installing a new cavity tray as a precaution.

Giving advice will always depend on the circumstances so you will have to use your professional judgement.

Insulated cavity walls

Although the building regulations introduced thermal performance requirements for walls in 1965 (maximum U value of 1.7 W/m2C), it was the energy crisis in the early 1970s that accelerated the tightening of the regulations and in 1975 the maximum U value for walls was reduced to 1.0 W/m2C. Currently, the maximum U value is around 0.17 W/m2C. Achieving such low U values can be very challenging and needs close attention to detail.

Since the mid-1960s, the house construction industry found the air space in a cavity was a convenient location for increasing the amount of thermal insulation. It was a relatively cheap method (especially if it is incorporated as the work proceeds); hidden from view and as long as the dwelling was properly built then the technical risks could be minimised. However, problems occurred when the cavities of older properties were filled with insulation retrospectively.

How cavity wall insulation is meant to work

A number of different materials are used to insulate existing cavity walls:

- Mineral wool, including glass and rock wool;
- Beads or granules made of expanded polystyrene (EPS), expanded polystyrene (EPS), and polyurethane (PUR);
- Foamed insulants, including urea-formaldehyde (UF) and polyurethane (PUR).

Most commonly used insulation materials are treated with some form of waterproofing agent so when the cavity is properly filled and the material compacted it should

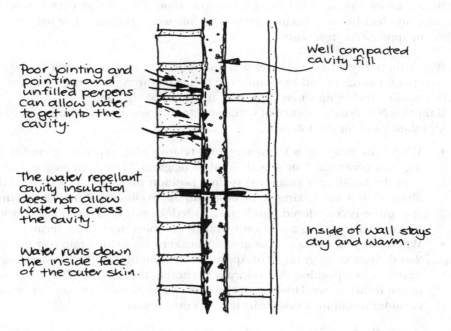

Figure 6.20 This cross section shows how a cavity wall retrospectively filled with insulation should resist moisture penetration (to be redrawn).

not allow moisture to soak through and affect the inside (see Figure 6 .20). To work properly, the cavity (and the insulation) should be at least 50mm thick: anything less will make the building vulnerable to dampness.

This will allow the water to run down the interface between the inner face of the outer skin of the cavity wall and the insulation and as long as it can pass down to the base of the cavity and drain away into the ground then it should not cause a problem.

Retrospectively insulating a cavity wall usually involves drilling holes in the outer leaf of the cavity wall and injecting a variety of insulation products under a slight pressure. The work is done from the outside and once the cavity is full, the holes are filled with mortar. This process is shown in Figure 6.22. Here we will focus on recognising when a cavity wall has been insulated.

Identifying retrofitted cavity wall insulation

The main visual indicator that the cavity has been insulated is by the mortar-filled drill holes in the outer skin of the wall. These can be difficult to spot especially if the wall has been recently repointed. However, here are a few tips:

* The holes are between 10–15mm in diameter (the size of a 10 pence piece) and are usually drilled through the mortar at the junction of the vertical and horizontal joints;

Figure 6.21 This is a drill hole for cavity wall insulation. Although it was positioned at the junction of the perpend and horizontal mortar joint, it has damaged two adjacent bricks. As the mortar used to repoint the hole is poorly matched, spotting that this wall has been retrofitted with cavity wall insulation is usually straightforward.

- No matter how careful the operative, the drill usually damages the adjacent bricks. Consequently, even if the wall is repointed, you should still be able to see this damage (see Figure 6.21);
- Most installers will follow a specific drilling pattern designed to make sure the cavity is properly filled (see Figure 6.22). A defining characteristic is a short row of drill holes just beneath window sills.

Although the insulation will be hidden in the cavity, you might be able to see evidence of the material in semi concealed spaces. For example:

- In roof spaces at eaves level, or to the flank wall where the insulation has 'spilled' out of holes left in the original mortar joints (see Figure 6.23);
- In meter boxes or beneath kitchen sinks; indeed, anywhere that you might expect to find a hole through the wall (see Figure 6.24);
- Behind the holes in air bricks where the cavity has not been sleeved through by the cavity insulation installer.

How cavity wall insulation can fail

In this section we have identified a number of factors that can result in problems.

Figure 6.22 The rendered outer skin of this cavity wall has not been repainted since the cavity was insulated. It shows the typical drilling pattern with the distinctive two or three holes below the window sills. However, if the render is redecorated and the pointing of the drill holes properly done then the evidence might not be so clear.

Figure 6.23 Although this looks unbelievable, this large pile of blown polystyrene beads ended up in the loft because the top of the cavity walls was not sealed.

Figure 6.24 Wherever cavities are not sealed, cavity wall insulation will find a way to escape. In this case, they nearly filled up the meter box. As this type of polystyrene bead is coated with a resin the beads can soon set solid.

Constructional faults

If the cavity insulation has not been properly installed and/or there are pre-existing constructional faults, voids in the cavity insulation can occur and these can redirect the water running down the inner face of the outside skin towards the inner skin causing dampness internally (see Figure 6.25).

Not only can these voids cause penetrating dampness they can result in uninsulated 'cold spots' on the inside face of the inner skin. Condensation and mould growth would soon follow (see Section 6.5).

Insulation less than 50 mm thick

To work properly, the cavity before insulation has to be at least 50 mm wide. However, cavities can often drop below this width because of:

- **Regional building practice**. In some areas of the country (especially the northwest), older cavity walls were built with 'finger' cavities that can be as narrow as 20 mm or less. These were often two skins of bricks in stretcher bond laid alongside each other without any form of structural connection. Identifying narrow cavities can usually be achieved by measuring the overall thickness of the wall. Anything below 250 to 260 mm may fall into this category.

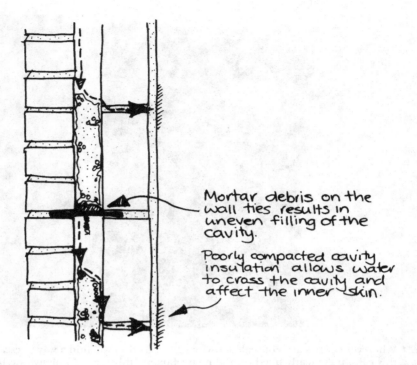

Mortar debris on the wall ties results in uneven filling of the cavity.

Poorly compacted cavity insulation allows water to cross the cavity and affect the inner skin.

Figure 6.25 Sketch showing how pre-existing constructional defects can result in problems with the cavity wall insulation.

- **Where natural stone has been used for the outer skin.** In many cases, although the front face of the stone was dressed, the rear face was left irregular and this reduced the width of the cavity. According to Historic England (2016, p. 5), older forms of 'hollow' stone walls were held together with 'through stones'. In such cases, these would not only provide a route for water to cross the cavity, but they would also result in a cold bridge.

Cold bridging, condensation and mould growth

Cavity walls of all ages will be peppered with numerous cold or thermal bridges. These are parts of a building where there is a direct connection between the inside and out-side by one or more elements or components that are more thermally conductive than the rest of the wall. Some examples are shown in Figure 6.29.

As a result, there will be wasteful heat transfer across the cold bridge, resulting in lower internal surface temperatures that will be different from adjacent, better insulated areas of the wall. If the surface temperature of the cold bridge drops below the dew point for the air in the room, condensation may form and this can result in mould growth.

In a poorly insulated property with an uninsulated cavity wall, the temperature difference between the 'cold bridge' and the rest of the internal wall surface may be

Figure 6.26 This property was built at the turn of the 19th century and the outer face is clearly built in stretcher bond. This suggests cavity wall construction that was in common use in this region at that time. But is it a cavity wall? See Figure 6.27.

smaller in relative terms and, depending on the internal environmental conditions, may not result in any problems. However, if the same cavity is insulated and these cold bridges have not been tackled, the difference in surface temperature between the cold bridge and the insulated wall may increase, resulting in condensation and mould growth where none existed before – a very frustrating problem for the building owners.

Consequently, most commentators recommend that potential cold bridges are identified and insulated before the cavities are insulated. For example, in their publication 'Cavity wall insulation in existing dwellings: A guide for specifiers and advisors',

Figure 6.27 As part of a larger repair scheme on the property shown in Figure 6.26, new air bricks were inserted into the wall at low level. This revealed that the wall consisted of two skins of brickwork with 'finger cavity' that were very narrow. The bricks were 110 mm wide, which explained the 250 mm overall width of the wall. If the wall is 268 mm or over, then it should not have a narrow cavity. Incidentally, the two separate skins were not connected with any form of cavity wall tie and were potentially unstable.

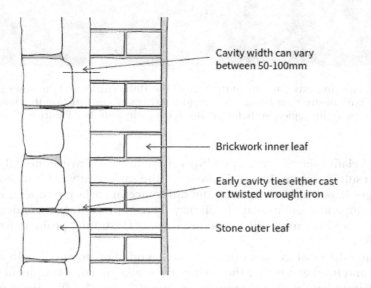

Cavity width can vary between 50-100mm

Brickwork inner leaf

Early cavity ties either cast or twisted wrought iron

Stone outer leaf

Figure 6.28 This shows how the cavity width can be reduced by the undressed face of the stones on the cavity side.

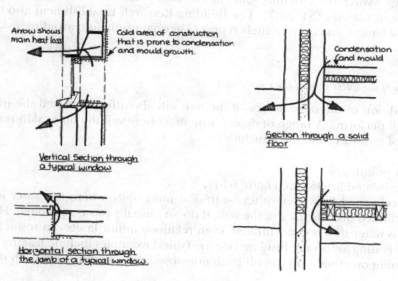

Figure 6.29 Typical locations of 'cold' or thermal bridges in domestic construction.

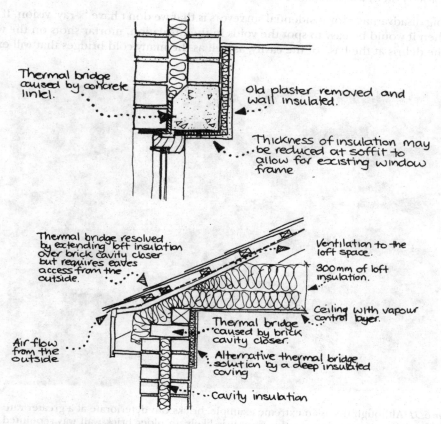

Figure 6.30a and *b*: Examples of how the thermal bridges can be resolved (taken from Thermal insulation: avoiding risks (BRE 2002b).

the Energy Saving Trust identify some of the most problematic thermal bridges that should be insulated (EST 2007). The Building Research Establishment also identify a range of typical situations in their report 'Thermal insulation: avoiding risks (BRE 2002b).

Condition of the outer skin of the cavity wall

The condition of the outside face of the wall will also affect how well the insulated cavity wall performs. A range of defects can affect how well the outer skin resists the passage of moisture, and these include:

*   Poor pointing;
*   Deteriorated bricks (see Figure 6.31);
*   Concentrated flow from other faults. For most walls, where rainwater is evenly distributed over the face of the wall, it doesn't usually cause a problem. However, if this water flow is concentrated, even relatively minor faults can result in moisture getting across the insulated cavity. Typical examples include leaking gutters, running overflows and run-off from non-absorbent cladding higher up the wall.

Assessment of cavity walls that have been insulated

One big disadvantage for residential surveyors is that we don't have 'x-ray' vision. If we did, then it would be easy to spot the voids in the insulation, mortar snots on the wall ties, the debris at the base of the cavity, as well as the many cold bridges that will exist

Figure 6.31 Although this is an extreme example, bricks can deteriorate at a greater rate than the mortar joints. In this case, it is likely an older brick wall was repointed with cement mortar that accelerated the deterioration of the bricks themselves.

in most dwellings. As most practitioners have not been blessed with these superpowers, clients are told in the terms of engagement that hidden parts of the building cannot be assessed. Although this is an important part of the contract, if this happens with too many elements in a building, clients will still be disappointed with the resulting report.

In an attempt to provide clients with more useful advice but without recklessly increasing our liability, we think it is important to try and put the client in what we call 'the right ballpark' – it is an approach we describe in a number of different chapters in this book. This involves collecting all the necessary information that you need to make a broad 'risk assessment' of the likelihood of problems caused by any particular feature – in this case, the cavity wall insulation.

The cavity wall insulation risk assessment protocol

This protocol consists of a number of stages:

- Collecting the appropriate information;
- Use this information to complete the 'risk assessment' protocol; and
- Use the outputs to give balanced advice to the client.

Collecting the appropriate information

In the preceding section, we have identified the general problems that could result in moisture penetration through a cavity wall and you should be identifying these during your inspection. Where a cavity wall has been retrospectively insulated, you will have to collect some more information so you can make an appropriate assessment. This includes:

The exposure of the property to driving rain

The property's exposure to driving rain will play a significant role in determining whether the cavity walls were suitable for insulation in the first place and, if they are, what type of insulation should have been used. This rather detailed process is beyond the scope of any pre-purchase report, so the following approach has been adapted from that described in the BRE's Good Building Guide 44 part 2 (BRE 2001) for buildings up to 12 m high.

- Using the map shown in Figure 6.32, determine the wind driven rain zone. Those most at risk are zones 3 (severe) and 4 (very severe). This exposure zone map is also included in Approved Document C of the English building regulations (diagram 12);
- However, it is not only about these broad exposure zones. Local conditions, such as open hillsides or valleys, can create problems by funnelling winds onto the wall and so increasing the likelihood of moisture penetration;
- Conversely, walls can also be well protected by trees and other buildings reducing the chance of problems.

Is the property of non-traditional, prefabricated or timber-framed construction?

Although many framed properties incorporate air cavities within their construction, filling these spaces is not recommended, as it will interrupt the behaviour of moisture within the cavities.

Exposure zones	Approximate wind driven rain *
1	Less than 33
2	33 to less than 56.5
3	56.5 to less than 100
4	100 or more

(litres/m² per spell)

Key to map

Figure 6.32 National exposure to wind-driven rain zones in the BRE's Good Building Guide 44, Part 2.

Is the property three or more stories high (more than 12 m)?

Some of the issues associated with cavity wall insulation become more critical when the wall is 12 m or higher and require insulation methods approved to be used at those heights.

Have the cavity walls of the property been subject to alterations, improvement or repair work?

Although it will always depend on the circumstances, some cavity wall insulation guarantees/warranties can be invalidated if the insulated cavity walls are disturbed in some way. Although many of the schemes are not clear about the type of 'disturbance' this constitutes, it is likely to include:

- Cavity wall tie repair/replacement;
- Physical replacement of the damp-proof course;
- Window replacement or the provision of new window/door openings;
- Provision of new openings through the wall (for example, new extensions and so on).

If you see evidence of significant disturbance to the cavity wall has occurred since the wall had been insulated, you should report this to your client and make clear recommendation that their legal adviser should check the cavity wall insulation/guarantee to see if it is invalidated by subsequent repair/improvement work and explain the implications.

The cavity wall insulation risk assessment protocol

Over the last 20 years, we have produced a number of straightforward protocols that aim to simplify the more complex processes described in the various authoritative publications. These protocols are designed to support your decision making and NOT to make the decision for you – the final judgement is based on your own professional view.

The cavity wall insulation assessment protocol poses a number of questions, each with a 'yes' or 'no' answer. This binary choice is admittedly crude, as in reality the choice is never black or white but shades of grey. Still, we think following this approach will put you in the right 'ballpark' and support your decision.

The other advantage of this protocol is that it provides a record of your decision. If someone questions your judgement at a later date, then this structured approach will enable you to clearly explain how you came to that view – a characteristic of a competent practitioner.

In addition to the foregoing issues, the questions in the protocol also include the following questions:

- **Was the property built before 1939?** Although this may appear arbitrary, we have assumed that the use of cavity walls became more common after World War II with a greater use of galvanised wall ties;
- **Weep holes** – if the wall has weep holes, then they are likely to be better draining and less susceptible to dampness problems;

- **Ten years left on the guarantee** – If guarantees have any value, then they should have a reasonable time to run. As the average homeowner moves every eight or nine years, then ten years feels the right amount of time;
- **Condition of the outside face of the wall** – condition clearly has an impact on the walls' ability to keep moisture out but so does the type of mortar joint. For example, a fully recessed mortar joint in a satisfactory condition could still leave the outer skin vulnerable to water penetration.

How to use the protocol

There are nine questions in this particular protocol and all have been drawn from the most important publications and guidance on cavity wall insulation. Carefully consider each question in turn and answer 'yes' or 'no'. Once you have responded to all the questions, you should make a professional judgement. Please note: the responses 'yes' or 'no' have been distributed between both the left- and right-hand sides.

Although this is not a mathematical formula, responses usually conform to three different profiles:

- The clear majority of the responses are on the left of the table (coded green). This suggests that the insulated cavity wall is less likely to be a problem as long as the walls are properly maintained in the future;
- The clear majority of the responses are on the right of the table (coded red). This indicates that the insulated cavity wall is more likely to be a problem and requires further investigation by an appropriately qualified person;

Less likely to be a problem	Assessing insulated cavity walls	More likely to be a problem
No	Was the property built before 1939?	Yes
Yes	Does it have weep holes over most openings?	No
No	Do natural stone blocks form the outer leaf?	Yes
Yes	Is the cavity wall insulation covered by an insurance backed guarantee with at least 10 years to run?	No
Yes	Is the outside face of the wall in a satisfactory condition and /or has flush joints?	No
No	Is the property in zone 3 or 4?	Yes
No	Is the property locally exposed (what ever the exposure zone)?	Yes
No	Is the dwelling a prefabricated property (concrete or steel) or timber framed?	Yes
No	Have the walls been disturbed by repairs and/or improvements since they were insulated??	Yes
Yes	Is the property less than three storeys?	No

Figure 6.33 The cavity wall insulation risk assessment protocol.

- The responses are a mixture of the two. Your own professional judgement will be very important in these marginal cases. We think exposure to driving rain should have the biggest influence on your choice.

Giving advice

Your report to the client should be built around the following framework:

When insulated cavity wall is LESS likely to be a problem

Inform the client:

- How the wall is constructed, that it is insulated and has a guarantee;
- Of the benefits of cavity wall insulation but outline the growing controversy;
- That you saw no evidence of problems and the likelihood of future problems are low as long as the wall is satisfactorily maintained;
- They should ask their legal adviser to check the validity of the guarantee and explain the implications.

When insulated cavity wall is MORE likely to be a problem

Inform the client:

- How the wall is constructed, that it is insulated and has a number of problems and doesn't have a guarantee or it is out of time;
- Of the benefits of cavity wall insulation but outline the growing controversy;
- Of the problems you identified and their implications;
- The need for further investigations by an appropriately qualified person and the possible implications.

The particular problems with UF foam cavity insulation

One of the first insulants to be used in the late 1960s and throughout the 1970s was UF foam, which involves the injection of a water-based chemicals system that produces an expanding insulating foam in the cavity. Some estimates suggest that up to 1 million homes were insulated with UF foam. One advantage is that as the foam expands, it can fill all the nooks and crannies within the cavity. However, a number of problems soon became apparent:

- When the foam was injected, it gave off a vapour that could result in an allergic reaction. If the internal leaf was not properly sealed, this vapour could affect the occupants. In the 1970s, a number of buildings were injected with UF foam and resulted in symptoms of nausea, sore eyes, skin rash and headaches being reported by occupants. As a result, the use of this insulant was included in the building regulations (see Approved Document D Toxic substances 2015);
- Although UF foam solidifies within a few hours, it will continue to 'cure' over the next few days. During this time, the foam can shrink back, resulting in cracks and gaps and these can redirect water to the inner skin, as illustrated in Figures 6.34 and 6.35.

Figure 6.34 This cavity-walled property was built in the 1970s. It has a dressed stone outer skin and concrete block inner skin. According to the owner (who designed the house), the cavity was insulated from the inside before the walls were plastered.

Figure 6.35 This is the view of the gable wall from inside the loft. A number of mortar joints were not properly sealed, allowing the foam to squeeze through. This evidence should initiate a 'follow the trail' process.

For these reasons, UF foam has developed a poor reputation. Despite this, it is still used to insulate cavity walls, especially those that are 'hard to treat'.

In existing properties, it is important for residential practitioners:

- To identify older installations of UF foam wherever possible;
- If UF foam has been used, carry out a more thorough inspection as there will be a greater likelihood of defects with this system; and
- Even if there are no visible problems, warn clients about the potential problems this system could pose in the future.

Identifying UF foam installations

It is very difficult to identify a material that is hidden. However, if the inner skin hasn't been properly sealed (and it never was), you might be able to spot the foam squeezing out of gaps. Typical locations include:

- Through any open mortar joints and around embedded timbers in roof spaces, understairs cupboards and within underfloor areas;
- In airbricks that haven't been sealed properly; and
- In meter boxes where the cavities have been left open.

Remedial work for all types of problematic insulated cavity walls

Where further investigations identify problems with the cavity wall insulation, the nature and extent of the remedial work will always depend on the circumstances but could include the following:

- Repairing the walling – this could include repointing, render repairs and cutting out and replacing deteriorated bricks and so on;
- Adding impervious cladding – where the wall is very exposed, additional protection maybe required especially on the higher storeys;
- Opening up the cavity wall to remove any debris/constructional faults;
- Install cavity trays and weep holes – this can be an expensive process where there are large numbers of openings;
- Insulation of cold bridges – increasing the thermal resistance on the internal faces of lintels, sills and other features can be disruptive and require redecoration internally.

In some cases, where the cavity wall insulation is found to be the cause of the problem, it may be appropriate to remove it. To do this, small patches of bricks are taken out along the top and base of the walls, so the existing insulation can be removed by a combination of air being blown into the upper holes and sucked out along the bottom. Any debris/constructional faults can be removed and the appropriately insulated either externally, internally or using a cavity insulant that is matched to the conditions.

6.4 Moisture-related defects at the base of walls

6.4.1 Introduction

'Rising dampness' is probably the best known of all types of moisture-related defects in this country. It is part of our cultural tradition and has even had a 1970s television

Figure 6.36 A brick has been removed from the rendered outer skin of this brick/cavity/brick
wall to a property built in the early 20th century. The base of the cavity is full of
debris.

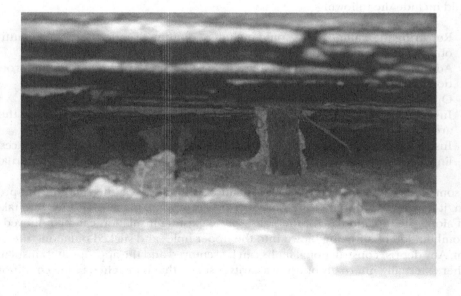

Figure 6.37 This photo shows the cavity of the wall shown in Figure 6.36. The underside of the
thick metal cavity wall ties with large amounts of mortar droppings can clearly be
seen.

Figure 6.38 This is the view on the inside of the same property. The visible damp patches correlate to the debris in the cavity. Resolving this problem will be a complex and costly operation in which the cavity will have to be opened up in numerous locations.

sitcom named after it. It can also be one of the most damaging forms of moisture problems and is difficult to resolve. This section will not focus on either the science or the solutions; instead, it will concentrate on how to recognise and report on the problem.

Traditionally, capillary action was seen as the main culprit allowing free water to rise from the damp subsoil into the drier wall. Many commentators claim that the height of rising moisture depends on a number of factors:

- **The amount of moisture in the soil** – Some sites can be wetter than others. The height of the water table is very influential;
- **Weather conditions** – 'Rising dampness' does not react immediately to changes in the amount of precipitation but over the medium term it can become worse during the wetter months;
- **Internal and external finishes** – Some finishes can prevent the wall from drying out. These typically include renders, tiling, wall boarding, vinyl-based wallpaper and paint and so on. These features can force the moisture up to levels higher than would normally be expected;
- **High internal humidities** – If the internal air is very moist then the drying effect of the wall will be reduced. As a consequence, it can rise higher and affect a larger area.

It is a commonly held view that 'rising dampness' rarely gets higher than 1000–1500 mm.

6.4.2 Is rising dampness a myth?

A debate has developed in the property and construction sectors over the last 20 years about the mechanisms of 'rising dampness'. When an academic at South Bank University tried to recreate this phenomenon in the laboratory, he failed to do so despite using a combination of materials and construction techniques (Howell 2008).

Writing in the Independent in November 1997, Jeff Howell interviewed Mike Parrett, then manager of the housing disrepair team for Lewisham Council. Parrett's team used calcium carbide meters (often called 'Speedy meters') to more accurately check moisture levels in walls and test for the presence of ground salts. According to Parrett, they did not find a genuine case of moisture being drawn up from the foundations.

Further fuel was thrown on the fire when the BBC released a programme on rising dampness as part of a series called 'Raising the roof' in 1999. This six-part series investigated the 'murkier' activities of construction professionals. The 'fly-on-the-wall' programme filmed a number of specialist contractors giving inappropriate advice to homeowners.

Since that time, a number of high-profile commentators and organisations have contributed to the debate. We have included some short headline quotes that may help give you a flavour:

> Stephen Boniface is a Chartered Building Surveyor with many years' experience of assessing, repairing and renovating older buildings. In a longer opinion piece on his company's website, he concluded, '. . . rising damp may exist in some situations, but it is rare. Often the cause of what seems to be rising damp can be found to be something that has happened to that building – what has been done to bring about a change to the building? I am not saying that there will never be a need for specialist intervention but in my experience, it is extremely rare'.
>
> (Boniface 2019)

Yet, despite these views, the leading authorities in the construction and property sectors continue to recognise 'rising dampness' as just one of the mechanisms by which moisture can affect buildings. For example:

- BS 6576:2005 (plus A1 2012) Code of practice for diagnosis of rising damp in walls of buildings and installation of chemical damp-proof courses (BSI 2012);
- Building Research Establishment BRE Digest 245 (2007 edition). Rising damp in walls: diagnosis and treatment (BRE 2007);
- Property Care Association: Investigation and control of dampness in buildings. March 2017 (PCA 2017).

Changing terminology

Faced with this debate in the industry, it is a challenge for residential practitioners to objectively diagnose the cause of 'rising dampness' and advise their clients in a balanced way. In keeping with our approach described in Section 6.1 of this chapter, we have attempted to develop a more objective approach to the diagnostic process. Rather than continue to use the controversial term 'rising dampness', we will refer to 'moisture-related defects at the base of walls'. Although some may think this is indulging in pointless semantics, we think it is the first step in preventing practitioners from jumping to early conclusions.

6.4.3 The causes of moisture-related defects at the base of walls

A number of different causes can be identified:

Failure of existing DPCs

Over the years, designers and builders have been aware of the need to provide an impervious layer within the thickness of walls to prevent moisture problems. Since the Public Health Acts in 1875, physical damp-proof courses (DPCs) have become common in walls and materials have included:

- Slate;
- Lead/copper;
- Bitumen impregnated felts;
- Non-absorbent engineering bricks (usually blue in colour);
- Specially manufactured ceramic bricks.

Figure 6.39 This shows the remnants of a hessian-based DPC that had been soaked in bitumen before being incorporated in a solid wall built in 1915. The DPC had been damaged by the demolition of the wall above. Although this DPC was brittle, it was still performing satisfactorily – the brickwork below was very moist, but the course above it was satisfactory.

Although some of these original DPCs may have deteriorated over time, according to the BRE, if the building has a DPC, it is unlikely to have failed, as most DPC materials have a very long life (BRE 2007). However, slate DPCs can crack following movement in the walls, metals can corrode and bitumen-based products will become brittle. Therefore, where older buildings have an older DPC, you should not automatically assume it is 100% effective. Even where a DPC has physically 'failed', it doesn't mean moisture will rise up the wall – this will still depend on those factors identified in Section 6.4.1.

Bridging of the DPC

A major cause of low-level moisture problems is through physical bridging or the bypassing of the DPC (if there is one). Many of these examples are associated with post-construction alterations and repairs, especially those of a DIY nature and typically include (BRE 2007):

- External paths and/or ground levels have been increased so they bridge the DPC. Where moisture levels in the ground are high, large amounts of moisture can get around the DPC, affect the wall and damage the plaster and decorative surfaces inside. In these cases, the moisture gets into the wall laterally, which is a very different mechanism to rising upwards;
- External render and/or internal plaster coatings have been taken down to the outside ground or floor level and so bridge the DPC. Although the bridging pathway will be relatively small, these finishes will also prevent the base of the wall from drying out, especially if the render and plaster materials are impermeable. This will increase moisture levels towards the base of the wall;
- Pointing over of the DPC. Although this does provide a physical pathway for moisture, as the depth of the repointing is usually less than 10 mm or thereabouts, it is unlikely to cause a serious problem by itself. In these cases, you should look for another contributing factor;
- Lack of continuity between the DPC in the wall and the damp-proof membrane (DPM) in the floor. To properly exclude moisture, it is vital that these two damp-proofing elements are positively linked. It is common to find a complete absence of DPMs in solid floors in older properties (especially those built before WW2). Even when DPMs became more commonplace, they were often laid on the surface of the slab, making it difficult to provide an effective DPC/DPM link. This can provide a large bridging pathway where much of the moisture travels laterally;
- The base of cavity walls filled with rubble. The base of many cavities will be filled with debris from a number of sources, including the original construction (mortar 'snots', for example) or from subsequent work (dust and masonry shards from cavity wall tie repairs and cavity wall insulation). Where it is compact, this debris can provide a route for moisture to bypass the DPC (see Section 6.3.3 also);
- It is rare that vertical DPCs have been provided where the boundary and other garden walls meet the main external wall, and this can allow moisture to bridge over.

It is our view that many cases of 'rising dampness' are clearly assisted by these bridging mechanisms.

Moisture problems caused by hygroscopic salts

In Section 6.2.2 of this chapter, we mention the problems that hygroscopic salts can pose when using a moisture meter. Although these salts are usually associated with moisture problems at the base of walls, they can cause problems by themselves, even where moisture levels in the walls have reduced. Consequently, it is important to understand what they are.

All building materials and the ground beneath a structure will contain a range of soluble salts and these can be taken into solution by moisture/water moving through the capillaries or pores of these materials. This moisture will migrate to the surface of the material and evaporate, leaving behind the salts that crystalize and remain on or near the surface.

When this occurs during the early drying-out stages of a new building, it is called efflorescence and is usually a temporary and harmless phenomenon. However, where the moisture that affects the building originates in the ground, it can become contaminated by the salts of nitrates and chlorides that are hygroscopic. Hygroscopic salts readily absorb moisture from the air and will collect along the drying edge of the moisture-affected area, often forming a narrow band of furry white salts.

Hygroscopic salts can have the following effects:

- They can disrupt plaster and decorative finishes. For example, the salt crystals can collect behind wallpaper and paint films, causing unsightly blisters and blebs. If these are pressed gently, the scrunching sound of the salt crystals can be heard;
- The hygroscopic salts can readily absorb moisture from the air in the building even where the source of the moisture in the wall has been dealt with.

Because of these effects, whatever the cause of moisture problems, most commentators recommend that salt-contaminated plaster be replaced to avoid continuing moisture problems.

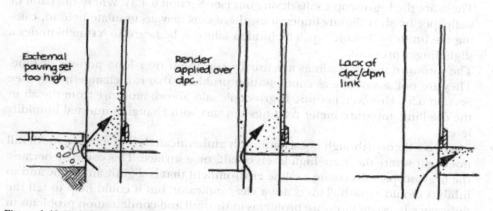

Figure 6.40 Examples of moisture problems at the base of walls that have been caused by the physical bridging of the damp-proof course.

6.4.4 The signs and symptoms of moisture-related problems at the base of walls

The classic symptoms

As discussed in Section 6.2, during the survey, you should look for both visual signs as well as using your moisture meter.

Signs and symptoms

The first step is to make sure that the problems have not been caused by moisture from another source. The BRE (2015) suggest the following sources are considered:

- Are there any leaking gutters or downpipes close by?
- Is there a problem with rainwater run-off from window sills, copings and so on?
- Is the house less than 5 years old? If so, it could be drying-out moisture. This could apply to houses that have been extensively refurbished;
- Are there any leaks from plumbing, wastes pipes, washing machines, condensate pipes from boilers and so on?
- Could it be condensation? Persistent condensation can often 'fool' the moisture meter and result in a misdiagnosis (see Section 6.5.4).

Signs and symptoms typical of moisture problems at the base of walls include:

- Persistently high moisture meter readings from the base of the skirting up to the limit of the affected area (see Figure 6.41). Although many commentators state that moisture problems of this type rarely extend beyond 1 metre above the floor level, they can occur in bands of any height below that level. Readings will gradually decrease higher up the wall but increase again at the boundary of the affected area. This is because hygroscopic salts are normally active at this level and will register higher readings than on adjacent unaffected surfaces;
- The decorations will be spoilt and the plaster finish may be deteriorating. There will be a characteristic 'tide' mark towards the base of the wall. There may be a band of furry white crystal growth or salts at the extreme edge of the affected area. These are the hygroscopic salts drying out (see Section 6.4.3). Where the paint or wallpaper finish is slightly impervious, these salts may accumulate behind, causing the finish to 'bubble' up. The hidden salts can be heard to 'crunch' under a slight finger pressure;
- The moisture meter readings are usually consistent over long periods of time. They are not as variable as condensation problems that can change hourly (see Section 6.5). However, because hygroscopic salts absorb moisture from the air in the dwelling, moisture meter readings can vary with changing internal humidity levels;
- There is a theory (though not scientifically authenticated) that mould growth will not occur where there are high levels of salts on a surface. This could be because the hygroscopic salts create a saline environment that is slightly antiseptic and so inhibits mould growth. This is not a clear indicator, but could help to tell the difference between moisture problems in the wall and condensation problems in some cases.

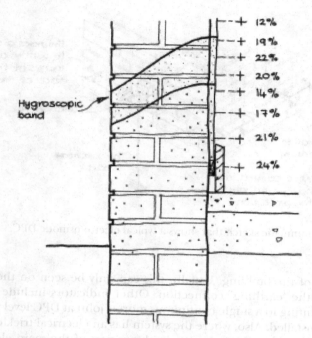

Figure 6.41 Section through a typical solid wall that has been affected by 'rising dampness'. This shows the variation in moisture meter readings that can be expected. Note the higher values in the hygroscopic salts band.

6.4.5 *Moisture problems to the base of walls that have been treated for 'rising dampness'*

Since the 1960s, the rise in popularity of a variety of remedial treatment systems has meant that a significant proportion of houses have had some form of treatment in the past. We have seen one property that has had four separate treatments over the years.

Even where these treatments may have been appropriate, many systems are approaching the end or even exceeding their guarantee period. On the other hand, some systems may have been installed following a misdiagnosis and never worked in the first place. In these situations, special care must be taken in diagnosing and reporting on the problem. The typical range of remedial treatment systems could include:

Injected chemical damp-proof course – A row of holes can be seen a few brick courses above external ground level. These may be pointed up with mortar or plugged with specially made plastic inserts. Some systems involved drilling and injecting the bricks themselves, whereas others relied on injecting the mortar bed joints.

Electro-osmotic damp-proof course – These could be difficult to spot. A continuous copper or titanium wire is embedded in a horizontal mortar joint around

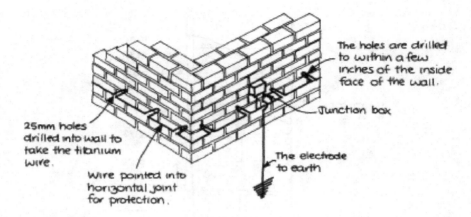

The holes are drilled to within a few inches of the inside face of the wall.

Junction box

25mm holes drilled into wall to take the titanium wire.

The electrode to earth

Wire pointed into horizontal joint for protection.

Figure 6.42 A diagrammatic sketch that shows a typical electro osmotic DPC.

the whole of the building. Usually, this can only be seen on the wall where it joins with the 'earthing' connection. Other indicators include relatively new mortar pointing to a single continuous mortar joint at DPC level where the wire has been installed. Also, where the system has an electrical trickle charge, there may be a transformer connection in the region of the main electrical switchboard (Figure 6.42).

- **'Dalton' drying tubes** – These are large diameter (25–30 mm) ceramic tubes that are fitted into larger holes that have been 'cored' out of the wall. They are usually set at 150–250 mm centres. The open ends of these tubes are often covered by a small grid (Figure 6.43). Although these are often used in older buildings to help the walls dry out more naturally, the pores of the tubes can soon become clogged with salts left behind as the moisture evaporates, rendering them ineffective.
- **The Schrijver system** – This is a relatively new product that has been used in the Netherlands for a number of years. With this system, a number of small, manufactured ventilators are installed in the outside face of the wall. According to the manufacturer, these are positioned just above the internal skirting board level. Apparently, the outside air flows across the small openings in the ventilators and natural airflow transports the moisture to the outside. In some cases, we have seen Schrijver ventilators fitted at first floor level (Schrijver 2021). This is a controversial system, and writing in the Daily Telegraph in 2010, Jeff Howell reported: '. . . I was first contacted by Frank Schrijver in 1997, when he asked me to help publicise his so-called "invention". I asked him then to provide me with independent scientific evidence to support his theory and now, 13 years later, I am still waiting.' An example of the Schrijver system is shown in Figure 6.44.
- **New physical DPC** – depending on the method used this can usually be recognised by evidence of new brickwork and/or mortar just above ground level. Some contractors may have replaced one or two courses of brickwork when they inserted the new DPC, while others may have 'cut out' just a single mortar joint in a more surgical fashion.

Figure 6.43 This photograph shows ceramic 'drying tubes' that have been installed in a brick wall. The smaller filled holes close by suggests that a chemical damp-proof course may also have been installed.

Figure 6.44 This shows the base of a single-storey bay window to a terraced house built in the late 19th century. The walls are solid brick with an ornamental stone plinth just above ground level. The Schrijver ventilating components have been fitted in the second course above the top of the plinth. Keen-eyed readers will also see evidence of two chemical injection DPCs: one in the brick course above the stone plinth and the other through the stone plinth itself. This is clearly a property with unresolved moisture-related problems.

It is difficult for surveyors to advise clearly on these matters because there is such a wide range of opinions within the industry itself. Our only advice is to follow an objective diagnostic process so you can evaluate whether the DPC system is working satisfactorily.

6.4.6 Replastering associated with remedial work

Whatever 'damp-proofing' system has been used, according to the Property Care Association, 'Water rising from the ground often introduces contaminating salts into the walls and plaster coats. This contamination will often result in a need for the plaster to be removed and replaced using specially formulated salt resistant plasters' (PCA 2020).

According to the BRE (2007), several different replastering systems can be used:

- Cement and sand plasters: mixed and applied properly, cement-based plasters are effective moisture barriers in their own right. However, their vapour permeability is low and their strength is very high, so they may not be suitable for all buildings, especially those built of softer masonry and lime-based mortars. They are also very dense and can have high thermal conductivity values. In some cases, they can result in surface condensation;
- Renovating plasters are available but according to the BRE these should be used only if the cause of the dampness has been removed. They also recommend that a product should be used only if it has been assessed as suitable and covered by third-party certification (for example, the British Board of Agrement certificate or BBA);
- Dry linings: in conjunction with a new DPC the BRE advise that there could be circumstances in which linings may be more suitable for a directly applied plaster system. Typical examples include walls heavily contaminated with hygroscopic salts, most likely where an agricultural building is being converted or where flooding has occurred. Many contractors now use proprietary moulded plastic membranes with plaster finishes or linings with plasterboard on metal tracking systems (see Figure 6.45).

6.4.7 Chemical DPC installations, guarantees and call backs

Anyone who has been closely associated with the remedial treatment of 'rising dampness' will know that a 20–30-year guarantee does not automatically mean that is the end of any problems. Where a property has been treated and dampness re-occurs, there could be a number of difficulties getting the guarantors to honour their commitments:

- The company might not be in existence anymore. If there is no insurance-backed guarantee, then there is no easy remedy;
- If the company is still trading then they will wish to investigate the problem for themselves. In many cases, when reputable contractors discover that the chemical DPC has indeed failed, they will carry out remedial work. Sadly, the industry is characterised by less professional operators who use a whole variety of excuses why the new dampness is not their fault. In our experience, these can include:

  - 'It's not rising dampness, it's condensation. Open your windows more';
  - 'Somebody used the wrong sort of plaster on this wall';

- 'It's not the DPC, it's a leaking gutter outside';
- 'Sorry but we didn't inject this bit of the wall. Our guarantee only covers the bits we did'.

These can be appropriate explanations for why moisture problems have returned to a property. Yet the experience of many commentators and surveyors has resulted in a more cynical view. This is because even where a contractor accepts responsibility, the terms of the original guarantee may only cover re-injection and not any of the replastering and making good of decorations that occur as a result.

When a contractor is prepared to carry out remedial work, there is some doubt whether it is appropriate to re-inject a wall that has already been treated. Concerns include:

- Will the new chemical be compatible with the original?
- Will any re-drilling damage the masonry causing cracking within the heart of the wall? This could lead to a loss of injection fluid, which might result in a poor-quality DPC.

6.4.8 Reporting on moisture problems at the base of walls

Stopping moisture rising up an existing wall where the original DPC is either missing or defective is a very complicated and skilled operation. There are so many possible

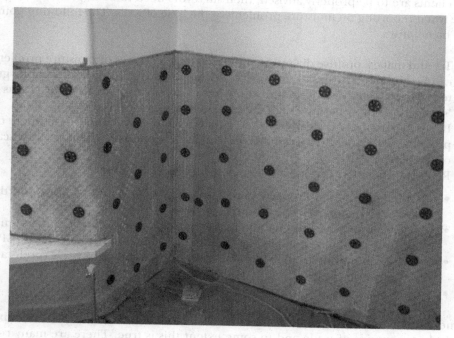

Figure 6.45 This shows a proprietary moulded plastic membrane used to stop residual moisture and hygroscopic salts from affecting the final plaster finish.

solutions that the BRE has acknowledged this difficulty. In *Digest 245* 'Rising dampness' (BRE 2007), they state:

> 'The methods for installing a new DPC are described generally as 'traditional' (the insertion of a physical damp-proof course) or 'non-traditional'. It is strongly recommended that non-traditional methods should be considered only if they have been awarded a third-party certificate. The only method currently satisfying this requirement is chemical injection. This is the only method which BRE considers suitable where insertion of a physical DPC is not possible'.

This clear view suggests that only two forms of DPC installations meet the BRE's requirements: a new physical DPC or a chemical injection.

Most commentators and reputable contractors also recognise that the insertion of a new DPC is not enough on its own and state that any remedial work should be seen as a system that includes the following components:

- The chemical injection or physical DPC itself;
- Replastering with a salt-retardant; waterproof plaster; or suitable dry lining;
- Building repairs and future management such as keeping the property in good order.

If all of these are not addressed properly, then the treatment scheme is unlikely to be effective.

If clients are to be properly advised, then surveyors need to be as proactive as possible when reporting on moisture problems at the base of walls. The following points may be helpful:

- Try and make a positive diagnosis of the defect. Many cases of moisture problems can be identified by straightforward repair of obvious defects. For example, gutters, window sills, overflows and so on. Yet many surveyors refer their clients to 'specialists' simply because high meter readings are registered in a dwelling;
- If 'rising dampness' is clearly suspected then try and give an indication where the problems were noted. This can help give an indication to the client as to the scale of the defect;
- Refer the client to a reputable damp-proofing specialist who:
  - Is a member of the Property Care Association (PCA) and employs staff that have been on appropriate training courses;
  - Issue insurance-backed guarantees that are approved by the PCA. This is usually those products issued by the Guarantee Protection Insurance Ltd (GPI);
  - Have been trading in their present name for a number of years (preferably ten years or more);
  - Are prepared to give references from jobs recently completed.

Some commentators might be concerned that referring clients to PCA members might be a restraint of trade and to some extent this is true. There are many competent damp-proofing specialists who are not part of this trade association, but the problem remains of how to help clients choose from a broad and unknown field.

Membership of the PCA is not an automatic guarantee for good quality, but it does represent a benchmark and provide your client with some form of recourse.

Where a particularly complex problem or potential dispute exists, consider referring your client to a building surveyor or other impartial professional who has a proven deep understanding of moisture-related problems.

6.5 Condensation problems

6.5.1 Introduction

Traditionally, condensation used to be associated with older 'public' housing, especially those prefabricated systems built after the Second World War. However, this type of defect can be found in all different housing types, especially as the current building regulations have increased insulation levels and reduced air infiltration rates.

One of the great myths surrounding condensation is that it is closely associated with the lifestyles of the occupants. When it is observed in a dwelling, we have come across a remarkable range of advice from surveyors, including:

- Open the windows and turn up the heat;
- Lifestyle changes including (and these are all direct quotes):

  - Don't take too many baths;
  - Don't let your kettle boil too long;
  - Don't dry clothes indoors;
  - Put lids on your saucepans; and
  - Get rid of your pet fish!

- Wash the walls down with bleach;
- Don't put furniture up against the external walls.

If this sort of advice is given following a carefully considered diagnostic process, then it could help in resolving the problems. In practice, it often results from a misunderstanding of what condensation is and why it occurs. The surveying profession has a culture of blaming the occupants of a dwelling for this type of defect and we all need to take a broader view.

The task facing the surveyor carrying out a pre-purchase inspection or survey is to objectively assess the true cause of the problem. Then the client can be advised whether the defect:

- Is indeed transient and will disappear with a change of living conditions; or
- Is a result of inherent constructional inadequacies that require significant financial expenditure.

6.5.2 What causes condensation?

All air contains water vapour. This is moisture in the form of an invisible gas. The warmer the air, the more water vapour it can hold. When warm air hits a colder surface, it is cooled and its ability to hold water vapour is reduced. If it is cooled enough, then some of that water vapour will 'condense', forming tiny droplets of water on the cold surface. This is what is known as 'condensation'. The temperature at which this occurs is called the 'dew point'. This dew point temperature will vary depending on the amount of water vapour in the air. If there is a lot (for example, the air is saturated, high humidities), then a slight drop in temperature will cause condensation on the coldest surfaces. If the air doesn't contain much vapour, then the air temperature would have to drop by a lot before condensation begins. The measure of how much water vapour the air is holding is called 'relative humidity' and occurs within a range of 0–100%. Figure 6.44 gives an explanation of how relative humidity varies with temperature.

6.5.3 What is condensation?

This section will not provide an in-depth review of the fundamental science of condensation. Instead, the main principles will be quickly reviewed.

One of the biggest challenges for the surveyor is to decide when condensation becomes an identifiable defect. For example, after cooking a big meal, taking a bath or hosting a house party, the high levels of moisture produced in any household will always result in condensation on window panes, mirrors and so on. In most cases, this will disappear after an hour or two. Condensation becomes a defect when condensation is persistent and a regular feature during the colder months. High levels of moisture can discolour decorations, cause mould to grow and in the worst cases lead to dry rot, as well as affect the health of the occupants.

Variable nature of condensation

There are no firm rules about when condensation will and will not occur. One set of occupants can live in a house that is not affected, while next door, a similar-sized

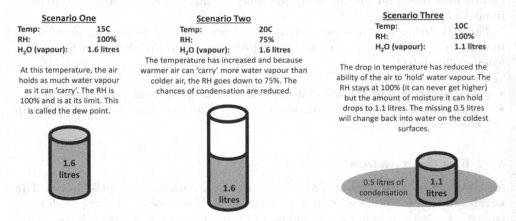

The relationship between relative humidity (RH) and temperature

Imagine an airtight room. Nothings gets in or out. Assume the internal air holds 1.6 litres of water vapour (an invisible gas). Please note: these arbitrary figures have been chosen for illustration purposes.

Scenario One

Temp:	15C
RH:	100%
H$_2$O (vapour):	1.6 litres

At this temperature, the air holds as much water vapour as it can 'carry'. The RH is 100% and is at its limit. This is called the dew point.

1.6 litres

Scenario Two

Temp:	20C
RH:	75%
H$_2$O (vapour):	1.6 litres

The temperature has increased and because warmer air can 'carry' more water vapour than colder air, the RH goes down to 75%. The chances of condensation are reduced.

1.6 litres

Scenario Three

Temp:	10C
RH:	100%
H$_2$O (vapour):	1.1 litres

The drop in temperature has reduced the ability of the air to 'hold' water vapour. The RH stays at 100% (it can never get higher) but the amount of moisture it can hold drops to 1.1 litres. The missing 0.5 litres will change back into water on the coldest surfaces.

0.5 litres of condensation   1.1 litres

Figure 6.46 Explanation of the relationship between relative humidity and temperature.

family could suffer from steaming windows, mould growth and ruined decorations. This is because the level of condensation in most homes is influenced by a complex interaction of factors. The British Standard code of practice for control of condensation in buildings (BS 5250:2011) identified a broad range of causes (BSI 2011). In most cases, the following are the most influential:

- The level of insulation in the dwelling – Low levels of insulation in a dwelling can give rise to:
  - High rates of heat loss;
  - Low temperatures – especially on the surface of windows, walls and ceilings;
  - Expensive bills.

It is essential that the dwelling should have a sufficient level of insulation to prevent this. The current Building Regulations are a reasonable measure of what is currently acceptable, but the vast majority of the housing stock is well below this level. Figure 6.47 compares the insulation values of typical constructions over

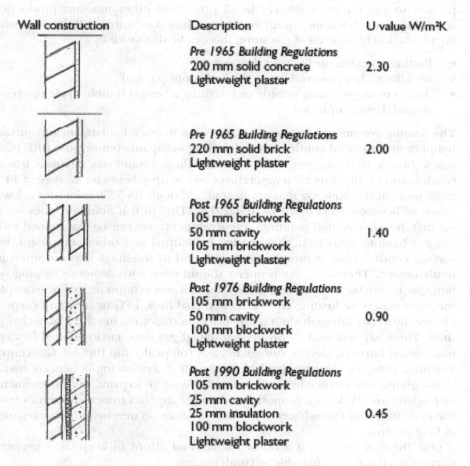

Wall construction	Description	U value W/m²K
	Pre 1965 Building Regulations 200 mm solid concrete Lightweight plaster	2.30
	Pre 1965 Building Regulations 220 mm solid brick Lightweight plaster	2.00
	Post 1965 Building Regulations 105 mm brickwork 50 mm cavity 105 mm brickwork Lightweight plaster	1.40
	Post 1976 Building Regulations 105 mm brickwork 50 mm cavity 100 mm blockwork Lightweight plaster	0.90
	Post 1990 Building Regulations 105 mm brickwork 25 mm cavity 25 mm insulation 100 mm blockwork Lightweight plaster	0.45

Figure 6.47 The 'U' values of some common wall constructions.

time. This tightening of the thermal regulations has continued. For example, in 2007, maximum U values for walls had dropped to 0.3W/m2K and at the time of writing (2022), the maximum value can be as low as 0.16W/m2K.

- **Moisture production levels of the household** – It is important to understand what are 'normal living activities'. For example, figures for specific domestic activities in BS 5250 can help in estimating what can be expected for a particular household (drying clothes, washing and so on) (BSI 2011). Although the total moisture output will vary with the nature of the household, an average daily total of 10–12 litres is not unusual. The dwelling must be able to cope with this. Although the way people live can affect moisture levels, the following activities are the ones that have the biggest impact:

  - Cooking;
  - Drying clothes indoors (for example, over radiators, on clothes' dryers, unvented tumble dryers, and so on);
  - Liquified petroleum gas heaters (LPG).

  Looking at other aspects of lifestyle will be largely irrelevant because, in comparison to the moisture sources listed previously, other moisture-production activities are much less significant and can be classed as 'normal'. If the occupants are producing high levels of moisture, this can be dealt with by:

  - Putting an extractor fan over the cooker;
  - Installing a low powered fan in the bathroom (s); and
  - Always drying washing outside or installing a vented tumble dryer or even a vented drying cupboard.

- **The heating system and its use** – Any dwelling needs adequate air and surface temperatures to avoid condensation. Ignoring occupant comfort, the BRE (Garrett & Nowak 1991) suggested that most dwellings should stay relatively free of condensation if the internal temperatures do not drop below an average of 14°C in all parts of the building that are heated, although BS 5250 mentions a lower range of between 10°C to 12°C (BSI 2011). This British Standard goes on to say that in well-insulated buildings, these temperatures can be maintained without any heating apart from that created by lighting and other equipment, but surface condensation is most commonly found in dwellings that are intermittently heated. Therefore, any property should have a whole-house heating system that is capable of achieving these temperatures economically. For example, the most expensive heating types are open coal fires, LPG heaters and on-peak electric heaters (although the energy price increases during 2022 may change this). These are followed by properly installed gas fires, enclosed coal heaters, many wood burners, electric storage heaters (off-peak) and fuel oil. Gas central heating can be the most economical (especially if condensing boilers are used), although this may change in the near future. As we go to print, the Governments of England and Wales are proposing to ban gas supplies to new properties from the mid-2020s and this is likely to increase reliance on new heating sources such as heat pumps.

  One thing is for sure: if a household cannot afford to keep their property warm, then it will be vulnerable to condensation.

The ventilation system and its use – Ventilation in dwellings is essential to remove smells and indoor pollutants, to provide ventilation for combustion appliances where appropriate, and to remove excess moisture. Ventilation rates are measured by 'air changes per hour' (ach). Older houses are more 'leaky' than modern constructions. This difference became more marked in 2006, when air tightness requirements were first included in Part F of the building regulations. From that date, all new homes (or a sample on larger developments) had to undergo a whole house pressure test to demonstrate that the air permeability had achieved an acceptable maximum level.

The problem with ventilation is establishing the ideal rate. This will vary with property type and can be challenging with older 'leakier' buildings. Too little ventilation will result in a build-up of moisture, while too much ventilation will result in excessive heat loss as the house loses all the warm air. In the latter case, the lowering of internal temperatures would itself lead to an increase in condensation.

A wide range of ventilation methods and products have been used in homes and this is likely to increase as building standards change and evolve. The most common types include:

- Air bricks through the wall;
- Trickle vents in window frames;
- Extract fans, either intermittent or continuous. These can be manually operated, fitted with an over-run timer or controlled automatically with a humidistat;
- Whole house, heat exchange ventilation systems (these will described in more detail in Volume 3 of this series 'Assessing building services');
- Passive stack ventilation where a duct connects the 'wet rooms' with the outside air usually at roof level. Temperature differences and the natural buoyancy of air will drive the air change.

Some of these are more appropriate than others in helping a household to achieve an appropriate balance, as the variables between different properties are considerable.

6.5.4 *Diagnosis of condensation*

Like all forms of dampness, condensation is easy to confuse with other causes. The following signs and symptoms may help distinguish between this and other forms of moisture related problems:

- There will be considerable moisture on the windows especially those that are single glazed. There may be evidence of water pooling on window sills and damp staining at the base of the window reveals. In the worst cases, this staining may be present below the window sill level where the condensation has run off;
- In affected areas, moisture meter readings will be concentrated near to the surface of the wall. The deeper the probes go into the wall, the drier it becomes;
- Moisture meter readings will often vary across even mould-affected areas. This is because the level of condensation will change throughout the day and across the seasons. The amount will depend on how recent the moisture production activities were carried out. This can be contrasted with directly penetrating moisture

and problems towards the base of the wall, where more consistent moisture meter readings can be expected;

• Problems are worse in the winter. Even where dwellings suffer from serious condensation problems, the symptoms can virtually disappear during the summer months and are very hard to spot. The exception is where wall and ceiling surfaces have been affected by extensive mould growth. In most cases, it is impossible to remove the mould stains, even with powerful cleaning agents;

• Mould growth occurs on the coldest parts of the building first. This includes the external walls and especially at cold bridges positions;

• Mould growth will often be worse where air movement is restricted including behind furniture and in wardrobes and other cupboards;

Figure 6.48 Mould growth to a window reveal of a terraced property with an uninsulated cavity wall. Condensation begins on the coldest parts and shortly after mould growth begins to grow.

- Mould growth has no defined edge and will gradually peter out across the affected area;
- There are no hygroscopic salts.

6.5.5 Reporting on condensation problems

Because this type of defect is closely associated with the way people live in their dwelling, it is very difficult to give clear advice. Three different situations might be observed and the report could reflect this. These are summarised in Figure 6.49. This approach assumes that the cause of the dampness has been correctly attributed to condensation.

Although this advice is based on the physical evidence collected during the survey, there are occasions when you may want to warn your clients of potential future condensation problems. A typical example would be where the subject property has been previously under-occupied by typically elderly occupants. The moisture production rates of this type of household will be very different from those of a larger family with children. In these circumstances, you may want to advise the client to prioritise preventative work, such as additional ventilation in the wet areas, as well as adjusting other features so environmental conditions within the home can remain in balance.

This type of advice will have to be adapted to suit each particular situation and level of service. The details will depend on the surveyor's knowledge and experience, but this framework could form a useful basis for positive advice about a defect usually ignored by surveyors in the past.

6.6 Built-in moisture

6.6.1 The drying-out problem

Using traditional methods of construction, building a three-bedroom house can use up to over 8,000 litres of water. Most of this is used in the concrete, cement mortar and plaster finishes. Because contractors are very keen to speed up the sale process and make a return on their investment, many houses are far from dry when the first occupants take up residence. This is especially true during the winter. The drying-out time for a floor screed has been estimated as taking 1 month for each 25 mm of thickness. Therefore, a concrete floor slab with screed topping could take up to 6 months to fully dry. Combined with the normal moisture-production activities of the first occupants, water vapour levels can soon rise.

Key inspection indicators

- High levels of condensation on the window panes and frames even where double glazed;
- White 'furry' salts appearing on the surface of the plaster;
- Mould growth may occur where air movement is particularly restricted, for example beneath or behind cupboards;
- Decorations 'bubbling' up because of entrapped moisture, tiles becoming detached from walls and floors and so on.

These effects can be unrelated to weather conditions or the lifestyles of the occupants. The key is how recently the property was built. It could take at least a year for

Category	Description of condition	Advice to client
Good (satisfactory repair)	There is little evidence of damage from condensation. There may be some isolated mould growth and paint damage to bathroom and kitchen windows. There is no visual evidence of mould growth in typical cold bridge locations.	There is a small amount of condensation damage that has resulted in superficial damage. If care is taken to keep the dwelling up to adequate temperature standards, keep moisture production down to a minimum and keep the dwelling reasonably well ventilated, these problems should not give any cause for concern.
Fair (minor maintenance)	There is evidence of damage in a few locations. This could include mould growth to window frames in a number of rooms, limited mould growth to wall and ceiling finishes in cold bridge locations and behind some furniture.	Although the problems are not serious, some work is required to bring the internal environmental conditions back into a balance and avoid future problems. Remedial work could include some additional ventilation work, making good of damaged surfaces, improving the heat input into the dwelling and possibly some limited insulation work to the coldest surfaces, etc. To make sure that the remedial work is properly targeted and cost-effective, advice from a building surveyor or other professional experienced in this sort of work may be appropriate.
Poor (significant or urgent repair needed)	There is significant damage in a number of areas. Windows and adjacent reveals and sills are badly affected by mould and condensation run-off; wall and ceiling surfaces in cold bridge locations are densely covered with mould. There is considerable mould growth behind the furniture, which is beginning to affect the adjacent wall surfaces.	The dwelling is significantly affected by condensation-related defects. One or more of the key influences on condensation formation (i.e., heating, insulation, ventilation and moisture production) are inadequate and out of balance with the rest. It is likely that significant remedial work will be required to bring the internal environmental conditions back into balance and so avoid future damage and unpleasant living conditions. Advice from a building surveyor or other professional experienced in this sort of work will be required. The scale of work could include improved insulation, heating and ventilation as well as ensuring moisture production rates are sensibly managed. The cost of this work could be high.

Figure 6.49 Table linking condensation defects to client advice.

some houses to properly dry out and so condensation-type defects during this period may be closely associated with construction moisture. After this period, moisture from another source must be suspected.

Not only can new houses be affected by this problem but also recently refurbished dwellings. A major modernisation can use 1,000–2,000 litres of water and have similar effects.

Reducing the problems of residual moisture has motivated many manufacturers and builders to use more prefabricated components. Timber-framed walls in place of masonry and dry lining with plasterboard instead of two coat plaster will all help to reduce this problem.

Giving advice

If this type of problem is suspected, then the client should be advised that additional ventilation and heat should be provided to assist the drying process. Forced drying methods (for example, industrial heaters and dehumidifiers) are not recommended, as the building should be able to reach an equilibrium naturally. Forced drying can cause cracking and warping of timbers and so on. You should also advise that if the symptoms persist for more than a year after the home was constructed, further investigations need to be carried out, as there might be another source of moisture. Additionally, because most new and refurbished properties will be covered by some form of guarantee or warranty, you should tell your client to ask their legal adviser to investigate this matter and explain the implications.

6.7 Traumatic dampness

This is a general term given to moisture defects that are caused by water leaks or bursts from such things as water and waste pipes, tanks, cisterns, radiators, sinks, baths, basins and so on. These can be truly traumatic and associated with large volumes of water, collapsed ceilings and ruined decorations. In other cases, the leaks can be small and gradual and the symptoms mistaken for other forms of dampness. Here are some examples from real cases:

- A new bungalow was showing high moisture meter readings at the base of the internal and external walls. The paint to the wall surface began to flake off and a characteristic tide mark had developed. For many months, this was mistaken for problems associated with a DPC/DPM fault. Further investigations revealed that the central heating pipes had been buried in the floor screed and a joint in one of the copper pipes had developed a leak, allowing water to soak into the floor. This had become so saturated that the moisture had soaked up the wall, causing the same symptoms as 'rising dampness';
- A watery brown stain had developed on the ceiling of a bedroom to an upstairs flat. Because the roof covering was the original slate, a roofer was instructed to overhaul the roof slopes. The stain continued to grow and a loft space inspection revealed a water tank directly above. Even then, the cause was not that simple. An inspection of the tank failed to identify any plumbing leaks but did show a high level of condensation run-off from the underside that was dripping onto the ceiling below;

- A cold-water storage tank suddenly developed a large leak that resulted in 50 gallons of water flooding through the house. Ceilings collapsed and carpets and possessions were ruined.

These cases illustrate how difficult it can be to make a precise diagnosis, especially when it has to be based on a single visit. Despite this, typical characteristics can be associated with these types of defects:

- The dampness is usually in isolated areas and the level of dampness can be very high;
- Moisture meter readings show a sharp change from high to low;
- Not usually associated with the weather or the time of year;
- There are often drainage, plumbing, heating or sanitary fittings nearby;
- Hygroscopic salts are not usually present (apart from longer term leaks).

Giving advice on this type of defect is usually straightforward, as long as the diagnosis is accurately made.

6.8 Thermal insulation, energy conservation and concerns about the environment

6.8.1 Introduction

In the early part of this century, when the first edition of this book was published, we described how energy issues had first appeared on social and political agendas. Not only was the cost of fuel rising, but people wanted to be warm and comfortable in their home. We also pointed to a growing section of the population that was motivated by environmental concerns: wanting to help conserve non-renewable resources and contribute to the reduction of greenhouse gases. The last 20 years have seen these concerns grow. Although there will always be differences of opinion, most commentators now agree that there is a clear link between energy use and changing climates. Few would now deny that the melting ice sheets, rising sea levels and weather extremes have nothing to do with energy use on this planet.

Scope of surveys and inspections

Formally, before June 2020, RICS Home Survey products included little on energy matters. Although the RICS Building Survey had a section of the report on energy-related matters (Section K), in practice, few residential practitioners used the section properly. The Home Buyer report included even less, according to the RICS professional statement at the time:

> If the surveyor has seen the current EPC, he or she will present the energy-efficiency and environmental impact ratings in this report. The surveyor does not check the ratings and cannot comment on their accuracy.

To better reflect the environmental concerns of the nation, the RICS enhanced the role of its members and described a more active role in the Home Survey Standards (RICS 2019). For example:

> Concerns over climate change and legislative and commercial changes in the energy sector have created a demand for clear and objective guidance on energy matters. Consequently, energy advice will be of great value to many clients. The nature of this service will be influenced by a range of factors that may change over time, for example, global, regional and national legislation and practice; the nature of the subject property; and the competence and technical knowledge of the RICS member or RICS regulated firm (RICS 2019, p. 18).

The HSS also outline the expectations for each level of service. This is our interpretation:

- Level one – If not the vendor or their agent hasn't provided you with a copy of the EPC, you should download the most recent certificate and include the relevant energy and environmental rating in the report;
- Level two – This should include all that described for level 1 and in addition, you should check any obvious discrepancies between the EPC and the subject property. Typical examples would include where a report shows the property as having insulated cavity walls when in fact they are of solid construction, or the low efficiency controls on the heating system on the EPC when the actual ones are better. You should report these to your client and briefly explain the implications;
- Level three – In addition to those matters described previously, at this level you should give advice on the appropriateness of any energy improvements recommended by the EPC. This might typically include not externally insulating or double glazing a listed building or one in a conservation area. Depending on your own skills and knowledge, this could also extend to more refined advice, such as how to tackle cold bridging or making sure the cavity walls are suitable for thermal insulation.

In our view, this is one of the most important features of RICS's Home Survey Standard; it has the flexibility to allow practitioners to offer a range of services as long as they have the skills to deliver it, the client understands the scope of the service and it is properly incorporated in the terms of engagement.

6.9 References

Allergy UK (2021). 'House dust mite factsheet'. Available at www.allergyuk.org/information-and-advice. Accessed 15 September 2021.

Asthma UK (2020). 'Asthma facts and statistics'. Available at www.asthma.org.uk/. Accessed 20 July 2020.

Boniface (2019). 'The rising damp myth'. Available at www.wcp-architects.com/2019/04/the-rising-damp-myth/. Accessed 15 September 2021.

BRE (1985). 'External masonry walls insulated with mineral fibre cavity-width batts: Resisting rain penetration'. *Defect Action Sheet 17*. Building Research Establishment, Garston, Watford.

BRE (1987). 'Windows: Resisting rain penetration at perimeter joints'. *Defect Action Sheet 98*. Building Research Establishment, Garston, Watford.

BRE (1991). *Tackling Condensation*. Building Research Establishment, Garston, Watford.

BRE (2001). 'Insulating masonry walls'. *Good Building Guide 44 Part 2*. Building research establishment. Garston, Watford.

BRE (2002a). 'Assessing moisture content in building materials'. *Good Repair Guide 33 (Part 2)*. Building Research Establishment, Garston, Watford.

BRE (2002b). 'Thermal insulation: Avoiding risks'. *BR 262*, 3rd edition. Building Research Establishment, Garston, Watford.

BRE (2007). 'Rising dampness in walls: Diagnosis and treatment'. *Digest 245*. Building Research Establishment, Garston, Watford.

BRE (2015). 'Diagnosing the cause of dampness'. *Good Repair Number 5 (Revised)*. J Huston, Building Research Establishment, Garston, Watford.

BSI (2011). *Code of Practice for the Control of Condensation in Buildings* (BS 5250:2011). British Standards Institution, HMSO, London.

BSI (2012). *Code of Practice for Diagnosis of Rising Damp in Walls of Buildings and Installation of Chemical Damp-Proof Courses* (BS 6576:2005+A1:2012). British Standards Institution, HMSO, London.

DCLG (2017). *Advice for Building Owners: External Wall Insulation (EWI) Systems with a Render or Brick-Slip Finish*. Advice Note 13. Department of Communities and Local Government, London.

DECC (2012). *Review of the Number of Cavity Walls in Great Britain Methodology*. Department of Energy and Climate Change, London.

Department for Business, Energy and Industrial Strategy (2018). *Household Energy Efficiency. Headline Release*. DBEIS, London.

EST (2006). *External Insulation for Dwellings, GPG293*. Energy Saving Trust, London.

EST (2007). *Cavity Wall Insulation in Existing Dwellings: A Guide for Specifiers and Advisors'*. Energy Saving Trust, London.

Fryer v. *Bunney* (1982). 263 EG 158.

Garrett, J. and Nowak, F. (1991). *Tackling Condensation. A Guide to the Causes of, and Remedies for, Surface Condensation and Mould Growth in Traditional Housing*. Building Research Establishment, Garston, Watford.

GEsensing (2013). *Surveymaster® Protimeter Dual-Function Moisture Meter Manual*. GE Sensing. Niskayuna, NY.

Historic England (2016). *Energy Efficiency and Historic Buildings: Insulating Early Cavity Walls*. HEAG083. English Heritage, London

Howell, J. (2008). *The Rising Damp Myth*. Nosecone Publications, London.

INCA (2015a). *INCA Technical Guide 01 – Fire Protection for EWI Systems*. Insulated Render and Cladding Association, London.

INCA (2015b). *Best Practice Guide: External Wall Insulation*. Insulated Render and Cladding Association, London.

INCA (2021). 'About INCA'. Available at www.inca-ltd.org.uk. Accessed 15 September 2021.

MHCLG (2019). *English Housing Survey Headline Report, 2017–18*. Ministry of Housing, Communities and Local Government, London.

Murdoch, J. and Murrells, P. (1995). *Law of Surveys and Valuations*. Estate Gazette, London.

Ormandy, D. and Burridge, R. (1988). *Environmental Health Standards in Housing*. Sweet & Maxwell, London.

Oxley, T.A. and Roberts, E.G. (1987). *Dampness in Buildings: Diagnosis; Treatment; Instruments*. Butterworths, London.

Oxley, T.A. and Gobert, E.G. (1994). *Dampness in Buildings*, 2nd edition. Butterworth Heinemann, London.

Parnham, P. and Russen, L. (2008). *Domestic Energy Assessors' Handbook.* RICS, London.

PCA (2016). *White Paper: External Wall Insulation – Who Are We Kidding?* Property Care Association, Huntingdon.

PCA (2017). *Investigation and Control of Dampness in Buildings.* Property Care Association, Huntingdon.

PCA (2020). *Guidance Document: The Use of Moisture Meters to Diagnose Dampness in Buildings.* Property Care Association, Huntingdon.

Protimeter (2020). 'Frequently asked questions'. Available at www.Protimeter.com/faq.html. Accessed 6 August 2020.

RICS (2019). *Home Survey Standard, RICS Professional Statement.* RICS, London.

Ridout, B. and McCaig, I (2016). *Measuring Moisture Content in Historic Building Materials.* Historic England, London.

Schrijver (2021). 'Web page of Schrijver Systeem'. Available at https://schrijversysteem.nl/. Accessed 7 January 2021.

SPAB (2020). 'Knowledge base'. Available at www.spab.org.uk/advice/knowledge-base. Accessed 6 January 2021.

TRADA (1999). *Wood Information Sheet No. 14.* Timber Research and Development Council, High Wycombe.

Which? (2021). 'Home grants'. Available at www.which.co.uk/reviews/home-grants/article/home-grants/the-green-deal-afMJp3S8hrgc. Accessed 6 January 2021.

7 Wood rot, wood-boring insects, pests and troublesome plants

Contents

7.1 Introduction

These topics have a few features in common:

- They are all living things;
- They are loosely-related to moisture related defects in buildings;
- They can quickly cause problems for a building owner. This can, in turn, affect the value and the saleability of the property.

DOI: 10.1201/9781003253105-7

Therefore, they have been included in the same chapter.

7.2 Wood rot

7.2.1 Introduction

Wood rotting fungi are the chief cause of wood decay in this country. Essentially, the fungus grows on the timber, extracting the sugars out of the cellulose, causing the material to breakdown. There are many different species of wood rotting fungi, but we will focus on the two main generic types:

- Dry rot (Serpula lacrymans);
- Wet rot (Coniophora puteana in particular).

7.2.2 Dry rot

This is the most destructive of all the different types of wood rot (PCA 2016, p. 2). If left undetected, it can seriously weaken timber structural components and, in the worst cases, lead to collapse. The eradication of an outbreak can involve a complex procedure of repair and treatment that could cost thousands of pounds. To properly detect an actual or potential outbreak, understanding a few fundamental principles is useful (Singh 1994, p. 36):

- To properly grow, dry rot needs a source of moisture, oxygen and cellulose (usually from the timber).
- The ideal conditions for growth are:

  - Moisture content of timber between 20–40%;
  - Adequate temperatures, best 18–23°C;
  - Poorly ventilated stagnant conditions.

Dry rot is a very delicate fungus. It does not like varying conditions and ironically it can very easily be killed if these ambient conditions are altered. When conditions are favourable or the fungus is threatened, an outbreak can produce 'fruiting bodies' that usually appear in more obvious parts of the building. These widely distribute the very characteristic brick-coloured orange spores.

The main indications of an outbreak of dry rot include:

- There is a silky cotton wool-like mass or growth that may be of considerable size. The fungus can be very 'fresh' and 'fleshy' and white in colour at the extreme edge of the growth. Behind this, long, thin, vein-like hyphae develop that connect the growing front with the source of food at the original outbreak. These hyphae can be several millimetres in diameter. When dry, they can be snapped. According to the Property Care Association (PCA), water droplets can form on the surface of the mycelium, giving the fungus its name 'lacrymans' (Latin for 'tears'). Lilac tinges are more common, especially in less humid situations where the surface mycelium is reduced to a thin silken grey skin (PCA 2017, p. 3);
- The affected timber is gradually broken down and shrinks in size. It becomes darker in colour and will often crack both across and along the grain. This

produces 'cuboidal' cracking that is so typical of dry rot. Small cubes of very light timber can be easily broken away;

- The surface of the timber will become uneven. The rot can develop behind a paint film and produce a 'wavy' effect on the surface. When prodded with a blunt probe (a car key is the ideal instrument), the probe will quickly penetrate deep into the timber;
- When conditions are favourable, fruiting bodies will appear in the vicinity of the main outbreak. These look like fleshy growths or 'brackets' that appear relatively quickly (a matter of weeks) and are soon covered with tiny, bright brick-red particles or spores. These spores can spread across a wide area and, if undisturbed, can literally cover adjacent surfaces (see Figure 7.1).

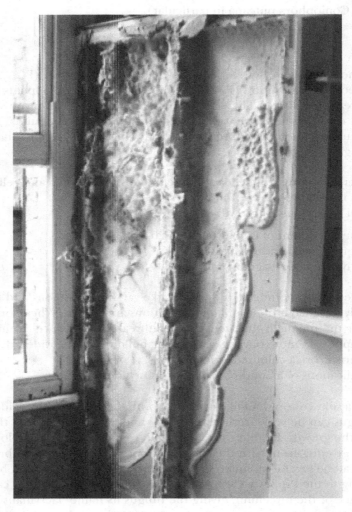

Figure 7.1 Photograph of dry rot within a timber partition. The plasterboard to the wall has been removed to reveal a luxurious dry rot growth. The outbreak began in the vertical timber stud, which is fixed to the inside face of the external wall (to the LHS of the picture). The source of moisture and the cause of the outbreak was a leaking gutter that saturated the wall.

Location is the key for identifying dry rot. Because it requires stagnant and stable environments, the main growths are usually below suspended timber floors, in under-stairs cupboards, within the thickness of timber partitions and so on. Only the fruiting bodies tend to be noticed when they appear in the more open areas of the dwelling. Inspection of basements and subfloor voids is important in this respect.

7.2.3 Wet rot

In the opinion of the PCA (2017, p. 3), many different species of fungi could be classed under the 'wet rot' heading, but the most common are *Coniophora puteana* and *Fibroporia vaillantii*.

Wet rot occurs when timber is excessively wet for most of the time. The moisture content would be in the region of 28–40% and wet rot has a preference for the wetter end of this scale. Wet rot turns the timber very dark brown and causes longitudinal fibrous-type cracking. In the final stages of attack, the timber becomes very brittle and is readily powdered. Wet rot can affect the heart of timber sections and so when it becomes noticeable on the surface, the middle of the timber is often completely destroyed.

Where the timber is painted, a surface waviness is a typical sign. Wet and dry rot in the early stages can be hard to tell apart. For mortgage valuations and level one and two services, this is not critical because if any type of rot is discovered, then a refer-ral to a specialist would be required. Despite this, location is often the distinction between the two. Wet rot can be found in the wetter location:

- Fully exposed window and door frames especially at the joints;
- Joist ends where they are built into very damp walls;
- Dry rot is not common in roof spaces but wet rot can affect the ends of rafters, wall plates, chimney trimmers, and so on;
- Timber that is dampened by regular leaking from piped water services. Leaks from WCs are typical locations.

7.2.4 Inspecting for rot

The main controlling conditions are the presence of moisture and timber products. Consequently, a moisture meter is vital if potential outbreaks of dry rot are to be iden-tified. Berry (1994) also recommends other items of equipment that may be useful:

- Depending on the level of service and the specific terms of engagement agreed with clients, you may need various 'opening-up' tools to take up floorboards, drill inspection holes, and so on;
- A surveyor's mirror or a digital camera/smartphone on a telescopic handle ('selfie stick'); and
- A borescope with a light source and flexible light guide. At the time of writ-ing, miniature endoscopes that wirelessly connect to apps on smartphones have become available. Their small size means they can fit through very small gaps (for example, through large gaps in poorly fitted floorboards, through airbricks and so on). These allow a partial view of a space that would otherwise be concealed. However, the performance of these emerging technologies varies, so ensure they are fully effective.

7.2.5 Case study: inspecting and reporting on a suspended timber floor

An older suspended timber floor can be very vulnerable to both wet and dry rots. Up to late Victorian times, poor ventilation, lack of isolating DPCs, earth or rubble sub-floors surfaces and poor workmanship generally can result in rot-infected timbers. According to Marshall et al. (2013, p. 91), the construction of raised timber floors improved during the early part of the 20th century when a concrete slab was commonly laid over the earth and floor joists were no longer built directly into the external wall but supported by its own sleeper wall and isolated with a DPC.

To properly assess the condition of a suspended timber floor, a full inspection would be required. This would include getting access to the floor void, either by lifting a number of floorboards around the room or literally crawling beneath the floor itself. Apart from being beyond the scope of most levels of service, the presence of fixed floor coverings and well-nailed floorboards would prevent this approach. Although the scope of inspection will depend on the terms of engagement, a typical service would include:

- Lift the corner of any loose floor coverings to identify the nature of the flooring itself;
- If this is possible then the surface of any timber components should be tested with a moisture meter as widely as possible. You should also look for signs of wood boring insects (see Section 7.3);
- If an access hatch or loose floorboards are conveniently positioned (for example, carpets or heavy furniture do not have to be moved) then these should be lifted and an inverted 'head and shoulders' inspection carried out with the aid of a strong torch;
- Where a 'trail of suspicion' presents itself (for example, high moisture meter readings to floor joists, signs of wood-boring insects or wood rot, unusual features, and so on) and the sub-floor void is easily accessible (say 1000 mm clear space without safety hazards such a loose wiring and uneven surfaces) it might be 'reasonable' for the sub-floor area to be inspected further. There cannot be any hard and fast rules for this – it will always be a matter of professional judgment and the terms of engagement previously agreed with the client.

If there is no hatch and the floorcoverings are fixed, it is still possible to make an assessment of the likely condition of the floor. Assuming that it is a typical suspended floor construction that would be found in a three-bedroom interwar semi-detached house, the following indicators should be noted:

Internally:

- Is the floor springy? The informal 'standard' test is the 'heel drop test'. You should rise up on your toes and drop down on your heels. Excessive rattling could suggest either undersized or deteriorated timbers (or an overweight surveyor);
- Is there evidence of high moisture meter readings to the walls, skirtings and floorboards around the edge of the room? Are there any flight holes to exposed timbers?
- Can you see any 'waviness' in the skirting boards or even fruiting bodies?

Externally:

- Is there enough ventilation to the sub-floor void from the outside? On average, there should be one 225 × 150 mm air brick every 1.5 m run of external wall. These should be clear of all debris and extend right through the wall (the standard test is to poke through a piece of rigid wire, for example, an unwound wire coat hanger);
- Even if there are sufficient air bricks, are they well distributed around the floor? The classic situation would be the mid-terrace house, where the rear extension floor is solid while that under the main house is suspended (see Figure 7.2). In these cases, large areas of the underfloor area may not have any through ventilation and would remain stagnant, allowing moisture levels to rise. Rot could easily develop;
- Is there a DPC in the wall and is it at least 150 mm above the external ground level? If not, then joist ends could be damp and possibly rotten;
- Is the internal floor level well above the external ground level? If not, then it might be very difficult to achieve an adequately ventilated floor void;
- Have external paths or flower borders been built up against the wall, possibly bridging the DPC?

If there are no adverse signs, then further investigations may not be necessary, but if there are any indications of potential problems, then further investigations will be warranted. This is the classic case of the 'trail of suspicion' that must be followed.

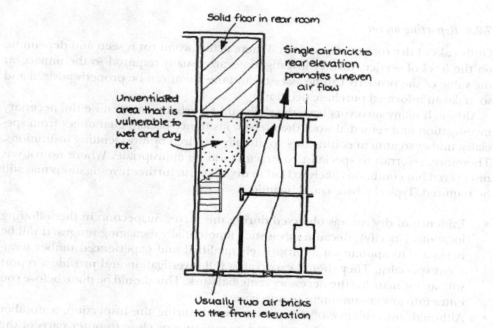

Figure 7.2 Plan of a typical suspended timber floor showing inadequate ventilation because of a lack of through flow of air.

Typical locations that should be inspected include:

- Skirting boards – These can act as a 'wick' for dry rot and are often the first visible element affected. The moisture readings of the timber and adjacent wall are very important;
- The base of door frames and linings where they are adjacent to a solid floor;
- Woodblock flooring or timber laminate flooring finishes especially where a DPM is missing or defective;
- The rear faces of built-in cupboards or heavy furniture that has not been moved for some time;
- Timber ground-floor joists where they have been built into adjacent brick walls or have not been properly isolated from damp masonry by a DPC;
- Timber joists of upper floors, fixing grounds or wall plates are often built into the thickness of older walls. High levels of rain can penetrate the walls on exposed elevations, leading to rot in the timbers close to the outside surface. As climate changes and weather extremes become more common, this is another example where you should focus on checking for moisture on those parts of the building that are most exposed to the prevailing weather;
- Timber close to water services that could leak. Examples could include rising water mains, radiators, bath wastes, toilet connections and so on;
- Timber sections in roofs that are constantly damp such as valley beams of 'butterfly' roofs, timber trimming pieces around chimney stacks, especially the secret gutters at the rear.

7.2.6 Reporting on rot

Outbreaks of dry rot are very serious. Where actual wood rot is seen and depending on the level of service, further investigations are usually required so the impact on the value of the property can be assessed and the client can be properly advised and so make an informed purchase decision.

Although many surveyors have the skills and knowledge to organise the necessary investigation and remedial work themselves, insurance-backed guarantees from specialist timber treatment contractors are usually required by most lending institutions. Therefore, referrals to specialist contractors will be appropriate. Where no rot was discovered but conditions likely to lead to dry rot exist, further investigations may still be required. Typical advice could include:

- Evidence of dry rot was observed during the survey/inspection in the following locations (specify). Because this is such a potentially damaging fungus, it will be necessary to appoint an appropriately qualified and experienced timber treatment specialist. They should carry out a full investigation and provide a report with an estimate for the necessary remedial work. This should be done before you enter into any commitment to purchase;
- Although no evidence of dry rot was noted during the inspection, a number of timber components were affected by moisture or close to other parts of the building that are affected by moisture defects (specify location and extent). Under these conditions, there is a possibility that wood-rotting fungi could develop in the foreseeable future. To ensure this problem is rectified, an

appropriately qualified and experienced damp-proofing and timber specialist should be asked to inspect the property and provide a report together with any appropriate estimates. This should be done before you enter into any commitment to purchase.

When referring clients to timber specialists, precautions similar to those outlined in Chapter 6.4.8 should be taken. A particular word of caution should be made at this point: in a pre-sale situation, it is unlikely that vendors will allow any type of inspector/surveyor to take up floor coverings, laminates or floorboards. Consequently, even specialist reports can reach only provisional conclusions. Although there is little a residential practitioner can do about this situation, at least we can tell the client that further investigations will need access to concealed areas and this should be discussed with the vendor.

7.2.7 Inspecting unimproved basements and cellars

To ensure consistency, it is important to be clear on the difference between a cellar and a basement. In our view:

- A cellar is a room under the ground floor of a house that is usually used for storage. In older properties, these were used for coal storage and were not habitable;
- A basement is part of a building that is either partially or completely below ground level. The building regulations define a 'basement storey' as 'a storey with a floor which at some point is more than 1200 mm below the highest level of ground adjacent to the outside walls' (Department for Levelling Up, Housing and Communities and   Ministry of Housing, Communities & Local Government, 2010). Basements were often intended to be used for habitable purposes, although older examples may not meet current requirements.

Where part or all of the ground floor has a cellar beneath, then the whole floor structure can often be fully inspected. In many other cases, these will be 'underdrawn' or have an old lath and plaster ceiling beneath. Not only does this limit the extent of the survey, but it can restrict free ventilation and increase the risk of timber defects. In these cases, consider advising the client to remove the underdrawings to protect the timber. However, the impact of removing a cellar ceiling may affect fire protection to upper floors, especially for buildings with two or more storeys.

Despite the assertions of many owners that their cellars are 'dry', many older cellars and basement walls that are in full contact with the adjacent ground will invariably be affected by moisture-related problems. Special care should be taken to test joist ends, as they are often built into damp brickwork. Other trails of suspicion could include:

- Signs of standing or running water. This could be caused by excessive groundwater or leaking drains. This can be seasonal, so you should look for evidence of 'tide marks' or water staining. In some regions, there may even be a drainage gulley in the cellar floor – a clear indication of regular water flow;
- Large coal chutes or other openings that could allow water to enter the space;
- Large amounts of stored timber, cardboard and even paper that could allow rot to develop.

7.3 Insect attack

7.3.1 Introduction

According to Bowyer (1977) '. . . infestation of timbers is probably, to the layman (sic), the prime reason for a . . . survey'. He goes on to explain that this public interest is probably due to '. . . the extensive publicity given by commercial bodies interested in the treatment and eradication of the scourge'. This identifies the two key problems in assessing properties for insect attack:

- Owners are very concerned when 'woodworm' is discovered in their home; and
- Many remedial treatment firms often carry out blanket treatments of whole dwellings when the presence of insects has been noted.

Both reactions are often out of proportion to the scale of the original infestation.

Before this is discussed any further, it will be useful to review the nature of wood-boring insects. Many that use wood as a food source prefer the damp timber of standing trees, freshly felled logs or decaying matter. Few will attack dry timber in buildings. Yet if the conditions are right, wood-boring insects can badly damage timber to the extent that structural integrity can be threatened and the value of the dwelling reduced. Their life cycle begins when the eggs of the beetles are laid on the surface or in cracks in the timber. Small grubs or larvae hatch out and bore into the wood, feeding as they go and forming a network of tunnels. As the larva moves along, it excretes bore dust or frass (see Figure 7.4). After a number of years of feeding within the timber, the larva pupates close to the surface. When the adult emerges, it bites through, leaving the 'flight hole' that is such a typical symptom. The adults do not cause any further damage. After mating, the females will lay their eggs in suitable timber and so continue the infestation.

7.3.2 Main types of wood-boring insects

There are a large number of different species of wood-boring insects. Although the Property Care Association includes a more complete review, most surveyors will only normally encounter the following five types (PCA 2017):

Common furniture beetle (Anobium punctatum)
> *Appearance:* The adult beetles are 2.5 mm–5 mm in length and are reddish to blackish brown in colour. The white larvae are about 6 mm long.
> *Shape and size of flight hole:* circular 1–2 mm
> *Bore dust* (under a 10× lens): cream, granular, lemon-shaped pellets

According to the PCA, the common furniture beetle naturally inhabits dead stumps and fallen branches in woods and hedgerows but is more abundant in building timbers and furniture. It is the most common of the wood destroying insects found in buildings in the UK and most damage is found in timber which has been in use for 10 years or more.

This type is the one with which most people are familiar and wrongly call 'wood-worm'. The adult beetles are capable of flight and this enables them to travel and infest other timbers. In the period from March to September, adults can be found on the window ledges of houses containing infested timber.

Characteristics include:

- It is found in the sapwood of all softwood and European hardwood timbers;
- It can infest timbers with moisture contents that are typically found in ventilated roofs and suspended floors. In such an environment, the extent of the damage will be moderate and the activity of the insect low. A heavy infestation, on the other hand, is a sure sign that a moisture problem will almost certainly exist as well;
- Where it affects joinery, staircases and intermediate floors in centrally heated environments the beetle tends quickly to die out;
- Heavy infestations can cause structural weakening of timber where timber replacement may be necessary.

House Longhorn Beetle (Hylotrupes bajulus)

Appearance: The House Longhorn beetle is somewhat flattened, measures from 8 mm to 25 mm in length and is brown or black in colour. When fully grown, the larvae are commonly 18 mm long but may attain a length of about 30 mm and are most active in the sapwood of softwoods. The larvae feed for a relatively long time (an average of six to seven years), but this varies with temperature and the moisture content of the wood.

Shape and size of flight hole: oval 6–10 mm

Bore dust (under a 10× lens): cream powder, chips and cylindrical pellets

Generally limited to Surrey and West London and measures have been included in the building regulations to try and control the insect.

Characteristics include:

- Only found in parts of Surrey and west London;
- Found mainly in roof timbers and because of the size of the larvae it can cause structural weakening. The damage can be so extensive that only a thin outer veneer of sound timber is left. According to the PCA (2016), there is very little external evidence of infestation, except that sometimes the outline of the larval borings may be detected by an unevenness of the wood over them;
- Not encouraged by damp conditions.

DeathWatch beetle (Xestobium rufovillosum)

Appearance: The DeathWatch beetle measures from 5 mm–8 mm in length. Its colour is dark chocolate brown with patches of short yellowish hairs, which give the insect a variegated appearance. In old specimens, these hairs may have been rubbed off, in which case the mottled appearance is less obvious. The larvae are curved and white, covered with long fine yellowish hairs and are over 6 mm in length. Probably the most famous wood-boring insect about which a great deal of mythology has developed;

Shape and size of flight hole: circular 2–3 mm;

Bore dust (under a 10× lens): brown bun-shaped pellets in the bore dust or frass.

Characteristics include:

- Distribution – common in the southern half of the UK; less frequent in the north of England and virtually unknown in Scotland;
- Only found in buildings more than 100 years old; mainly affecting hardwoods;
- Requires moist conditions and often associated with fungal attack. Therefore found in wall plates, truss ends, bonding timbers, panelling, and so on;
- May cause severe damage so the implications have to be considered carefully.

Lyctus powder-post beetle (Lyctus brunneus)

According to the PCA, Lyctus beetles are found in unseasoned or recently dried hardwood timbers. Oak, ash, elm and sweet chestnut, as well as some imported timbers, are commonly infested. The life cycle is short and completed in a 12-month period, with emergence between May and September. Only the sapwood is attacked, as it is the starch in this portion of the wood that provides the food for the larvae. In timber which has been cut for some time, the starch may be so depleted that the larvae are unable to feed. Hence, these beetles are never found in old wood, but may be encountered in timber yards, fencing and also in comparatively new furniture if any sapwood has been included;

Shape and size of flight hole: circular 1–2 mm;

Bore dust (under a 10× lens): a fine flour-like bore dust (frass). This differs from the common furniture beetle where the frass is coarser, lemon shaped and gritty.

This beetle is found all over the world and can be very destructive. It can reduce the sapwood to a powdered mass within a few years. The larvae are curved and white with a yellowish head and dark brown jaws and, when fully grown, measure approximately 6 mm in length.

Characteristics:

- Attacks the sapwood of hardwood timbers;
- Does not attack softwoods so most modern dwellings are safe;
- Usually infests timber before it is delivered to site.

WOOD-BORING WEEVIL

Appearance: Adults are 2.5–5 mm in length, reddish brown to blackish, with a snout (rostrum) and long flattish body. Life cycle is short at around 12 months and unusually, the adult beetle also lives for around 12 months. Adults therefore feed and leave characteristic 'striations' (channels) on the timber surface;

Shape and size of flight hole: ragged 1 mm;

Bore dust (under a 10× lens): fine brown and angular;

Very closely associated with very damp and partially decayed timber. Often found behind skirting boards and decaying wall plates when fungal decay is present. They are normally brought under control by the measures taken to deal with outbreaks of wet rot.

Characteristics include:

- Attacks only damp and decaying timber;
- It does not cause timber deterioration but it can speed it up;
- Adults feed and leave characteristic 'striations' (channels) on the timber surface (see Figure 7.3);
- Found in rot-affected joist ends, backs of skirtings on damp walls, and so on.

7.3.3 Locating infected timbers

The BRE (1998) identify a number of key locations in a dwelling where insect attack is most common, especially if the timber is excessively moist. These are illustrated in Figure 7.5. Exposed timbers should be inspected with a strong torchlight, looking for

Figure 7.3 This shows a replacement floor joist over a very wet cellar of a terraced house built in 1890. The timber was moist enough to allow a community of wood-boring weevils to move in. They have left characteristic 'striations' or channels across the timber surface. The frass can clearly be seen. Lower moisture levels in the cellar will be key to getting rid of the weevil problem.

evidence of the flight exit holes. A hand-held magnifying glass will help in this process. Indicators of activity include:

- Freshly cut holes, which tend to have sharp and well-defined edges. Where the inside walls of the tunnel can be seen, they tend to be light in colour when compared with those holes that have been open for some time. Where holes are observed, they should be measured so the type of beetle can be identified. A short metal engineer's ruler can help with this process;
- On adjacent surfaces beneath the affected timber, small piles of freshly ejected bore dust or frass can often be seen. Again, the more recent the dust, the lighter the colour (see Figure 7.4);
- In the worst cases, sometimes the actual larvae can be found by probing the damaged timber with a bradawl or penknife. The BRE point out that this can be very rare in practice.

7.3.4 Insect attack and standard surveys

Although there are a number of key locations where insect attack can be expected, in occupied premises few of these areas will be fully exposed. Fixed floor coverings, bath

Figure 7.4 This shows a floor joist over a shallow sub-floor void that has an earth surface. The whole floor had been infested with a variety of wood-boring insects, including the death watch beetle. As the adults emerged from the underside of the joist, the frass fell to form small conical piles on the surface below. A hand-held mirror will enable you to assess the extent of the infestation.

panels, internal linings and inaccessible roof and floor voids may all prevent access. When exposed timber surfaces can be seen, every opportunity should be taken to look for evidence of beetle attack. The flight holes of all the common types can easily be seen by the naked eye and torchlight and then measured. A magnifying glass is a useful addition. A sharp implement (such as a bradawl or a stout knife) can help to assess how bad the attack could be. However, many types of survey products are non-destructive and so this may go beyond the scope of the inspection agreed in the terms of engagement.

In a dwelling where a lot of timbers are found to be affected by excess moisture, then the likelihood of infestation is higher and so special attention should be paid. This is a typical 'trail of suspicion' that should be either followed or at least emphasised in the report.

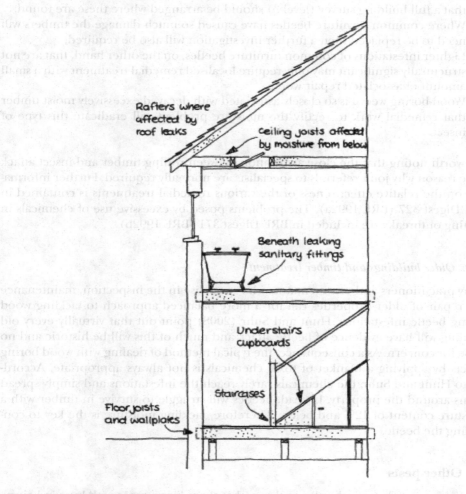

Rafters where affected by roof leaks

Ceiling joists affected by moisture from below

Beneath leaking sanitary fittings

Understairs cupboards

Staircases

Floor joists and wallplates

Figure 7.5 Common locations of insect attack in a domestic dwelling. The risk of insect attack is higher where the moisture levels in the timber are greater.

7.3.5 Assessing the extent of the damage and reporting

Where insect attack in timber is found, it is beyond the scope of most surveys to precisely identify the type of attack and recommend treatment. The BRE (1998) identify several categories of damage in timbers in an attempt to limit unnecessary work. Where House Longhorn, Lyctus powder-post and DeathWatch beetles and extensive infestations of common furniture beetles are found to be active, further remedial treatment is usually needed. Consequently, referral to a specialist timber treatment contractor is justified, who then should identify the type, extent and treatment required. In addition to this referral, additional advice can usefully be given to the client to indicate the seriousness of the infestation. For example:

- If the timber has been attacked by House Longhorn, DeathWatch or Lyctus then the damage could be considerable. In addition to remedial treatment, the replacement of structural timbers is often required. The BRE (1998) recommend that a full building survey (level 3) should be arranged where these are found;
- Where common furniture beetles have caused so much damage the timbers will need to be replaced then a further investigation will also be required;
- Lighter infestations of common furniture beetles, on the other hand, that are not structurally significant may only require localised remedial treatment with a small amount of associated repair work;
- Wood-boring weevil is so closely associated with decayed excessively moist timber that remedial work to rectify the moisture problem will eradicate this type of insect.

It is worth noting that the close association between rotting timber and insect attack is the reason why joint referrals to specialists are normally required. Further information on the relative effectiveness of the various remedial treatments is contained in BRE Digest 327 (BRE 1992a). The problems posed by excessive use of chemicals in treating outbreaks are included in BRE Digest 371 (BRE 1992b).

7.3.6 Older buildings and timber treatment

Many practitioners and commentators who specialise in the inspection, maintenance and repair of older properties call for a more balanced approach to tackling wood boring beetle infestations. Hunt and Suhr (2008) point out that virtually every old building will have evidence of beetle attack and much of this will be historic and no cause for concern. As a consequence, the typical method of dealing with wood boring insects by applying a blanket of toxic chemicals is not always appropriate. According to Hunt and Suhr, the chemicals rarely reach the infestations and simply spread toxins around the property. Instead, beetles will struggle to survive in timber with a moisture content of 12% and below. Therefore, tackling moisture is the key to controlling the beetle.

7.4 Other pests

There are a number of other living things that share living space with human beings. Not all of these will affect the condition of the dwelling, but some might be very expensive to eradicate and can threaten the health of the occupants. If they affect the dwelling, then most clients will want to know.

On their website, the British Pest Control Association (BPCA 2020) state that pests have the potential to:

- Contaminate homes and spread disease;
- Damage possessions and get into foodstuffs;
- Damage property and in some circumstances can cause fires (electrical faults) and traumatic plumbing leaks (pipe leaks).

The main types are described as follows.

7.4.1 Pigeons and other birds

Birds can indirectly damage buildings:

- Nests can block drains, gutters and flues. This can lead to water damage and sometimes dangerous heating appliances when the proper operation of flues is prevented;
- Pigeons have discovered that the space beneath solar panels provide good places to roost. Not only can the droppings reduce the efficiency of the panels, but they can also dislodge electrical connections;
- Droppings can disfigure surfaces of buildings and encourage other biological growths including lichens and algae. Limestones and sandstones can be badly affected by the high acid content released from the droppings;
- Dead birds can affect the operation of appliances and fittings especially where they were roosting in the roof space;
- Birds can introduce parasitic insects that can cause nuisance, damage and health risks especially where food is stored or prepared.

Pigeons are the most common culprits, but sparrows and starlings can also be a problem in large enough numbers.

Identifying bird nuisance

In many cases, the problem can be self-evident. Sometimes dozens of birds can be seen roosting on ledges, roofs, telephone wires, guttering, etc. Where they have gained access to roof spaces, they can be seen on purlins, struts, etc. There are few surveying experiences to compare with unexpectedly disturbing a dozen flapping and scared pigeons in a dark confined roof space of an empty property. Where the birds are not evident, accumulations of droppings are a key indicator. These may be fouling the ground some distance below. Internally, nests and dead bodies, as well as droppings, may be evident.

Giving advice

As with all other pests, the Environmental Health Department of the local authority may be able to offer help, advice and in some cases appropriate treatments (although many authorities do not provide this service any longer). There are also a range of pest control contractors. In general, the approach will include:

- Removal of all droppings, dead bodies, nests, etc.;
- Before the areas are inhabited, access must be denied. This can include:
  - Blocking up any gaps or holes in the fabric;
  - Protective measures to roosting points including (see Figure 7.6):
    - Steel-sprung wires; metal or plastic vertical spikes;
    - Netting across roosting sites;
    - Various nontoxic gels that repel the birds through a number of mechanisms.

Removing the source of food can also discourage roosting in a particular area. Depending on the extent of the problem, some of these measures can prove expensive.

Figure 7.6 This shows part of a solar panel array on the hip slope of a pitched roof. The amount of stainless steel 'pigeon spikes' around the panels and along the hip tiles indicates a problem with roosting pigeons and/or other types of birds. Although these can provide an effective deterrent, it is important that they are regularly checked and repaired/replaced.

7.4.2 Rats, mice and squirrels

Although rats are the most troublesome, all rodents can damage the fabric and fittings of a building. To wear down their continually growing teeth, they regularly gnaw hard materials and will shred softer items to form nests. Rats in particular can:

* Gnaw through plastic pipes and electrical cables;
* Chew through copper pipes and cause water damage; and
* Cause localised subsidence through extensive tunnelling.

The main concern about rats is the spread of disease. Salmonella and Weil's disease can be deadly. Rats can also have a deep psychological impact on the occupiers. The rodent has never got over the public relations disaster of the Black Death and will always be associated with unhealthy conditions. The irony is that in many urban areas, rats either pass by or through most dwellings.

Key inspection indicators

To the untrained eye, spotting rodent infestations is not easy.

* Look for small trays or pots of poisoned bait (usually a grain that has been dyed blue) installed by pest control agencies. These can usually be found in roof spaces, semi-concealed ducts, behind heavy furniture, etc.;
* Neat circular entrance holes to a network of rat burrows in external areas. These can go down to 750 mm and extend horizontally for great distances;
* Gnawed and damaged building components and finishes;
* Nests of neatly piled shredded material in underfloor areas, lofts, and so on;
* With squirrels, trees that have branches within 3 m of the building and gaps between components especially at eaves and fascia board junctions;
* Large amounts of pellet-shaped dung (see Figure 7.7 for more details).

Giving advice

Where these indicators are noted, the matter should be reported to the client. The Environmental Health Department or independent pest control contractors should be asked to identify the true nature and extent of the infestation. The remedial work will depend on the circumstances but could include putting down poison baits, removal of dead rodents and possibly the blocking of holes, blocking off disused drainage connections, etc. and other disruptive ancillary building work. Further advice on how to reduce the risk of pest infestation in buildings is outlined in BRE Digest 415 (BRE 1996a).

7.4.3 Bats

These mammals are different to the other living things described in this section. Although some people may see them as pests, they are protected by the law. The

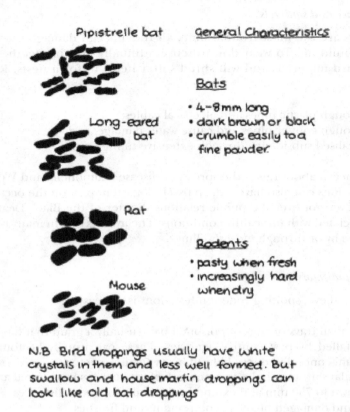

Figure 7.7 Comparison of bat droppings to those of other animals (reproduced courtesy of English Nature 1995).

Wildlife and Countryside Act 1981 (as amended) and the Conservation of Habitats and Species Regulations (2017) (as amended) make it an offence to:

- Kill, injure, catch or keep bats;
- Damage, destroy or obstruct bat roosts;
- Disturb bats by entering known roosts or hibernation sites;
- Sell, barter or exchange bats, alive or dead.

It is unlikely that surveyors will intentionally be involved in harming bats directly. In normal circumstances, it is the contractor rather than the surveyor who will do the damage. Surveyors can get involved indirectly by being part of a repair or remedial work 'chain of events'. For example, the surveyor may notice a serious outbreak of wood-boring beetle in a loft and recommend that it should be treated. A specialist contractor might then follow and spray the affected area, inadvertently killing the bat colony. Should the law be contravened in this way, the surveyor will be judged by applying the test of 'reasonableness'. Should a reasonably competent surveyor have noticed the bats during the inspection or survey? As with building defects, if there was not an obvious 'trail of suspicion' then the surveyor cannot be expected to have spotted these very small mammals.

Figure 7.8 This shows a recently replaced inspection chamber to a combined drainage system to an early 20th century terraced property. The large cylindrical droppings are clear evidence of rodents.

Precautions before the survey is undertaken

It may be an offence under the Acts to enter known roosts or the hibernation sites of bats. This is because if they are disturbed, their chances of successfully breeding or surviving the winter may be affected. This poses two problems for the surveyor inspecting a dwelling:

- If the owner knows that there is a bat colony in the property and informs the surveyor before the inspection a special mitigation licence may have to be obtained from Natural England before the inspection can go ahead;
- If the colony is undiscovered but detected by the surveyor during the inspection, (s)he may have to withdraw from the roost and obtain a licence as described previously.

In both cases, it is likely that the survey will not be completed. One way to avoid this occurring is to ask the owners if, to their knowledge, they have any bats in their property when the inspection appointment is being arranged.

Recognising bats in buildings

There are two main issues to consider:

- How to recognise the presence of bats; and
- Once they are known about, what advice should be given?

Spotting the signs

Most bats are seasonal visitors forming small colonies during May or June and leaving for more sheltered roosts in August. They do not migrate but stay in this country to hibernate through the winter. The following summary may help to identify bat roosts:

- Bats have adapted to live in all kinds of buildings including houses, churches, farms, ancient monuments and all sorts of industrial buildings. They can be found in urban, suburban and rural environments.
- Bats choose their roosts carefully:
  - During the summer they prefer sites that are warmed by the sun and are most often found on the south and west sides of a building.
  - Most types prefer small roosting spaces. The pipistrelle prefers places outside, such as in the eaves, behind tile and timber claddings, etc. and has been known to colonise recently built houses. The brown long-eared bat, on the other hand, is usually found just below the ridge in the loft spaces of older buildings.
- Although it may be possible to see the bats themselves, bat droppings are usually the clearest indicator of any roosts. They are roughly the same size as mouse droppings (see Figure 7.9), but crumble to a powder when dry. They are generally found stuck to walls near the roost exit or in small piles beneath where the bats hang. As colonies can contain between 50–1,000 bats, some of this evidence might be significant.

Giving advice

The Conservation of Habitats and Species Regulations 2017 states that any building owner must contact English Nature before anything is done that is likely to affect bats or their roosts. This could include:

- Alteration and maintenance work including thermal insulation;
- Re-roofing works;
- Getting rid of unwanted bat colonies;
- Re-wiring and plumbing works in roofs;
- Timber treatment for beetle and rot infestations;
- Treatments for wasps, bees, cluster flies and rodents.

Figure 7.9 This distinctive pile of bat droppings was directly below the ridge of a trussed rafter roof where the bats roosted.

Many of these activities may be recommended by a surveyor following an inspection. Therefore, where bats are found in a building, clients should be advised that:

- A possible bat roosting or hibernation site has been identified in the property; give the precise location.
- If the discovery of the bats has limited the inspection, clearly identify the restrictions and explain the possible consequences of not carrying out a full appraisal.
- Outline the implications of having bats in a property. This could include:

  - The need to contact Natural England to gain approval for any work to be carried out;
  - Possible restrictions on how any planned work might be carried out. For example:

    - Work might be restricted to certain times of the year;
    - Contractors may be allowed to only use specific types of materials;
    - Bats might have to be carefully moved before work can start.

These restrictions could add to the cost of the work.

On a more positive note, the client could be told that sharing their home with bats will present very few practical problems and be no threat to their family's health. Bats are considered by many to be remarkable animals with some amazing features. Home owners have a crucial role to play in the conservation of bats in this country and this can be enhanced by clear, objective advice from the surveying profession.

7.4.4 Masonry bees

'Masonry bees' have occasionally hit the headlines when they have been discovered burrowing through the brickwork of a house. These tabloid headlines tend to conjure an image of a spreading swarm of house-eating bees. In reality, *Osmia rufa* (or masonry bees) usually inhabit earth banks and soft rocks (BRE 1996b). On the odd occasion the bees choose the soft mortar of older buildings as a suitable nesting site, usually on the south facing sunny side. The damage is usually done by the female bees, who burrow into the masonry to form galleries and tunnels to house the pupal cells of the next generation. Although they act in a solitary capacity, a number of different bees can inhabit the same area if the conditions are right. Sometimes the bees may overwinter within the masonry and enlarge the tunnels. This can result in extensive damage over a number of years. In a few cases, this has led to isolated rebuilding of the wall.

Key inspection indicators

- Numbers of large holes and linked tunnels in the mortar joints of older and softer brickwork and masonry;
- Broken off pieces of brick or stone on the ground beneath;
- Evidence of frequent bee flights during the summer months.

Giving advice

Remedial work will usually involve:

- In some cases, the Society for the Protection of Ancient Buildings (SPAB 2020) state that temporary fine netting can be placed over the walls;
- Raking out and repointing the affected areas with a mortar that is not too hard for the bricks but too hard for the bees. Usually best done in late summer or autumn;
- Where soft masonry has been significantly affected then a render coat may have to be considered;
- Many building owners provide artificial nesting boxes for bees that are commercially available.

The BRE (1998) point out that the use of pesticides and other injection treatments are unlikely to be effective on their own and SPAB (2020) point out that not only can chemicals stain masonry, but their use is also ecologically undesirable as the bees are effective pollinators.

7.5 Troublesome plants

Trees and shrubs have been discussed in Chapter 5 and can undermine the very foundations of the dwelling. There are a number of other plants and organic growths that, although they will not lead to structural failure, can adversely affect the fabric of the building. The main ones are listed in the following section.

Figure 7.10 The tunnels of masonry bees can be clearly seen in the soft sandy mortar of this interwar bungalow.

7.5.1 Creepers and climbing plants

These can include a whole variety of plants that are encouraged by the occupants to climb up the dwellings. Few of the fashionable gardening programmes on TV or radio point out the less desirable effects!

- Walls and roofs may be disturbed by the plant's roots.
- Gutters and downpipes can be blocked leading to water damage.
- The wall is kept damp for most of the time and is unable to be properly inspected for defects. This may lead to neglect.
- It is difficult to repair or paint the wall satisfactorily.
- Root suckers can disfigure the wall surface.

The main types of plants are:

- Ivy and other creepers:

  - Ivy has small roots that grow in search of moisture and darkness. They can gain a foothold in open or cracked joints, forcing them apart even more. As the growth thickens, the damage can become quite serious. On balance, unless the building is in very good condition and can be kept under strict observation, then ivy is best removed.
  - Creepers – these tend to attach themselves to the building via small surface suckers. Although these are not as damaging as ivy roots, they can secrete small amounts of acid that can pit susceptible stonework.

Figure 7.11 This buddleia plant is growing in the gap between the wall and the lead waste pipe coming out from the building. The ability of this plant to thrive in such adverse conditions is admirable.

- Other climbing plants – plants such as roses, jasmines and honeysuckle all need to be supported by a framework of trellis or wires attached to the walls. These frameworks should be appropriately fastened to the wall. Ideally, the whole plant and supports should be able to be bent forward to allow maintenance and repair of the wall beneath.
- Plants on or near buildings – the BRE point out a number of other situations where plants can cause damage:

  - Ground-cover plants should be kept clear of ventilation bricks;
  - Woody plants that have root systems that can penetrate masonry should be cut down and the stumps poisoned;
  - Seeds from trees can lead to plants growing almost anywhere on a building. Flat roofs, secret gutters, valley gutters, etc., are all at risk. Where this growth occurs, all plant growth should be removed.

7.5.2 Giving advice

At the very least, the climbing plants should be kept well trimmed back from the eaves and gutters and clear of door and window frames. On balance, it is better to recommend the complete removal of ivy plants. This is normally done by cutting the main stem and digging up or poisoning the stump to kill it off. The remainder of the branches and leaves should be left in place until it completely dies back and dries out. Only at this time should the plant be removed and the roots carefully dug out. Repointing is usually required.

References

Berry, R.W. (1994). *Remedial Treatment of Wood Rot and Insect Attack in Buildings.* Building Research Establishment, Garston, Watford.

Bowyer, J. (1977). *Guide to Domestic Building Surveys.* The Architectural Press, London.

BPCA (2020). *Domestic Pest Control Advice.* British Pest Control Association. Available at https://bpca.org.uk/pest-advice/domestic. Accessed 7 January 2022.

BRE (1992a). 'Insecticidal treatments against wood boring insects'. *Digest 327.* Building Research Establishment, Garston, Watford.

BRE (1992b). 'Remedial wood preservatives: Use them wisely'. *Digest 371.* Building Research Establishment, Garston, Watford.

BRE (1996a). 'Reducing the risk of pest infestation'. *Digest 415.* Building Research Establishment, Garston, Watford.

BRE (1996b). 'Bird, bee and plant damage to buildings'. *Digest 418.* Building Research Establishment, Garston, Watford.

BRE (1998). 'Wood boring insects: identifying and assessing damage'. *Good Repair Guide 13,* part 1. Building Research Establishment, Garston, Watford.

The Conservation of Habitats and Species Regulations 2017. Available at https://www.legislation.gov.uk/uksi/2017/1012/contents/made.

Department for Levelling Up, Housing and Communities and Ministry of Housing, Communities & Local Government (2010). *Approved Document B: Fire Safety.* Volume 1, Dwellings, HMSO, London.

English Nature (1995). *Bats in Roofs – a Guide for Surveyors.* Nature Conservancy Council for England, Peterborough.

Hunt, R. and Suhr, M. (2008). *Old House Handbook: A Practical Guide to Care and Repair.* Frances Lincoln, London.

Marshall, D. et al. (2013). *Understanding Housing Defects.* By Duncan Marshall, Derek Worthing, Roger Heath and Nigel Dann. Routledge, Abingdon.

PCA (2016). *Best Practice Guidance. Wood Destroying Insects in Buildings.* Property Care Association, Huntingdon.

Property Care Association (2017). *Guidance Note: Fungal Decay in Buildings.* PCA, Huntingdon.

Singh, J. (1994). *Building Mycology: Management of Decay and Health in Buildings.* E&FN Spon, London.

SPAB (2020). 'Masonry bees'. Available at www.spab.org.uk/advice/masonry-bees. Accessed 7 January 2022.

8 Roofs and associated parts

Contents

DOI: 10.1201/9781003253105-8

8.1 Introduction

The roof covering is the property's 'hat' and it should fulfil 'its function as a weather shield . . . supported by a suitable structural system', the roof structure, which often 'provides some of the best evidence . . . to date a building' (Watt 2007, pp. 168–169). Furthermore, the roof space inspection can tell a surveyor so much about the original standard of construction because it usually remains in its original condition.

8.2 Extent of inspections

What is expected of the surveyor when inspecting a roof? One of the best sources of information is the standard guidance from professional institutions. This varies with the different types of inspections and surveys and will also be modified by the specific requirements of any lending organisation. The main elements are summarised in the following sections.

8.2.1 Level one service inspections

The guidance on undertaking roof void inspections has changed since this book was originally published and two publications have brought this about:

- Although we do not cover valuations of residential property, it is important to review how the 'Red Book' has affected how roof spaces are handled. A limitation has been included in the 2017 version of the Red Book (RICS 2019a). It makes provision that 'Roof voids . . . are not to be inspected' and this aimed to reinforce the message that the mortgage valuation was not a 'survey'. However, many Lenders continued to include within their guidance the need for a '*head and shoulders*' inspection.

- For condition reports at level one, the Home Survey Standard (RICS 2019b, p. 25) outlines the scope of roof space inspection. The first part applies to all levels of inspection and states:
  The RICS member will carry out an inspection of roof space that is not more than three metres above floor level, using a ladder if it is safe and reasonable to do so. Energy efficiency initiatives have resulted in thick layers of thermal insulation

in many roof spaces. Usually, it is not safe to move across this material as it conceals joist positions, water and drainage pipes, wiring and other fittings. This may restrict the extent of the inspection and the scope of the report. Consequently, this matter should be discussed with the client at the earliest stage.

The HSS goes on to describe a typical level one scope of inspection:

The RICS member will not remove secured access panels and/or lift insulation material, stored goods or other contents. The RICS member will visually inspect the parts of the roof structure and other features that can be seen from the access hatch.

So what does this mean? Here is our advice:

Subject to reasonable accessibility, the roof space is inspected only to the extent visible from the access hatch without entering it. The first problem is the access hatch. If there is not one, this should be clearly stated as it limits the judgements that can be made. Care must be taken looking for one. It will normally be on a landing above the stairs, but sometimes it could be tucked away in a small cupboard or above a tall wardrobe. An access hatch might still be considered inaccessible if:

- The ladder cannot be erected safely and easily without moving heavy furniture;
- The hatch is fixed or painted shut so that it could not be opened without significant force and potential damage to the finishes.

If it is openable, the ladder should be assembled and the loft hatch opened. A brief torchlight inspection should be carried out from that position. If an obvious defect is observed, should the surveyor venture further into the roof space? Does the surveyor always have to follow the trail of suspicion? This is difficult to define and will depend on the weight of the visual evidence.

In Smith v. Bush (1990) the issue of loft space inspections was discussed. The case famously involved a surveyor failing to note an unsupported chimney breast in the loft. This later collapsed into the bedroom below. The fact that the chimney stacks were observable externally and the chimney breast below was absent created a 'trail of suspicion'. This should have led the surveyor to look in the loft, even if it was not standard practice at the time. Applying this principle, a surveyor could identify a trail to be followed if the following defects are clearly observable by torchlight:

- Significant structural problems with the party walls, chimney breasts, roof timbers, and so on;
- The ingress of water is clearly evident, for example, staining, daylight showing through the covering, and so on;
- Visible wood rot or insect damage to the roof timbers;
- External signs that suggest potential problems, for example. sagging roof slopes, bowing ceilings, extensive damp staining to the ceiling below, etc.

This is not an exhaustive list but a few examples that may help you make the necessary judgement in marginal cases. This does not mean that you **have** to enter the roof space. At level one, it may be appropriate to specify the action required for the trail of suspicion to be followed so it can be established whether there is a problem or not. If there is, what action is then required for it to be rectified. In some cases, this may result in a referral for a further investigation.

8.2.2 *Level two service inspections*

The Home Survey Standard describes the level two inspection benchmark as follows:

> In addition to that described for level one, the RICS member will enter the roof space and visually inspect the roof structure with attention paid to those parts vulnerable to deterioration and damage.
>
> (RICS 2019b, p. 25)

As we have said before, the HSS is not prescriptive and it is up to the surveyor to interpret these benchmarks. Here is our view of the level two roof space inspection:

- The surveyor will get off the ladder and inspect the roof space where hatches are no more than 3 m above floor level. The extent of the inspection will depend on safety issues (such as the 'three points of contact rule' – walking on the tops of visible ceiling joists and holding on to the rafters/purlins above) and the number of stored items (see Chapter 3 for more information about safe inspection of lofts).
- The focus of the inspection should be confined to details of design, construction, general condition, hazards and legal matters. The surveyor should not move occupant possessions or lift insulation unless there is a reason to and it can be safely and easily done. The principle of 'follow the trail of suspicion' should be applied.
- Moisture meter readings should be taken to a limited and random number of timber components '. . . vulnerable to deterioration' (RICS 2019b, p. 25). This would typically include rafters close to party walls, purlin ends, underside of valley boards and timbers close to where walls and chimneys penetrate the roof coverings. Roof timbers in these locations should be thoroughly checked, as they are particularly vulnerable to damage. Moisture readings should be taken on associated accessible parts of the property, for example, chimney masonry and gable walls.
- Depending on the number of stored possessions and general safety issues, the surveyor should spend a significant amount of time in the loft on a level two inspection. Assuming a typical total inspection time of around 2.5 hours, we think 20–25 minutes will usually be spent in the loft space.

8.2.3 *Level three services (building surveys)*

In the introduction to this volume, we made it clear that the content is focused on level one and two services. However, understanding the scope of the level three roof inspection can help 'fix' the scope of level two. The HSS states the following:

> The RICS member will enter the roof space and visually inspect the roof structure, with attention paid to those parts vulnerable to deterioration and damage.

Although thermal insulation is not moved, small corners should be lifted so its thickness and type, and the nature of the underlying ceiling can be identified (if the RICS member considers it safe to do so).

Where permission has been granted and it is safe, a small number of light-weight possessions should be repositioned so a more thorough inspection can take place.

(RICS 2019, p. 25)

This suggests a more thorough inspection where things are moved. We think the interior of accessible roof voids should be inspected in detail and accessible connecting voids from the nearest vantage points to the extent practicable with the equipment available. All timbers should be checked for damage and it is important to record the moisture content of the timbers, sample timber sizes and design of the roof structure, particularly the pitch of the roof, which will help in reporting on the suitability of the covering. During a building survey, the surveyor should be prepared to spend a considerable amount of time inspecting all timbers, crawling down to the eaves, moving occupant possessions and generally getting very dirty.

Although it is important to define the limits of any inspection, no surveyor has ever been sued for going beyond the parameters of professional guidance. In fact, courts will expect this where a trail of suspicion exists.

8.3 Different types of pitched roof structures

A surveyor will meet a wide variety of roof structure types, including those described as follows.

8.3.1 Common traditional roof types

Some common types are listed here and two are illustrated in Figure 8.1.

- **Single lean-to roof** – Spans of up to 2.5 m, usually used above back additions of older terraced property;
- **Double lean-to roof** – Common on Georgian town houses with parapet wall to front and rear. Consists of a central valley beam supporting rafters bearing onto the party walls;
- **Collar roof** – The introduction of a collar between rafters allowed savings on brickwork as a room was formed partly in the roof space. The collar was sometimes so far up the rafters that it was not very effective at tying rafters together. This was associated with cheap work.
- **Couple roof** – Very simple roof consisting of rafters on wall plate with no provision for ceiling tie. This roof type suffers from a fundamental defect – lack of 'triangulation' (see Figure 8.2), a principle of most satisfactory pitched roofs, whereby the lower part of each rafter is prevented from spreading outwards by the ceiling joists, thereby forming a triangle;
- **Close-couple roof** – Ceiling tie (joist) introduced between rafter feet and as long as it has been constructed correctly can be very stable. Limited to spans of 4 m;
- **Close-couple roof with purlin** – To achieve wider spans, purlins were introduced to keep rafter sizes to reasonable dimensions. It is usual to have one or two purlins

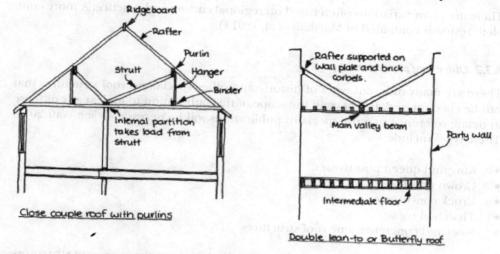

Figure 8.1 Two common types of traditional pitched (sloping) roof structures.

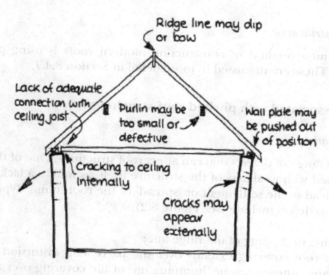

Figure 8.2 Sketch showing the typical signs of roof spread.

on each slope, and at a right angle to rafters. Struts should support the purlins and must be taken down to internal load-bearing walls. In many narrow-fronted houses, purlins can span from party wall to party wall. Collars are often found at purlin positions. A variation occurs when a room in the roof is provided. The floor level to the second storey is different to that of the rafter feet, so triangulation problems can occur;

- **'A' frame roof** – Lack of triangulation becomes more of an issue with this roof type, so named since it looks like a capital 'A' – see Section 8.4.

There are many variations often based on regional custom and practice. A more complete review is contained in Marshall et al. (2014).

8.3.2 Older roof structures

There are many different types of historic and other specialised roof structures that can be encountered. It is beyond the scope of this publication to look at any of these in detail; reference to more specialist publications will be necessary – see Watt 2011. These types include:

- King and queen post trusses;
- Crown post truss;
- Cruck construction;
- Thatched roofs;
- Steel and concrete frame roof structures.

Specialist knowledge and experience are required to assess these types of structures. Any surveyor must carefully consider whether he or she has the right background to make an assessment. Ignorance of a roofing type will be no defence in court.

8.3.3 New roof structures

The most common method of constructing modern roofs is using prefabricated trussed rafters. These are discussed in more detail in Section 8.4.7.

8.4 Defects associated with pitched roof structures

8.4.1 Roof spread

There is a wide range of defects that can affect roof structures. One of the most common is associated with weakness of the structure itself, as well as a lack of triangulation. This can lead to the settlement or 'spread' of the roof frame. Typical signs and symptoms of this defect include (see Figure 8.2):

- A sag, bowing or dipping of the ridge line;
- Where the roof covering extends over the party wall, distortion in the structure can cause unevenness or 'humping up' of the covering over the party wall positions;
- The feet of the rafters pushing outwards can cause the top of the supporting walls to distort, for example, bulge centrally, in some cases displacing the wall plates;
- Horizontal cracking to mortar courses just below eaves level where the top parts of the masonry are pushed outwards by rafters;
- Physical distortion of the roof components internally, in extreme cases causing splitting of timbers.

Lack of triangulation is even more of an issue with an 'A' frame roof. Lower sections of the rafters (above sloping sections of ceiling), situated beneath the horizontal collars (in effect ceiling joists), tend to thrust outwards when they are subjected to loads such as wind and snow (the 'live' loads) and the weight of the tiles and of the roof timbers (the 'dead' loads). However, if the collar is situated in the lower third of the roof, and the roof is otherwise fairly well constructed and properly nailed, whilst some movement can occur and this can result in some distortion in the roof structure and the walls, the roof should usually perform reasonably satisfactorily.

If the collar is in the middle or top third of the roof, the resultant lack of restraint to the lower sections of rafters means that the possible effects on the walls beneath can be significant and in some instances catastrophic. Modern 'A' frames prevent this issue by using a single steel beam at ridge level or two situated where the collar meets the rafters, spanning from wall to wall.

In addition to the 'signs and symptoms' above, in most cases of roof spread, there are also likely to be:

- Vertical gaps or cracks where internal walls meet external walls as those outer walls are pushed outwards;
- Gaps between floorboards and outer walls, due to the same cause.

Roof spread is often caused by structural deficiencies, as described in the following section.

8.4.2 Adequacy of timber sections – rafters

Before the national building regulations came into force in 1966, the sizes of timbers in many traditional roofs varied according to local bye-laws and construction practice. Additionally, research has shown that wind and snow loadings have increased over the last 20–25 years and following some collapses of roofs due to snow loads in mainland Europe, standard roof timber sizes were reviewed in 2004 and factors of safety increased. Consequently, when compared to current standards, it is common to find that older roof structures do not match up. But that does not mean they should be condemned, as you must also review all aspects of structural integrity (for example, lateral stability, triangulation and so on) so other serious deficiencies are not overlooked.

In the first edition of this book, we presented a number of traditional 'rules of thumb' that could be used to give an approximation of a roof structure's adequacy. On reflection, assessing the adequacy of any roof structure is more complicated than applying 'rules of thumb' and so these have been omitted. To show how complicated the sizing of roof components can be, take a simply supported rafter. The size will depend on the following:

- The 'dead' loads such as tiles, sheathing felt and tile battens. In some cases, this may include PV panels;
- 'Live' loads such as wind and snow. These will vary depending on the location and exposure of the site;
- The pitch of the roof: rafters on steeper pitches can span further than those on shallower slopes;
- The spacings between the rafters. Those set at 400 mm spacings can span further than rafters at 600 mm centres;
- The span of the rafter. For example, between the wall plate and the purlins, between the purlins and between the upper purlin and the ridge board;

Comfortable (lower risk)	Uncomfortable (higher risk)
• Lighter roof covering (see Section 8.5.2); • In a region where wind and snow loads are lower and/or the property is sheltered; • There are no PV or solar thermal panels; • A steeper pitch to the slopes; • Spacings of the rafters are closer to 400–450 mm; • The span of the rafters is between 1.75 mm and 2.0 m(maximum); • The rafter size is at least 50 x100 mm.	• Heavier type of roof covering (see Section 8.5.2) • In a region where wind and snow loads are higher and the property is exposed; • There are PV and/or solar thermal panels on the roof slopes; • The spacings of the rafters are closer to 600 mm; • The span of the rafters is over 2.0 m; • The rafters are less than 50 x100 mm.

Figure 8.3 Assessment scenarios for rafters.

- The strength of the timber. Under current building regulations, 'C24' graded rafters can span some 13–14% further than 'C16' graded timber.

It is not possible to reflect this level of complexity with a simple rule of thumb. However, we do not want to duck the issue. We think it is possible to develop a 'feel', based on experience over time, for what should be within the margins of acceptability. In Figure 8.3, we have described two scenarios: the 'comfortable' scenario outlines the features of a roof structure that is likely to perform satisfactorily and the 'uncomfortable' scenario describes a roof structure that may leave you with a sense of unease and require further consideration.

To avoid numerous referrals to roofing specialists or structural engineers, a sense of balance must be retained. If the roof structure sits within the 'comfortable' scenario above and there is no evidence of:

- Significant distortion, sagging or deflection;
- Deterioration of the timber components (for example, wood rot and wood boring insects), and
- The structure appears well triangulated;

Then no further work or referrals may be required.

8.4.3 Purlin problems

The purlin is usually the main supporting component to many roof structures and can suffer from a variety of problems.

Size of purlins

A similar approach to that described for rafters can be used for purlins (see Figure 8.3). The factors associated with dead and live loads, timber strength and pitch of the roof are the same as for rafters. The main variants are:

- The spacings of the purlins. This is usually the distance between the feet of the rafters and the first purlin; between the purlins (if there are more than one); and between the purlin and the ridge board;

- The clear span of the purlin – this is usually the distance between the supports to each end of the purlin. This could be a supporting wall or an intermediate strut.

Support to purlin ends

Typical faults include:

- Purlins should ideally be built into gable walls by at least 80mm, to help support the wall. Where purlins are built into exposed solid gable walls or chimney stacks, or through cavity walls, the ends are vulnerable to wood rot;
- Some purlins are supported on brick corbels that protrude out of the wall. These can break under load unless properly constructed;

Figure 8.4 Two examples of split rafters, in both cases weakened by cutting the 'birds-mouth' too far through them – see Section 8.4.6.

Figure 8.4 (Continued)

- Poor support from the wall on which they bearing (see Figure 8.6);
- Purlins should be in one length between supports. Where they are jointed, they should be properly 'scarfed' equally on each side of the support. If not, the purlin can rotate around this joint;
- Where purlins are supported at mid-span by a strut, this must bear onto an appropriately supported element (for example, an internal wall or a partition around a staircase). In some cases, this strut can be carrying a significant proportion of the weight of the roof slope. If it is not properly supported, deflection of the roof slope could result. Check the following features:

  - The struts bear directly onto the supporting wall below and not merely onto a timber plate positioned on the ceiling structure (see Figure 8.7);

Comfortable (lower risk)	Uncomfortable (higher risk)
• Lighter roof covering (see Section 8.5.2); • In a region where wind and snow loads are lower and/or the property is sheltered; • There are no PV or solar thermal panels; • A steeper pitch to the slopes; • Spacings of the purlins are closer to 2.1 m or less; • The clear span of the purlins is not more than 2.1 m; • The purlin is at least 75 x 225 mm but preferably greater.	• Heavier type of roof covering (see Section 8.5.2) • In a region where wind and snow loads are higher and/or the property is exposed; • There are PV and/or solar thermal panels; • A lower pitch to the roof; • Spacings of the purlins are more than 2.1 m; • The clear span of the purlins is not more than 2.1 m; • The purlin is less than 75 x 225 mm.

Figure 8.5 Assessment scenarios for purlins.

Figure 8.6 Inadequate support to purlins at the end bearing. These purlins are poorly supported by this partial party wall.

- The internal wall taking the strut matches up with walls on all the floors below, down to the foundations. The 'load path' from the top of the roof down into the foundations must be continuous. Special checks should be made where 'through' rooms extend the full width of the house, especially when these have been as a result of recent DIY alterations.

Ring purlins

The purlins to hipped roofs often have no visible support. This is because they act as an informal 'ring beam' and rely on the whole structure acting together as one. For a typical semi- detached property with a hipped roof, the purlins beneath the front and rear roof slopes will be supported by the party wall at one end and by the hip rafter at

Figure 8.7 Inadequate strut support to a purlin. This traditionally constructed roof was recovered with concrete tiles in place of original slates and had begun to deflect under the increased weight. The purlin was visibly deflecting and, trying to limit this problem, a contractor installed a small strut. This was poorly supported by a timber plate fixed to the ceiling joists at mid span, causing considerable deflection to the ceiling below.

the other. The hip roof slope should also have a purlin of its own and this should be accurately cut to form a well-formed junction with the other purlins.

To perform satisfactorily, a roof of this type should have the following characteristics:

- The purlins should be of substantial size (for example, 75 × 225 mm minimum);
- All timber sections should be cut closely together and properly spiked or nailed into position, especially at the junctions with the hip rafters and purlins;
- The rafters should be properly fixed to the purlins – for most roofs this means using two nails. Indeed, all nailed structural connections in a timber roof structure should be with two nails, not one, since otherwise the two timber pieces can rotate around a single nail connection;
- There should be adequate restraint to the feet of the rafters by either the ceiling joists, collars or other forms of restraint;
- The hip rafter should be of adequate size because this will be taking a considerable load from the purlins. It should be showing no significant distortion;
- There should always be a tie (sometimes called a 'dragon' or 'angle' 'tie') reinforcing the connection between the wall plates at the corner of the building below the hip rafter. Without this, the wall plates can move outwards, allowing the hip rafter to deflect (see Figures 8.8 and 8.9).

If a hipped roof has most of these characteristics and there is no evidence of distortion, no further action would be necessary.

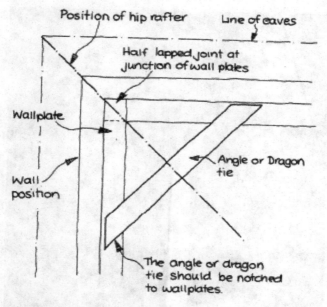

Figure 8.8 Sketch showing the typical arrangement for a 'dragon tie' that joins the adjacent wall plates to the side slope of a traditional hip roof.

8.4.4 Ceiling ties and joists

The ceiling structure has several functions:

- To support the ceiling finish;
- To help triangulate the roof structure by restraining the feet of the rafters.

Ceiling joists are often 'stiffened' by binders and hangers. Some of the most common ceiling joist problems include:

- The joists can be either too small or lack support at mid-span. If the joists are not big enough to span between their main supports, a binder with hangers should be provided;
- The joists are not at the same height as rafter feet and so do not offer any resistance to rafter spread (in other words the roof is not triangulated). In this case, look for typical signs and symptoms of roof spread (see Section 8.4.1);
- Poor nailing at junctions with other roof members. For example, single-nailing at the junctions with binders;
- Where joists do not restrain the rafters, check for timber collars. They are more effective if the collars are dovetailed to the rafters rather than just nailed onto the side;

Figure 8.9 The hip rafter to this roof was deflecting because the wall plates were spreading apart – there were no dragon ties to restrain them. A builder inserted a crude vertical strut to take the weight of the hip rafter. This was supported off the top of the ceiling joists, causing the ceiling to crack. Another unsatisfactory feature is the junction of the two purlins. These have been roughly cut to either side of the hip rafter and so will not provide a secure connection.

- Where the roof structure has a hip slope the feet of the hip rafters are often not tied to the ceiling joists as the latter often run parallel to the gable wall. Additional ties may be needed at this point to help prevent rafter spread.

8.4.5 Room in the roof

In some older housing, a room in the roof was provided at the time of construction. Characterised by a steep winding staircase up to a small room served by a dormer or skylight, these spaces can break the continuity of roof structure. Triangulation for rafters can be difficult to provide, as the floor structure and the wall plate are often

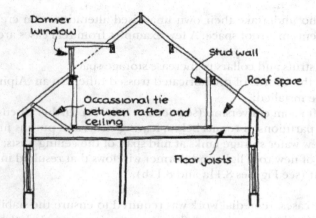

Figure 8.10 Room-in-the-roof structure – lack of connection between the floor or ceiling joists and rafter feet can often result in roof spread.

at different levels. In many cases where there are access hatches to the front and rear roof spaces, the true nature of the roof structure can be assessed. If these hatches are absent, care must be taken. Experience has shown that few of these types of roof structures are adequately triangulated (see Figure 8.10).

8.4.6 Other defects

Several other defects associated with the roof structure can be regularly encountered, including the following.

Over-cutting of joints

Where one timber section meets another, they are often 'birds-mouthed' together. This is where one is cut to fit tightly against the other. The depth of this birds-mouth should be no greater than one-third the depth of that timber – see Figure 8.4 for examples of where this was done incorrectly.

Weakening of structure through woodworm and rot

Dry rot is not common in roof spaces because the environmental conditions are so variable. Wet rot can affect timbers, especially where they are built into or in contact with excessively moist masonry. Typical locations include trimming timbers around chimney breasts, rafter feet and wall plates adjacent to leaking gutters, timbers beneath and around valley gutters and beneath and around horizontal and parapet gutters.

Woodworm can be found in most older roof spaces, but usually there are only low levels of infestation that pose no serious threat to structural stability. Where the infestations are particularly heavy, especially with DeathWatch and Longhorn Beetles, structural timbers can be considerably weakened (see Chapter 7 for more information).

Homeowners who undertake their own unadvised alterations can create a range of structural problems in a roof space. A few examples from real cases are as follows:

- Removal of struts and collars to increase storage space;
- Cutting out the chords of prefabricated trussed rafters so an 'Alpine' model railway could be installed;
- Informal loft room conversion (see Section 8.6.6 for further discussion);
- Removal of partitions in rooms below, leaving struts and purlins unsupported;
- Fitting of new water storage tanks at mid span of the ceiling joists;
- Installation of new roof light and dormer windows that resulted in the main purlin being cut (see Figures 8.11a and 8.11b).

In each of these cases, remedial work was required to ensure the stability of the roof structure. If any similar alterations to the roof structure are noted, the owner should be asked whether the appropriate building regulations approval has been obtained or if a competent contractor has carried out the work. If not, further investigations may be required, including noting the matter for the legal adviser.

Dormer roof structures

Many traditional houses have large dormer windows, where the front wall of the dormer is formed by the continuation of the external wall of the house above the roof line. Unless this wall is well restrained, it can become unstable. A classic sign is the wall leaning backwards out of vertical, putting pressure on the adjacent roof structure. Internal distortion of roof timbers and ceilings can often be seen.

Figure 8.11a The roof to this property has been affected by roof spread. Based on the external visual clues, the likely causes are heavier replacement concrete tiles and installation of roof windows.

Figure 8.11b An internal view of one of the new roof windows. The purlin has been cut through so the window could be fitted. The vertical posts supporting the ends of the purlins are supported off the existing floorboards and floor structure, with resultant distortion above and below. Also note this example of water penetration affecting the purlin end, caused by, as usual, the leaking parapet wall.

Dormers in many properties are often affected by several defects. These features need to be detailed very carefully if rainwater penetration is to be avoided (HAPM 1991, Section 4.3). Typical problems include:

- Surface and interstitial condensation – The dormer cheeks (sides) and ceiling detailing can make it difficult to achieve continuity of insulation and ventilation. Check for signs of mould;
- Inadequate trimming around the dormer opening – The top and bottom trimmers and the side supporting rafters will have to carry the truncated portion of the main roof that results from the dormer, for example the fact some rafters have been removed. If not properly designed, usually by at least 'doubling up' the rafters to each side and the joists beneath, timber members could deflect. Check around the dormers for signs of deflection and cracking to the finishes;
- Rainwater penetration – There are a number of vulnerable points around a dormer including the junction of the window frame and the dormer structure; the junction of the dormer and main roof covering and the apron flashing at the base of the dormer, usually under the window sill. Internal surfaces adjacent to these features should be checked for signs of water penetration. Matters can be made worse to the sides ('cheeks') if the dormer roof is pitched to either side and there are no gutters;
- Inadequate durability of the window in the dormer – Because this window is usually very exposed and is less accessible for maintenance than lower windows, it is often in poor condition. One solution is to have a low-maintenance PVCu or metal window, even if the rest are timber;

- Flat roof problems – Where the dormer has a flat roof, the covering is very exposed to wind, rain and sunlight. As it is very difficult to access, it is often 'forgotten' until it fails. This can be inspected externally using a selfie-stick or pole camera and the ceiling beneath carefully checked for signs of water leakage, providing use of the camera in this way has been clarified in the terms of engagement (see Chapter 3 on inspection for more information).

8.4.7 Trussed rafter roof structures

Most houses constructed since around 1965 have trussed rafter roofs, although there are significant regional variations. These prefabricated structures are designed in accordance with modern design codes (although they have changed over the years) and tend to be of a more reliable quality. However, because they are so well engineered, their reserves of strength can be less than those of traditional cut roofs. They are also more reliant on interactions with other parts of the building structure, such as the walls. Consequently, they are more vulnerable to damage caused by inappropriate installations and alterations. Typical *key survey indicators* include:

- Trussed rafter roofs must be adequately braced to have sufficient strength to resist lateral wind forces (see Figure 8.12). In early forms of this roof type (usually the 1970s), bracing was usually not provided; nowadays, they are sometimes incorrectly installed (for example, with one nail only) or cut through by operatives fixing flues, soil and vent pipes, and so on;
- Some early trusses have timber connectors at their joints, rather than metal plates – these can suffer from wood boring insect attack;
- Trusses should not usually be placed at greater spacings than 600 mm – any more than this could result in excessive deflection;
- The final three trusses should be fixed to the gable or flank wall with galvanised restraint straps, 'noggins' and 'packing pieces' to ensure the roof structure helps

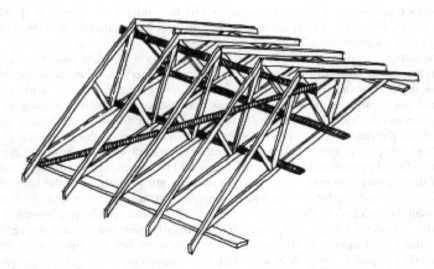

Figure 8.12 Typical diagonal and longitudinal bracing to a trussed rafter roof (reproduced courtesy of BRE).

support the wall against wind loadings. More information is given in Figure 5.15. An alternative is a 'gable ladder' with barge-board, relying on friction of the nog-gings passing through the supported wall;

- A trussed rafter roof is very sensitive to alterations – no part should be cut or removed without a properly engineered solution;
- Any water tanks need to be adequately supported with the weight spread across several trusses (see Figure 8.13);
- Trusses can be damaged and split where they are 'skew' nailed (nailed at an angle) to the wall plate – metal truss clips are a better solution;
- Surveyors should check the metal nail plates for signs of corrosion – important if high humidity levels are suspected;
- Earlier trussed rafter roofs will often have less ventilation than the current regula-tions require. Extra insulation added over the years might have reduced this even further. These factors can cause mould growth and damp staining.

8.5 Pitched roof coverings and their defects

8.5.1 Introduction

There are many different types of roof covering to pitched roofs. It is important to inspect them from above and below, if possible. The external inspection will depend to some extent on personal preference but will typically include:

- By binoculars from ground level;
- From adjacent windows that may provide a view over the roof surfaces;
- By the use of camera poles and drones.

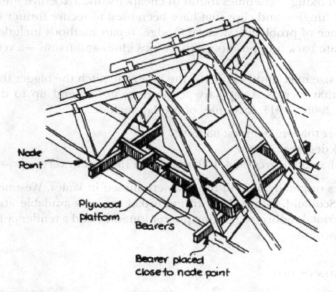

Figure 8.13 Support to water storage tanks in a trussed rafter roof (insulation, lid, overflow pipes etc. omitted for clarity) (reproduced courtesy of BRE).

Whatever methodology you adopt, it is important to include these matters in your terms of engagement and to explain the implications to your client.

8.5.2 Different types of pitched (sloping) roof coverings

Introduction

This section identifies the main types of roof covering and briefly outlines the most common defects. It is important that the surveyor can identify the different types of pitched roof coverings, especially between 'slates' and 'tiles', for they are not the same. Clients may view an otherwise perfectly prepared report with jaundiced eyes simply because 'slates' have been identified as 'tiles', or vice versa. This applies to all materials and components. Slates are a 'double-lap' covering and have an average thickness of around 5–6 mm. Tiles can be single (more usual) and double-lap, with a greater thickness, usually around 15 mm.

Natural slates

This is a very durable covering that has been used for hundreds of years. There is a great variety in both type and characteristics. Some are longer lasting than others; slates have been described as 'the most hardwearing of all roofing materials' (Rock 2006, p. 30). The main problems include:

- Sulphurous acids in rain can cause breakdown of carbonates in the slates. Capillary action can allow the rainwater to travel between slates and so deterioration, manifesting as delamination, can begin from the underside;
- Due to each slate moving in wind the nail hole can become enlarged, allowing the slate to slip out of position. This will especially affect slates that only have one centre nail fixing, sometimes found in cheaper work. Excessive numbers of lead or copper 'tingles' and clips that have been used to secure former slipped slates are evidence of problems. Modern 'bodge' repair methods include sticking the slipped slate back into position with various glues and foams – a very temporary solution;
- Incorrect size in relation to pitch – the lower the pitch the bigger the slate needs to be so that increased capillary action does not extend up to the next joint (Johnson 2006, p. 91). Examples of overlap include:

  - Steeper than 45 degrees: 63 mm
  - 30–45 degrees: 75 mm
  - Less than 30 degrees: 88–100 mm, up to 150 mm in very exposed areas.

- Most slates originally used in the UK were mined in Wales, Westmorland, Cornwall and Scotland. Other slates from abroad are now available and there have been anecdotal reports suggesting poorer longevity and a tendency for some varieties to discolour.

FIBRE-BASED 'SLATES' OR TILES

These were originally manufactured as asbestos cement slates with surface finish, so they resembled natural slates, often laid in a variety of patterns. These have now been replaced by non-asbestos fibres. Early types often warped, lost the surface finish and supported high levels of mould and moss growth (Johnson 2006, p. 102). Several different types are made by different manufacturers. It is important to become familiar with different systems. Regularly updated manufacturer literature will be a useful source of information. *Key inspection indicators* are:

- Fixing methods – especially the correct provision of copper discs or rivets that hold down the lower end of the slates;
- Evenness of the slates – newer slates have to be laid on a reasonably true and even supporting structure. If they are not properly supported from beneath, they can easily crack when walked over during normal maintenance work, which releases asbestos fibres;
- Accessories – for example, eaves and verge details, ridge and hip fittings. They should be from the same manufacturer's range as the slates themselves.

PLAIN CLAY TILES

These have been used for centuries, hand- and machine-made. They are a 'double-lap' covering, hence usually very watertight. The main problems are:

- If too porous can break down with frost, particularly close to eaves – most clay tiles absorb around 10% of their weight in water when it rains;
- Because of the double lap they weigh significantly more than most other pitched roof coverings;
- Some machine-made varieties can have variable durability;
- There can be a problem matching profiles, textures and colours when they need replacing;
- Check the nailing – for existing older roof coverings this will vary according to exposure:
  - Normal – every fourth or fifth course, every second course in some regions;
  - Exposed location – this can be every course;
  - Nailing can be checked from the loft by gently pushing up the tiles if there is no underlay;
  - In some older properties, tiles may be held in by wooden pegs – considerable doubt over durability of this type of fixing should be indicated to the client.

CLAY PANTILES

These are made of clay and so many of the problems identified with plain tiles also apply. Other issues include:

- Older varieties are usually never nailed, so are prone to wind damage;
- Because of their pronounced 'S' shape, they easily allow:
  - Driving rain to penetrate – a good underlay, dressed into gutters, is vital;
  - Birds to nest between underlay and tiles – which peck away at the hessian reinforcement in traditional underfelts to make nests, creating holes in the felt.

CONCRETE TILES

There has been large-scale use of concrete tiles since the 1970s. Many concrete tiles can be of poor quality. The main problems include:

- They can be much heavier than original roof coverings they replace and so cause deflection of the roof structure;
- The surface finish often weathers away and can contribute to the blocking of gutters and drains;
- More prone to water penetration than plain tiles or slates, since they have only a single lap, so must have a good underlay;
- Difficult to form properly weathered details especially at valleys, hips and abutments – this can often lead to poor 'extemporisation' on site;
- The fixings need to be in accordance with the manufacturer's recommendations to prevent uplift, especially at the edges, such as metal clips at verges.

STONE SLATES

Examples include sandstone in Yorkshire and around Northamptonshire ('Colleywestons'), or limestone in the Cotswolds. Because they are very heavy, the roof structure needs to be assessed with special care. Many older roofs may have become distorted over time because of this loading. If the building is listed or in a conservation area, new or reused stone slates will have to be used in any recovering or maintenance work. As this requires specialist knowledge and skills, the client should be warned about the possible cost consequences.

Other roof components

The surveyor must also make note of other parts of roofs. Typical *key inspection indicators* for some components include:

- Ridge and hip tiles are subjected to greater wind loads (sometimes causing uplift) than most other tiles and traditionally relied on lime:sand (latterly cement:sand) bedding mortar to prevent uplift – if there is insufficient mortar or it has eroded, tiles can become dislodged and present a safety hazard;
- Valleys:
  - Lead valleys are subject to warming and splitting in sunlight if the lead lengths are too long – with resultant water penetration;
  - Mortar fillets to adjoining tiles will often crumble and erode, blocking valleys and allowing water penetration; and
  - Minimum width for a valley is 100 mm – any less and there is a risk of leaves preventing rainwater run-off;
  - A close inspection beneath to confirm condition of the underlay and using a moisture meter is vital for any valley;
  - Modern valleys are usually in fibre-glass, easy to confuse with lead but with a much shorter life-span;
- Where roofs meet on a wall, mortar fillets are prone to cracking – a fillet with small embedded tile pieces will tend to suffer less shrinkage; and

- Mortar verge fillets can shrink, crack, erode and fall away, the verge tiles are then more liable to wind uplift – in 2014 it was reported 'nearly two thirds of all roofing claims against the Buildmark warranty . . . related to mortar failure' (Redland 2014, p. 4).

Protecting against wind damage – change in fixing standards

To prevent slipping and wind damage, each slate has generally always been nailed and many traditional clay tiles were nailed or pegged in place. For clay and concrete tiles, although there are local variations, until 2014, it was usually a requirement for tiles to be nailed:

- All eaves and verge courses of tiles;
- Every tile course beneath the ridge;
- Tiles around chimney stacks, roof-lights and adjoining valleys; and
- To all other areas of tiles, every 3rd, 4th, or 5th course (depending on manufacturer's recommendations).

In 2014, in view of increasingly volatile weather and apparent increases in wind speeds across the UK, BS 5534 (BSI 2014) was revised to improve fixing requirements:

- For single lap tiles, every tile should be nailed or secured with clips, usually in aluminum or stainless steel;
- The greater the pitch, the more fixings should be provided;
- Ridge and hip tiles must now all be secured with clips or similar, onto the roof structure. In other words, ridge and verge tiles should be 'dry fixed' and no longer bedded in mortar.

Given changing weather conditions, these increased fixing requirements should be considered as the modern 'benchmark' for properties of all ages. For properties likely to have lesser standards of fixing, you should consider warning the client that maintenance of the roof is likely to be more expensive than a modern roof, especially in exposed locations.

Problems with the underlay, or 'sarking felt' as it was formerly called

Newer or recovered roofs can often suffer from leaks due to faults in the underlay. For example, if the underlay has not been properly supported where it is dressed over the eaves into the gutter or over the edge of any bargeboard, water may build up in the resultant sag and leak internally. The same effect can occur where the underlay has not been fitted closely around soil and vent pipes that pass through the covering. Typical defects in underlay can include:

- Older underlays (for example, a type called 'kraft paper') become very brittle, break and leak;
- Moisture staining to the ceilings, inner leaf of the wall, or loft insulation;
- Poorly fitted underlays with inadequate laps, loose, punctured or partly missing sarking felt to the underside of the covering;

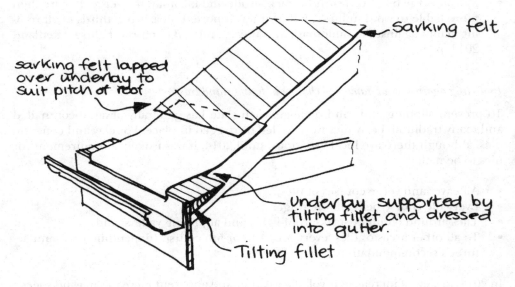

Figure 8.14 Recommended underlay support detail at eaves level.

- Condensation forming on the underside, particularly if ventilation is poor (see Section 8.6.5 for more discussion).

Where the underlay is dressed into the gutters, if it is the older standard type of felt (reinforced bitumen), it will degrade on exposure to sunlight, especially on the south slope. To prevent this, the BRE (1985a) recommended that a supplementary strip of DPC-quality material should be provided beneath the main underlay at the eaves and the verge. A *key inspection indicator* is to look for water stains to the fascia and soffit boards, and daylight behind the gutter, indicating a gap where there shouldn't be one (see Figure 8.14).

The underlay should not be:

- Stretched too tightly over the rafters because it can result in tears occurring when it is nailed and water running down the felt being trapped behind the batten;
- Too limp – there is a danger of water not draining away properly or ponding in any folds.

8.5.3 Assessing the condition of roof coverings

Slate and tile roof coverings should have a long life. Many natural slate roofs are still giving good service 80–100 years after they were first installed. Even where some slates have slipped or are broken, a competent roofing contractor can extend the roof's effective life even further. The problems begin when a high proportion of the slates or tiles are defective or have been repaired before. Roofs in this condition can be ravaged by high winds and storms. This can be expensive to repair, cause considerable damage to other parts of the property and be a safety hazard.

8.5.4 Temporary repairs to roof coverings

Many building owners carry out temporary repairs to roof coverings to save money. Rather than going to the expense of recovering the roof, cheaper proprietary repairs are often used. These can include:

- **Bitumen coatings to the roof slope** – sometimes called 'tunnerising', this involves the outside application of liquid bitumen products to the whole or part of the roof area. Occasionally, this is combined with hessian or glass fibre sheeting to reinforce the repair. Although this can keep the water out in the short term, the bitumen can soon break down, allowing water to penetrate again;
- **'Flashband' type materials** – an adhesive-backed aluminum or similar material. Very good at providing a patch repair, for example, to a lead valley or slates, but usually a clear indication that more fundamental repairs and replacements can be anticipated soon;
- **Spray foam adhesive coatings applied to the rear face of the covering** – this type of treatment has become increasingly common since the first edition of this book. Additionally, it is a controversial repair choice. Consequently, this has been considered in more detail in Section 8.5.5.

8.5.5 Spray foam insulation to the underside of roof coverings

Over 20 years ago, in the first edition of this book, we warned about the problems caused by what we described as 'rigid foam coatings to the rear face of the roof covering' (Parnham & Rispin 2001, p. 168). At that time, manufacturers and installers of polyurethane spray foam claimed the treatment extended the life of roof coverings by sticking the tiles/slates together, curing leaks and increasing the thermal insulation of the roof. We were not convinced. We had come across a number of projects where encapsulated roofing battens and rafters had rotted and roof space ventilation had been reduced to very low levels. We advised practitioners they should not recommend this treatment to their clients and where the foam was in place, make the client aware of the implications of buying a property with spray foam coatings which might require '. . . roof covering replacement in the near future'.

A fifth of a century later, polyurethane spray foam insulation is still being installed on the underside of existing roof coverings. Spurred on by a variety of government energy efficiency initiatives and grants, spray foam manufacturers and installers have been actively promoting the alleged energy benefits of applying spray foam to the underside of existing roof coverings and structures. It now comes in two varieties: the softer open cell that is claimed to be vapour permeable, and the closed cell, which is less permeable, more rigid and better suited to stabilisation functions. Many spray foam systems are supported by a number of British Board of Agrément certificates (BBA) and endorsements from organisations such as the Local Authority Building Control (LABC) and the Energy Saving Trust (EST endorsed product). So it must be OK to use – mustn't it?

Like most complex situations in our sector, the answer to this question is not straightforward. In our opinion, encapsulating timber, no matter what assurances the

Figure 8.15 This photo shows the slate roof covering of a terrace property built in the early
20th century. The original slate roof was in poor condition with many slates missing, cracked, loose and showing signs of previous repair. The owner reported that
the roof was leaking in at least three places. Figure 8.16 shows a photograph taken
in the roof space.

manufacturers give about vapour permeability, carries with it inherent risks. If timber
is not able to quickly and evenly reach a moisture equilibrium with the local atmos-
pheric conditions, then it will be vulnerable to wood rot and wood-boring insects.
This is especially true with roof coverings because of the high moisture environment
in which they exist.

This view has been confirmed through our own professional experience and that
of other residential practitioners who have reported numerous problems on forums
like the Surveyor Hub Community – an online community with over 3,000 members (www.facebook.com/groups/the.surveyor.hub). Although this is anecdotal, it
provides enough evidence to justify adopting a cautious process when assessing the
suitability of spray foam insulation. Figures 8.15 and 8.16 show just one of the many
examples that are causing problems for property owners and have influenced our
views.

Reactions in the marketplace

These concerns over the use of spray foam applications have had an impact on the
saleability of properties with this system installed. At the time of writing, a small number of lenders had a blanket policy of not lending on properties with spray foam insulation, while others left it up to the individual valuers to take a view. A 'nil' valuation is

Figure 8.16 This is a view of the roof space of the property shown in Figure 8.16. Rigid closed cell spray foam insulation had been applied directly to the underside of the slates in an attempt to secure the slates and stop the roof leaks. The foam comprehensively failed in its task. We advised the owner that stripping and recovering was the only option and this would be made more difficult and expensive because of the foam.

often returned. Although we think a 'nil' valuation is an overreaction, it is clear that the likely cost of remedial work would still have some impact on value.

Like Japanese knotweed (see Chapter 11), it is not surprising that many homeowners have discovered that their properties are unsellable. During the time it took to write this volume, the issue has been covered by a number of national newspapers and featured on a peak-time consumer programme on national television. The media have featured a number of stories in which owners had to have the spray foam removed or accept a sizeable reduction in the purchase price in order to sell their properties.

Certification

As mentioned previously, most of the major products have the support of a number of BBA and LABC certificates issued for spray foam. However, this does not provide a blanket clean bill of health. For example, the BBA certificate for one well-known spray foam product stipulates the following conditions must be met:

- Before application of the product '. . . existing constructions must be in a good state of repair with no evidence of rain penetration or damp. Defects must be made good prior to installation'. The certificate goes on to state, '. . . installation must not be carried out until the moisture content of any roof timber framing is less than 20%';

- In relation to tiled or slated pitched roofs, the product can be spray-applied directly to the underside of '. . . reinforced bitumen membranes, breathable roof tile underlays, or timber sarking boards between the rafters'. There is no mention of applying the foam directly to the slates or tiles where a sarking layer has not been used;
- To avoid interstitial condensation risk, '. . . it is essential that the roof design, construction and maintenance not only limit opportunities for vapour migration by diffusion but also by convection through gaps, cracks and laps in air and/or vapour control layers (VCL)s and through penetrations'. This is taken to mean that the ceiling above the habitable rooms must be well sealed against rising warm moist air (Isothane Ltd 2016).

The LABC certificate for the same product confirms many of these conditions, including:

- 'A well-sealed ceiling as described in BS 5250 is required';
- 'Roof spaces should be ventilated in accordance with BS 5250:2011';
- If the product is left exposed in non-habitable roof space, then '. . . fire warning labels are required'. However, no details about the nature of the labels are given.

The picture that is emerging is that the installation of spray foam insulation to current standards is a far more sophisticated process than that employed 20 or more years ago, when the foam was routinely sprayed directly to the underside of roof coverings. More recent certification requires a greater level of preparatory and ancillary work and in our view, this can be used to build a suitable protocol for residential practitioners that is described at the end of this section.

Upsetting the equilibrium

Before we put forward an assessment process, it is important to go back to basics. For most domestic residential properties with an uninhabited 'cold' loft space, the most cost-effective position for thermal insulation will be the rear face of the horizontal ceiling element. As long as the dwelling is well ventilated, the ceiling is well-sealed against air movement and the roof space is also satisfactorily ventilated, moisture problems in the loft space should be kept to a minimum. In these circumstances, if the property owner wants to increase the level of thermal insulation to match current standards, the most effective way of doing this is by adding extra insulation to the back of the ceiling to any that may already exist.

In comparison, adding a layer of spray foam to the back of the roof covering in an effort to increase thermal insulation levels does not make logical sense. Not only will it be more expensive, but it will also potentially upset the equilibrium of the roof space's environmental conditions.

This has been described by the USA-based Spray Foam Coalition in their publication titled 'Guidance on best practice for the installation of spray polyurethane foam' (American Chemistry Council 2012). According to the Coalition, buildings are complex systems in which all the major elements contribute to overall performance.

A change to one component can impact the other elements and can alter aspects such as thermal performance, indoor air quality and moisture levels, to name but a few. The Coalition used an example to illustrate this knock-on effect, where the 'thermal element' is moved from one area of the structure (say the back of the ceiling or loft floor) to another area (for example, under the roof covering between the rafters). If the implications of this change are not fully understood, then overall performance can be affected and any imbalance could result in excess moisture and possible wood rot.

Installing spray foam insulation without appropriate precautions may have a similar effect. For example, if the back of the ceiling below a roof space is already insulated to some extent, installing spray foam insulation below the roof covering will create two thermal elements: one at roof level and one at ceiling level, separated by an unheated roof space. Because the spray foam insulation is likely to dramatically reduce existing roof space ventilation, it will leave the remaining roof space more vulnerable to condensation, as warm moist air will continue to drift upwards from the dwelling below.

Assessing spray foam insulation

This leaves residential practitioners at the sharp end of the process (a position with which we are familiar). On the one hand, we do not want to cause an unreasonable overreaction against spray foam insulation, but on the other, we have to make our clients aware of the risks involved in buying a property with spray foam insulation on the underside of the roof.

Inspecting a spray foam roof

In many ways, this would be similar to the approach adopted for any roof space but with particular emphasis on those parts of the roof that would be vulnerable to deterioration in normal circumstances, such as:

- Around chimney stacks, beneath valley gutters and along the underside of the eaves construction;
- Looking for evidence of purpose provided ventilation; and
- Assessing how well the ceiling is sealed (for example, the loft access hatch, back of lights and any pipe and service penetrations).

In an effort to provide you with a balanced and objective approach to assessing spray foam insulation, you may want to consider placing the roof space in one of the following categories:

- **High risk.** The installations that carry the highest risk are those where the spray foam has been applied directly to the rear face of the tiles/slates (no sarking felt). In this case, you should be able to see the outline of the tiling battens on the surface of the foam (see Figure 8.16). As the foam will encapsulate the tiling battens and often the rafters, the risk of timber deterioration will be high, regardless of whether open- or closed-cell foam has been used.

  In these circumstances, we think it is important to warn your client about the high level of risk of timber deterioration and the likely impact on saleability.

Figure 8.17 This is a typical example of a late Victorian end of terrace property that was recovered in the late 20th century with heavier concrete tiles. This one image should establish a clear trail of suspicion to the roof space to see if the structure has been strengthened.

A referral to a suitably qualified person would be appropriate, with a broad indication of the scale of the likely repair works. In most cases, although this will not be known until a further report is received, it is likely the roof covering will require renewal in the short term to medium term.

- **Medium risk.** This scenario would typically include installations where the spray foam has been applied to the rear face of the sarking/roofing felt but does not encapsulate the rafters. In this case, the foam will often be evenly applied between the rafters, as the tiling battens on the outside of the sarking layer will be unaffected by the foam. Typically, the foam would have sealed the roof space and may have left it without purpose-provided ventilation. It would also have a 'leaky' ceiling – for example, loose fitting loft hatch, unsealed downlighters, and so on. This would leave the roof space vulnerable to moisture build-up and condensation.

In these circumstances, you may want to warn your client about the risk of timber deterioration in vulnerable locations (for example, around chimney stacks and below valley gutters and so on) as well as the likely impact on saleability even though the risk level is lower. A referral to a suitably qualified person would be appropriate but with a broad indication of the scale of the likely repair works. Although this will not be known until the further report is received, it is likely to be restricted to ancillary repairs such as sealing the ceiling (draught stripping the loft hatch and 'loft caps' over the back of light fittings) and providing roof space ventilation.

- **Low risk.** This scenario would typically include installations where the spray foam has been applied to the rear face of the sarking/roofing felt and does not encapsulate the rafters. There is likely to be evidence of work to seal the ceiling (for example, well fitted and draught stripped loft access hatches, 'loft caps' covering down-lighters and so on) and clearly visible purpose provided loft ventilation.

Although an automatic referral for a further investigation may not be justified, you may still want to warn the seller about the likely impact on saleability because of the controversy surrounding this insulation method.

Like all of the protocols in this book, the final judgement will be yours and must be based on the individual circumstances.

8.5.6 Recovering of roofs – statutory control and implications

Since 1985, the recovering of a roof may have required formal approval from the local Building Control Authority (Approved Document A, building regulations Section 4):

- If there will be a 'significant' increase in the loading (for example more than 15%), the existing roof structure and structure beneath (internal walls and so on) must be checked to ensure it can take any increased loading;
- Where the new roof covering will be significantly lighter, a check is required to ensure the roof is strong enough to prevent uplift in wind;
- When the checking shows strengthening works are necessary, those required 'material alterations' could include:

  - Replacement of defective members (for example, rafters or purlins), fixings and restraints;
  - Provision of additional structural members (for example, additional struts, or in extreme cases new steel beams) to enable new load to be supported;
  - Provision of additional restraint to prevent uplift, such as holding-down straps.

In most cases, the Building Control Authority will require calculations to justify the new proposals. Therefore, it is important to identify whether the roof covering has recently been replaced.

Alternatively, a contractor from a government-approved competent roofer scheme could be used as they can self-certify that their work conforms to the requirements of the building regulations. At the time of writing, this was being administered by the National Federation of Roofing Contractors (NFRC 2021).

Possibly the most usual circumstance a surveyor will meet is on a Victorian mid-terrace house where an original slate roof covering has been replaced with concrete tiles. The reason for using concrete tiles is usually because the cost is significantly less than putting natural slates back. However, slates (double lap but relatively thin) weigh on average around 35 kg/m^2; whereas single lap concrete tiles (relatively thick) weigh around 55 kg/m^2. The difference in applied loading is therefore around 60% (see Figure 8.17).

Some *key inspection indicators* include:

- Does the covering appear relatively new and un-weathered?

- Is it different from neighbouring properties or others in the road?
- Is there evidence of recent work visible in the roof space, such as structural repairs, new roofing felt, and so on?

If the evidence suggests renewal, the vendor should be asked for evidence that appropriate approvals have been obtained. In our experience, such a process has usually not been followed, which could be an indication of substandard work. If appropriate, the client should be clearly advised that the roof may require some remedial work. There are two options:

- Ask for retrospective approval to be obtained from the Building Control authority for which a fee will be payable and remedial work will almost certainly result (see Section 13.2);
- Appoint a chartered building surveyor, structural engineer or equivalent to assess the structure, recommend any remedial work and oversee the work including obtaining approvals. Once this has been completed, issue a certificate of structural adequacy or its equivalent.

Any loading calculation must take account of the self-weight of the structure and coverings, wind and snow loads and the water absorption of the covering itself. The latter can be as much as 10% of the original weight for some concrete and clay tiles.

In most cases, it is likely the legal adviser should be informed where the roof covering has been altered, in view of the building regulations requirement outlined previously, and possibly for planning purposes, especially if the property is listed or is in a conservation area.

8.6 Other features associated with pitched roofs

There are several other features associated with pitched roofs that may cause problems. These are outlined in the following sections.

8.6.1 Party walls

Party walls between terraced and semi-detached properties can either terminate below the roof line or extend above it. Each has its own problems:

- **Walls through roof line (for example, parapet wall)** – These are vulnerable at the junction of the roof covering and party wall. Flashings and soakers may be defective or totally absent. In some cases, the junction is waterproofed by a simple cement fillet, which is never very effective. The copings also present a weakness, allowing water to bypass the covering. Any one of these can lead to moisture penetration into the loft space or even the rooms below. The rafters close to the wall may become very vulnerable to rot;
- **Walls terminating below roof line** – These have separating security and fire functions. The walls can be completely missing or stop short of the underside of the roof covering. These partial party walls can offer inadequate support to purlins. In these cases, appropriate remedial work or further investigations should be recommended. Any lack of, or inadequate, party wall is a fire risk and therefore a safety hazard.

8.6.2 Gable walls

These can be very exposed to the weather, particularly if they have a parapet above. Problems include:

- Structural instability because the walls of older buildings can reduce in thickness at this height; making them very slender and vulnerable to lateral instability, bulging and bowing. This can be made worse by relatively heavy loadings from the roof structure and lack of adequate lateral restraint between the structure and the gable wall;
- Because of the reduced thickness, wind-driven rain can penetrate the mortar joints and cause damage internally, for example wood rot in purlin ends.

Because of these issues, where gable walls exist, they should form a key part of the inspection process.

8.6.3 Chimneys

These can be the source of many problems. Some of the main defects are as follows.

- Moisture penetration due to (see Figure 8.18):

  - Lack of, or inadequately situated, DPC – provision of a DPC is a relatively recent innovation;
  - Defective flashings, soakers and 'back' or 'secret' gutters at the rear of the chimney. The flashings to many newer properties may never have been properly wedged with lead into the mortar joint. Instead, they are held in position by mortar or sealant only. This can become easily dislodged in high winds;
  - Defective flaunching, chimney pots and flue terminals;

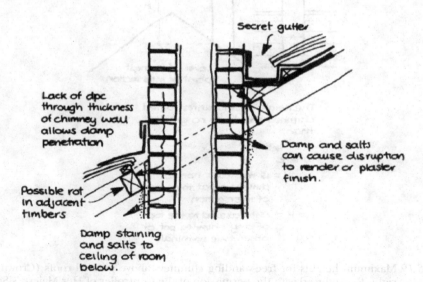

Figure 8.18 Possible routes for penetrating moisture around a chimney.

- Poor pointing to stack allowing water to by-pass the flashings;
- Cracking and deterioration caused by lack of satisfactory flue lining, allowing salts, condensates, tar, smoke and soot to contaminate plaster and masonry – provision of a flue lining only became a national Building Regulations requirement in 1966.

- Stability problems due to:

  - Slenderness ratio, the chimney being too high for its width. The height of the chimney should be no more than 4.5 times its width (see Figure 8.19) (OPSI 2013, p. 35). Chimneys to rear additions or offshoots are especially vulnerable, as they were often taller, so the combustion products avoided the main house;

- Sulphate attack may cause the mortar joints to expand (see Section 5.7.3). Maximum expansion normally occurs on the west side of an exposed chimney, causing the whole stack to bend over;
- Where the entire, or part of the, chimney breast has been removed internally, rendering the whole stack above unstable;
- The lack of bond between the chimney and the adjacent wall can result in the separation of the two. In the worst cases, this can lead to the escape of flue gases into the roof space, which is a safety hazard.

Because of this variety of problems, chimneys require special attention during any survey. The BRE have produced a useful guide on the repair and rebuilding of traditional chimneys (BRE 1990b).

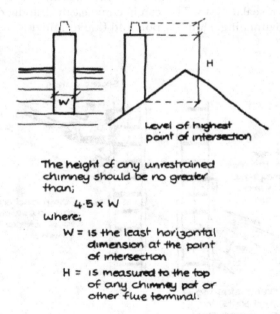

The height of any unrestrained chimney should be no greater than;

4·5 × W

where;

W = is the least horizontal dimension at the point of intersection

H = is measured to the top of any chimney pot or other flue terminal.

Figure 8.19 Maximum heights for free-standing chimneys above pitched roofs (Crown copyright. Reproduced with the permission of the Controller of Her Majesty's Stationery Office).

8.6.4 *Thermal insulation of pitched roofs*

Part L of the Building Regulations (OPSI 2016) now links the insulation values of the different elements of a building to that building's overall energy efficiency. The requirements are under continual review and surveyors must maintain their knowledge of what their clients expect. At the time of writing, the practical thermal insulation requirements for roofs are:

- 'Cold' pitched roofs:

  - To horizontal ceilings, mineral wool or similar insulation laid to a depth of around 300 mm, usually in two layers at right angles to each other, thereby helping to prevent heat loss (or gain); and thus covering the ceiling joist part of the truss; and
  - To sloping ceilings (for example, to an 'A' frame roof) 100–125 mm rigid insulation board tightly fixed between the rafters and 50–60 mm similar rigid insulation board above or below the rafters (plus bonded plasterboard if beneath), to help prevent 'cold bridging'.

- Flat roofs, 100–125 mm rigid board insulation between joists and 50–60 mm above or below joists as above.

In those roofs inspected by most surveyors, for example pitched roofs, the increasing depth of insulation required over the years has made it difficult and hazardous for surveyors, service engineers and homeowners to move around a roof space – some surveyors now recommend installing a permanent access walkway for this reason. Such roofs warrant a warning about this hazard for homeowners and the service engineers.

8.6.5 *Roof ventilation*

Part 'C' of the Building Regulations covers ventilation of roofs for 'cold deck roofs, i.e. those roofs where the moisture from the building can permeate the insulation' and other roof types (OPSI 2013a, p. 40). Without it, moisture levels in roof spaces can quickly build up and result in mould growth on the roof timbers and, in the worst cases, wood rot (BRE 1985b).

In all cases, guidance confirms that ceilings between living accommodation and roof spaces should be as airtight as possible (BRE 1985c):

- No gaps or holes in the ceiling around light fittings, water pipes and electricity cables;
- Doors and hatches into roof spaces should not be in rooms that develop significant quantities of water vapour, such as bath or shower rooms and kitchens; and
- Hatches into roof spaces should be well insulated and sealed with draught-proofing.

Additional requirements where the underlay is not vapour permeable, for example, type high water vapour resistant (type HR) felt, are:

- If the building has a width of up to 10 m:

  - Pitched roofs (15° pitch or more) – cross ventilation between each side equal to a 10 mm continuous strip at eaves level,
  - Less than 15° pitch – cross ventilation at eaves of 25 mm;

- For a building width more than 10 m and where the roof pitch is greater than 35°, there should be ventilation at ridge level equal to at least 5 mm wide;
- Lean-to roof of 15° or more – 10 mm wide strip at eaves with ventilation at high level equal to at least 5 mm wide strip along the whole length of the junction with the wall. This can also be provided by a series of individual ventilating slates or tiles fitted evenly along the roof slope;
- Lean-to roof of less than 15° – ventilation same as the foregoing, but the requirements are 25 mm at eaves and 5 mm at high level.

Ventilation of roof spaces is more difficult where the ceiling follows the underside of the rafters. Two main features are required:

- Ventilation at eaves equal to 25 mm wide strip *and* ridge level equal to 5 mm wide strip to promote flow of air above insulation and;
- Minimum air gap of 50 mm between the top of the insulation and the underside of the roof covering.

There are equivalent requirements where there are dormers on the roof, plus additionally:

- A 5-mm wide strip at dormer window sill level;
- For dormers with cold pitched roofs:
  - A 10 mm eaves vent gap, unless
  - The dormer roof is flat, when the equivalent gap should be 25 mm.

If there is no underlay, this requirement may be waived because the roof space would be so draughty anyway. For recovered roofs where a continuous strip at the eaves is not possible, proprietary roof tile vents can be used. For example, if 'Glide-vale' tile ventilators were used, the requirements would be:

- For standard loft space ventilation where there is no habitable space – one ventilating tile every 1.0 m;
- Where a 'room-in-the-roof' exists with a sloping ceiling to the underside of the rafters;
  - Eaves ventilation – one ventilating tile every 400 mm centre (i.e., every rafter spacing);
  - Ridge ventilation – one ventilating tile every 2.0 m centre.

For an average terraced property, this could result in a considerable number of ventilating tiles being fitted along the lower part of the roof. If there are only one or two (or none at all), this suggests that the work might not have building regulation approval. Appropriate advice should be given to the client (see Figure 8.20).

For roofs that have a 'vapour permeable' (type LR) underlay, condensation is less likely to occur, but only if 'the ceiling is well sealed and the eaves have a minimum continuous ventilation opening of 7 mm' (LABC 2017, p. 214). In practice, surveyors should be very careful with this type of underlay, since anecdotal evidence suggests either LR type underlays are not fully effective or some new properties are being

Figure 8.20 Ventilating tiles to a pitched roof. The regulations have changed considerably over the last 20 years. When this roof was recovered (over 20 years ago), where continuous ventilation is not provided at fascia level, several special ventilating tiles had to be provided near the eaves. At the time of writing, in some circumstances, permeable roofing felt has removed the need for ventilating tiles completely.

completed with gaps in ceilings. They should be particularly careful on old properties with replacement coverings; where it is likely LR underlay has been used, there are likely to be ceiling gaps, in which case LABC suggests an eaves gap of at least 7 mm.

For further information, see the LABC Warranty Technical Manual (LABC 2017).

Around 2000, the use of some plastic underlays became usual for a time, but these were found to suffer significant condensation. Since then, fitting a 'breathable' underlay has become the norm. These are 'vapour-permeable' (incorporating very small perforations that allow vapour to escape from the roof space but prevent rain water from getting in). Typical problems with these modern underlays include:

- They can be prone to leaking where they are torn during installation and at nail holes (older bitumen felts can also leak at nail holes but the bitumen can often be self-sealing around nails when the covering becomes hot under sunlight);
- Condensation, despite being 'vapour-permeable' – many building control authorities insist on draught-stripping to loft hatches, sealing of other holes in ceilings (recessed lights, holes for service pipes and cables) and ventilation strips at eaves; plus, rooms below that develop water vapour (for example, kitchens and bathrooms) should have mechanical extract ventilation or similar;
- If stretched too tight, the eaves underlay can vibrate annoyingly with a drumming sound in the wind;
- 'Ballooning', where underlay pushes upwards on the underside of the tiles, due to wind pressure, resulting in even properly clipped and fixed tiles becoming

dislodged. Remedies include securing the underlay laps (through which the wind blows) with battens or gluing the underlay edges together with special adhesive strips (Redland 2014, p. 7).

8.6.6 Loft conversions and rooms-in-the-roof

A significant number of dwellings have had their lofts converted into habitable rooms. This can have both positive and negative effects:

- It can enhance the value of the property;
- If not carried out properly it can lead to structural instability;
- Usability and amenity – where the new space is poorly designed, difficult access and low head heights can make the space uncomfortable to use;
- Safety – a poorly designed conversion can compromise safety, especially in relation to escape in the case of fire.

It is essential that surveyors are aware of the issues which give rise to these effects so clients can be properly advised. Some of the key issues are outlined as follows.

Older loft conversions and original room-in-the-roof spaces

In many properties, a room-in-the-roof may have always existed. It is helpful to use the current building regulation standards as a benchmark to assess the room's suitability; these can be especially helpful in advising clients about repairs or safety improvements, but it is likely to be unreasonable to call for upgrading to meet the full scope of the current building regulations. Three questions can help in evaluating existing 'rooms-in-the-roof':

- Is the room accessible enough to be considered a habitable room? Are the access stairs so small and winding that the room should really be considered (and valued) as a storeroom?
- Does the space have adequate precautions in the event of fire or would the safety of any occupant be seriously compromised?
- Has the structural integrity of the roof been undermined by the room, for example, is there evidence of roof spread or is the structure well triangulated?

Consideration of these factors can help the appraisal and ensure the client receives balanced advice.

Loft conversions since 1985

Loft conversions have been specifically included in the Building Regulations since 1985. If the conversion does not have the necessary approvals, a Regularisation Certificate can be obtained.

During a standard survey, it will be impossible to assess standards of the conversion on inspection without an in-depth knowledge of the regulations and time-consuming exploratory and opening-up works. Therefore, you should look for the following *key inspection indicators*:

- Is there a loft or attic room? This may sound obvious, but they have been missed in the past;
- If yes, has it been created since 1985 and therefore should be covered by Building Regulations? The vendor may know; otherwise, look for key indicators below that might indicate it has been converted to appropriate standards, including:

  - Doors to the staircase fitted with self-closing devices down to the exit doors, although self-closers are not always required by the current building regulations. This makes the assessment even more challenging;
  - Partitioning made of robust standard that can resist the passage of fire (for example, plasterboard attached to a good frame and not hardboard); and
  - Rooflights or windows that conform to Approved document 'B' escape requirements.

Advice to the client

If the loft conversion has been completed since 1985, the client should be advised that planning permission and Building Regulations approvals should have been obtained. Even if this has been granted, the surveyor should recommend that the legal adviser ensure there is a 'completion certificate' confirming the work has been carried out in accordance with the original approval. If these conditions are not met, then there are several options:

- Ask for building regulation and planning permissions to be obtained in retrospect;
- Ask for full inspection by a chartered building surveyor, structural engineer or other suitably qualified person to assess the loft room in relation to current standards and confirm any necessary remedial work.

The client's legal adviser should be specifically instructed on what to ask for.

8.6.7 Pests in roof spaces

Living things can often be found in loft spaces. Not only can these be nuisances, but they should all be flagged as safety hazards. Most of these have been described in Chapter 7 but are listed here for convenience:

- Pigeons, swallows and swifts – these can gain access through missing slates or tiles or gaps at eaves and verges. Pigeons can pose a particular safety hazard;
- Bats – these are a protected species. Their presence must be reported, since their presence can increase costs of work and delay repairs, alterations and timber-treatment regimens – the matter should be flagged up for the legal advisor;
- Squirrels – these can be a nuisance, especially if trees are close by. They can cause considerable damage to timbers and cabled services by gnawing – an electrical and fire hazard;
- Rats – these have been known to climb up boxing around soil and vent pipes and can cause similar damage to squirrels;
- Wasps and bees – dangerous (and painful) if the nest is disturbed;
- Flies – with increasing ventilation requirements, resultant gaps make it easy for flies to become a nuisance in the roof space.

8.6.8 Services in loft spaces

Services in domestic dwellings are covered in Volume 3 of this publication, but because loft space inspections can give such a good insight into the condition of these elements, a brief reminder has been included at this point. The main elements are:

- Cold water storage and feed and expansion tanks;
- Hot water cylinders and boilers;
- Electric cabling, ceiling lights and TV aerials.

Some service engineers insist on proper access walkways in roof spaces to get safe access to installations such as boilers and water systems. If such access is not present, this should be noted and confirmed in the report; otherwise, your client may face an unexpected cost.

For a full checklist, see Volume 3.

8.6.9 Eaves roof spaces

Some roof spaces will consist of small eaves spaces, typically in properties with rooms in the roof. These are accessible through doors or hatches located in timber stud-work partitions between the roof space and living accommodation – see Section 8.6.6. These spaces usually give good opportunities to check areas such as triangulation of the roof structure, the top of the external wall, the depth of the joists supporting the upper floor and the ventilation (if any) to sloping ceilings. It is also common to see soil and waste pipes to upper floor sanitary fittings and other service pipes and cables. These areas should be inspected in the same way and for the same reasons described previously, but *key inspection indicators* include:

- The absence or otherwise of insulation to the timber studwork partitions between the roof space and living accommodation; and
- Whether the hatch is suitably insulated and draught-sealed, and absence of holes in the partition, to prevent warm, moist, air getting into the roof and causing condensation.

8.6.10 Roof spaces above flats

If the surveyor is inspecting a flat in a large block or conversion:

- For level two surveys an inspection of the roof space is excluded, unless the void is accessible from the flat itself;

If there is indeed an accessible roof space, all the previous comments regarding construction, condition, safety and other matters apply. The surveyor should pay particular attention to the following *key inspection indicators* for roofs above flats, including confirming:

- The footprint of the roof space aligns with the floor of the flat – otherwise there may be boundary, trespass and possibly fire precaution issues;

- There is satisfactory separation against fire spread between the roof space of the flat and adjoining roof spaces of other flats and any common areas – for example, a masonry partition;
- The space may be part of a common roof space, where other tenants and maintenance personnel may have access rights (especially if there are common services) and where there may be asbestos-containing materials – the surveyor should always try to locate and read the asbestos register for common areas, in order to:

  - Protect themselves during inspection, and
  - Provide appropriate advice to the client.

8.6.11 Roof space hazards

Several hazards have been mentioned in this section. Other specific hazards the surveyor may encounter in roof spaces include:

- If insulation is over ceiling joists, meaning joists are not visible, moving around the roof space safely and easily is difficult;
- If there are loose storage and access boards in the roof; the client or trades-people carrying out repairs or maintenance could trip or fall – indeed, some trades-people will refuse to work in a roof unless such boards are properly fixed down;
- There may be asbestos-containing materials – such as old water storage tanks, or insulation; and
- The access hatch may be situated in a potentially hazardous position, for example directly over a stairwell.

The report should include appropriate comments, together with any necessary remedial actions.

8.7 Flat roofs

8.7.1 Introduction

Inspection of flat roofs is different to pitched coverings as it is rare to be able to view the structure. Therefore, any report must assess the condition of the roof covering and speculate about the structure hidden beneath. This must be done by judging the condition of what cannot be seen by that which can. On domestic properties, flat roofs can be found in several different locations:

- Over the whole property – flat roofs were very fashionable during the 1960s and early 1970s;
- Over back additions or 'offshots';
- New extensions added since the house was constructed;
- Flat roofs over front and rear bay windows, balconies, and so on;
- Linking front and rear roof slopes of the main roof construction, sometimes at upper floor level, which can complicate an inspection.

The two main elements are now considered.

8.7.2 *The roof structure*

A surveyor is likely to encounter two different types of roof structures:

- **Suspended construction** – Usually comprising joists, decking (timber, profiled metal sheet) and covering;
- **Solid construction** – Reinforced concrete, cast concrete between filler joists, hollow clay pots between concrete beams, etc.

One method of telling the difference is to knock on the surface; timber will give a hollow sound. If surveyors decide climbing onto the roof can be done safely, they must beware of using the 'drop-heel' test on the roof surface – if the structure has been weakened by rot, damage and injury could result. Because timber roofs are more vulnerable to defects, they need greater consideration during the survey, so this initial distinction is important.

However, over the last 20 years, many flat roofs are of the 'warm sandwich' type, so tapping the surface won't help confirm the deck construction. If the flat roof meets a pitched roof, inspecting from within the main roof where the two roofs meet can be instructive.

During the inspection, the *key inspection indicators* are:

- The roof surface should have a slight, even slope or fall;
- No signs of ponding of water – even in dry weather there should be some staining, moss or lichen around the edges of former puddles;
- Rooms beneath that produce water vapour, such as bathrooms or kitchens, increase the risk of problems caused by condensation;
- Ventilation, if a 'cold-deck' design, should be provided by:

  - Ventilation strips around all sides of the roof under the eaves behind the fascia board – equal to a 25 mm wide continuous strip on two opposite sides of the roof, or
  - Ventilation fittings set into the roof structure, extending through the covering,
  - Roof space ventilation must also be well distributed around the roof. If there is no cross-ventilation, parts of the flat roof void may become stagnant and more likely to suffer condensation. This could lead to wood rot development even though the overall level of ventilation may be acceptable (see Figure 8.21).

- The roof types described previously, with ventilation, are known as 'cold' roofs. If there is no ventilation, it is possible the roof is a 'warm sandwich' or 'inverted' type roof.
- Evidence of leaks to the rooms or spaces below – hidden timbers may have been adversely affected.
- Use of the flat roof – some occupants may use them as breakfast balconies or roof gardens, possibly exceeding design loadings. Indications of this misuse might include informal access (for example, through a window rather than a proper door), poor quality and inadequate detailing and workmanship, evidence of excessive defection (i.e., cracking, bulging and so on) to roof surface or ceiling below.

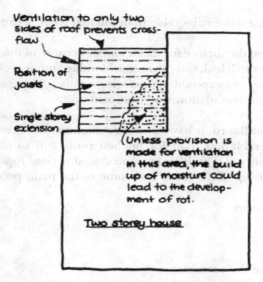

Ventilation to only two sides of roof prevents cross-flow

Position of joists

Single storey extension

(Unless provision is made for ventilation in this area, the build up of moisture could lead to the development of rot.

Two storey house

Figure 8.21 Diagrammatic sketch showing inadequate ventilation to flat roof because of lack of through ventilation as a result of infill nature of extension (flat roofs are ideal for this).

If a significant number of these adverse signs are noted, making a recommendation for a further investigation would be appropriate.

8.7.3 Roof coverings

The main types of flat roof coverings include:

- **Asphalt** – A combination of limestone aggregate and bitumen, laid to a 20 mm thickness Usually black when new, becomes light grey after exposure to sunlight. It is important to protect asphalt from the sun by either mineral chippings or solar reflective paint (usually white or silver);
- **Metal roof coverings** – Including lead, zinc and copper and other alloys. These will have features such as rolls and cappings. The installation and repair of these roof types is a specialised operation;
- **Bitumen-based roofing felts** – Fabric impregnated felts laid in wide sheets or strips bedded in hot bitumen. Originally based on asbestos (if suspected, worthy of note in the report as a hazard) or jute fabrics, modern versions are more durable, relying on polyesters and glass fibres. Although traditionally protected with mineral chippings, many felts are now self-finished with coloured mineral chippings or even reflective metal foils;
- **Glass fibre, resin and single ply plastic coverings** – The flat roofing industry has responded to customer concerns regarding the durability of felt roofs with several modern innovations;
- **'Green' roofs** – Still relatively few in number but will be encountered by most surveyors.

Leaks to flat roofs

Flat roof coverings are more vulnerable than pitched roofs because:

- Rainwater stays on the surface for much longer because of the shallow slope or fall of the roof – any small leak can cause considerable amounts of water penetration;
- A flat roof can be very exposed to damaging winds, a wide range of temperature changes and solar degradation, and occupant misuse.

It is unsurprising that flat roofs have a poor reputation. The life expectancy of most flat roofs is considerably less than for pitched roofs and so must be viewed with suspicion during a survey. The BRE have produced a *Good Repair Guide* on how to assess flat roof coverings (BRE 1998) and some of the main points are summarised here.

Key inspection indicators:

MAIN ROOF AREAS

- Splits or cracks in the covering;
- Bubbles, blisters or bumps in the surface;
- Evidence of ponding or deflected areas to the main roof area;
- Areas of the roof without solar protection, e.g.:

  - Lack of stone chippings;
  - Lack of reflective paint or other finish.

VERGES AND ABUTMENTS DETAILS

Many leaks from flat roofs result from inadequately designed or installed 'edge detailing' where the flat roof meets other parts of the building, including:

- Lack of, or defective, cover flashings, usually in lead, to external walls of the main building;
- Upstands to party, parapet walls and rooflights – the vertical parts of the roof covering can slip down, tear and split;
- Junctions with the main pitched roof slopes – sometimes the flat roof covering does not extend sufficiently beneath the pitched roof covering, including the underlay – see Figure 8.22;
- The eaves and verges of the roof – exposed to sunlight and possibly damaged during installation, this detail often splits or cracks later;
- Drainage fittings, outlets and any internal guttering – often poorly made;
- Pipes, flues or handrail supports penetrating the roof covering can allow water penetration;
- Paved areas on fire escape routes, maintenance access, etc. where the slabs can cause damage.

If faults are noted, an internal check should be made for evidence of roof leaks; but on a flat roof, leaks in one location can 'travel' and may only appear sometimes considerable distances away. Even if the defects have not yet resulted in roof leaks, the client should still be made aware of the faults and future implications.

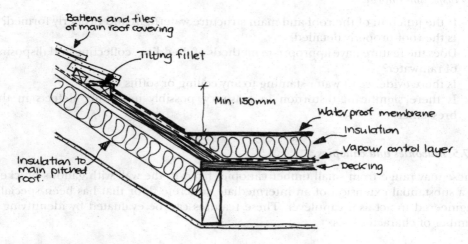

Figure 8.22 Detail of flat and pitched roof abutment.

TEMPORARY ROOF REPAIRS

Because flat roofs are so vulnerable to faults, it is highly likely older ones will show evidence of temporary repairs, including:

- Waterproof adhesive tapes and proprietary products applied over splits and cracks;
- Black bitumen poured over the surface, sometimes with reinforcing matting beneath;
- Patch repairs where areas of felts and asphalt have been cut out and replaced.

Temporary roof repairs are evidence that the existing covering is at or near the end of its useful life. Where there are a number of these repairs, immediate renewal should be recommended or a warning given that this could be required soon.

8.7.4 Bay and porch roofs

The roofs of porches and bays are often covered to a lesser standard than a roof to the main dwelling. Roofs over these features often include simple zinc on timber, bitumen on primitive concrete roofs and small-tiled pitched roofs with multiple hips, valleys and abutments. The junction with the main roof is often poorly waterproofed. These defects become even more critical because in many older properties, there is often a large timber 'bressumer' or beam spanning the width of the bay opening only just protected by the roof construction. Water leakage can lead to rot and consequent structural failure. Rainwater disposal may be blocked, inadequate or absent, allowing rainwater to discharge over the bay or porch below, causing an increased rate of deterioration in timber components. Where the roofs of these features cannot be inspected from ground level with binoculars or off a ladder, they can often be seen from the windows of adjacent upstairs rooms.

Key inspection indicators

- Is the junction of the roof and main structure watertight and properly formed?
- Is the roof properly detailed?
- Does the feature have appropriate methods of shedding, collecting and disposing of rainwater?
- Is there evidence of water staining to any ceiling or soffits below?
- Is there significant distortion in masonry, possibly indicating defects in the bressumer?

8.7.5 Balconies and canopies

These may range from small timber canopies fixed to the wall with gallows brackets to a substantial extension of an intermediate concrete floor that has been specially engineered to act as a cantilever. These features can be evaluated by identifying a number of characteristics:

Hazards

Safety is a major consideration on balconies for users and those walking beneath

Key inspection indicators

- A balustrade:

    - Strong enough and securely fixed;
    - High enough with no unacceptable gaps or horizontal rails that will allow or even encourage children to fall through or climb over – the building regulation rules (OPSI 2013c, p. 24) include:

        - Handrail at least 1100 mm above the balcony;
        - A 100 mm ball should not be able to pass between any component; and
        - No horizontal or similar rails (popular in the 1970s).

- A kerb to stop objects rolling off the balcony;
- People walking below a balcony should not be able to hit their head on any handrail supports that extend below the soffit of the balcony irself;
- The balcony may be a fire escape; indeed, some flats were designed with these as a vital part of the escape route – the surveyor should confirm its functionality and it has not been informally barred;
- Any timber component on a flat or similar balcony should be confirmed as a fire hazard.

Condition

- With concrete balconies, inspect:

    - For evidence suggesting the structure is under stress, such as cracking or local crushing around the bearings in the main structure, cracking to the top surface near the junction with the wall suggesting lack of adequate reinforcement, etc. – in these cases further investigation might be needed;

- For evidence of spalling or cracked concrete to the underside or exposed edges of the balcony, or even exposed and rusted reinforcement. This could be caused by corrosion of reinforcement too close to the surface or carbonation of the concrete itself, which has allowed further deterioration of the material. All loose material may need to be hacked back, the concrete reformed and the surface sealed;
- The junction of the main balcony with the junction of the external wall construction for signs of damp penetration including:
  - Signs of dampness internally – look for staining below the balcony and within the adjacent rooms and spaces;
  - Obvious deficiencies with the waterproofing layer to the balcony, including lack of a proper upstand, splits or slumps in the waterproofing layer, etc.;
  - Lack of a properly positioned cavity tray in a cavity wall and a link with that cavity tray – this should include an upstand and open vertical mortar joints above that drain the tray (NHBC, chapter 6.1, p. 18), (see Figure 8.23);
- With timber or metal structures:
  - Evidence of rot or corrosion to the decking might suggest current or future problems; and

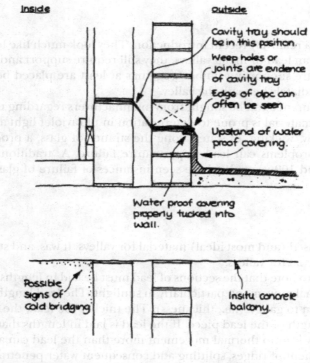

Figure 8.23 Link between a concrete balcony and a cavity wall. It is essential that the balcony has a waterproof covering and upstand linked with the DPC and cavity tray of the wall.

- Carefully check the support to the balcony, especially for:

  - Bent, twisted or out of plumb columns, which often have inadequate foundations and readily suffer from settlement;
  - Gallows brackets insecurely fixed to the wall, resulting in them 'pulling' out the top fixing and buckling or crushing at the bottom;

- Are the waterproofing materials in a sound condition?

The replacement of these features can be very expensive and so have an impact on value.

8.8 Valleys

8.8.1 Introduction

Where two sloping roofs meet, it is necessary to form a 'valley'. Alternatively, valleys can exist where a roof slope meets a wall on another building or the subject property, or where two roofs meet, for example, in the centre of a large property. Such a valley is usually near horizontal. Names for horizontal valleys include 'parapet gutter' or 'concealed' or 'secret valley gutter'.

8.8.2 Some different materials for valleys

Glass fibre

These valleys are a relatively recent introduction. They look much like lead. Although they weigh less than lead or metal valleys, they still require support and it is generally necessary to ensure supporting timber noggings at least are placed between rafters and trusses immediately beneath the valley.

Although reassurance can be sought from manufacturers regarding the life of glass fibre valleys, the material is prone to deterioration in ultraviolet light and water penetration caused by 'wicking' of water along the strands of glass, a process known as 'osmosis'. Such problems can lead to premature failure. A 'traditional' lead valley should last around 100 years. We have seen instances of failure of glass fibre valleys after 10 years.

Lead

This is the most used (and most ideal) material for valleys. It was, and still is, the most 'traditional' material for valleys.

It is important to note that the sections of lead must be laid in lengths short enough to allow for thermal movement, particularly in sunlight. The lead lengths must be calculated in relation to gauge (i.e., thickness). The thicker the lead, the longer can be the width and length of the lead piece. If the lead is laid in lengths that are too long, this will inevitably lead to thermal movement more than the lead can accommodate, causing development of ridges, splitting and consequent water penetration.

Lead valleys should never be laid on bitumen or mineral felt, as the heat developed within the lead can cause the lead to stick to the felt. This prevents the lead from moving, leading to splitting.

Sloping valleys

On sloping valleys, it is always necessary to ensure that there is a good distance of at least 100 mm between the raking or cut tiles and slates on either side of the valley so that water can run down the valley. If that distance is inadequate, leaves or moss can accumulate, causing blockage.

Flat valleys

Where a roof slope terminates at the building's edge, there is often a parapet wall. Water is then collected in a parapet gutter with a slight slope. Where lead meets the parapet wall or abutment, the lead gutter lining should be dressed up to the wall, with a further lead cover flashing dressed into the parapet and down over the upstand.

Alternatively, where 2 roofs meet in a flat 'V' section, water must be directed away from the centre of the building to the edge. The detailing of such a valley gutter must be very carefully considered as such an area is potentially prone to blockage by leaves, debris and snow and if water penetration does occur through such a detail, it can sometimes be catastrophic if significant water penetration occurs at one time or just as catastrophic in the long term if a slow leak occurs, which in turn leads to wood rot timbers. In general, such a valley is best laid in lead, with 40–50 mm 'steps' to allow for thermal movement. The lead lengths between steps are usually around 2–3 m. Underlay from the sloping roof must lap over onto the lead.

Key inspection indicators

These include:

- In the roof space, carefully inspect the timbers beneath any valley, including taking moisture readings at regular intervals. Take particular care under horizontal valleys – a favourite position for wood rot;
- Be even more careful under parapet gutters with the additional problem of a possible leaking parapet wall – most parapet walls leak, particularly older ones, since detailing any DPC (if present) is so difficult in practice;
- Ensure any underlay is dressed over and into the valley. Get out onto a horizontal gutter if access allows, subject to safety requirements;
- The gutter outlet through a parapet wall must be properly detailed in lead with discharge to a hopper to accommodate the sometimes-significant amount of rainwater which can run down such a valley;
- It is possible to mistake glass fibre valleys for lead – this can result in an incorrect anticipated life span, plus it can sow seeds of doubt in the client's mind about a surveyor's ability. The surveyor should always inspect with binoculars and look at the end of the valley to see the difference.

References

American Chemistry Council (2012). *Guidance on Best Practices for the Installation of Spray Polyurethane Foam.* Centre for the Polyurethanes Industry. (no location given)

BRE (1985a). 'Pitched roofs: Sarking felt underlay – drainage from the roof'. *Defect Action Sheet 9.* Building Research Establishment, Garston, Watford.

BRE (1985b). 'Slated or tiled pitched roofs: Ventilation to outside air'. *Defect Action Sheet 1.* Building Research Establishment, Garston, Watford.

BRE (1985c). 'Slated or tiled pitched roofs: Restricting the entry of water vapour from the house'. *Defect Action Sheet 3.* Building Research Establishment, Garston, Watford.

BRE (1990b). 'Repairing or rebuilding masonry chimneys'. *Good Building Guide 4.* Building Research Establishment, Garston, Watford.

BRE (1998). 'Flat roofs: Assessing bitumen felt and mastic asphalt roofs for repair'. *Good Repair Guide 16, Part 1.* Building Research Establishment, Garston, Watford.

BSI (2014). *BS 5534, Code of Practice for Slating and Tiling for Pitched Roofs and Vertical Cladding.* British Standards Institution, Kitemark Court, Davy Avenue, Knowlhill, Milton Keynes, MK5 8PP.

HAPM (1991). *Defects Avoidance Manual – New Build.* Housing Association Property Mutual, London.

Isothane Ltd. (2016). *Duratherm OS Roof Insulation, Agrément Certificate 10/4771, Product Sheet 1.* Isothane Ltd, Newhouse Road, Huncoat Business Park, Accrington, Lancashire BB5 6NT.

Johnson, Alan (2006). *Understanding the Edwardian and Inter-War House – A Historical, Architectural and Practical Guide.* The Crowood Press Ltd, Ramsbury, Marlborough, Wiltshire SN8 2HR.

LABC Warranty (2017). *LABC Warranty Technical Manual, version 8*, 2 Shore Lines Building. Shore Road, Birkenhead, Wirral CH41 1AU.

Marshall, Duncan, Worthing, Derek, Heath, Roger and Dann, Nigel (2014). *Understanding Housing Defects*, 4th edition. Routledge, Abingdon, Oxon OX14 4RN.

NFRC (2021). 'Competent roofer scheme'. Available at https://nfrccps.com/. Accessed 18 September 2021.

Office of Public Sector Information (OPSI) (2013). Building Regulations approved document 'A'. *Structure.* OPSI, Richmond, Surrey.

Office of Public Sector Information (OPSI) (2013a). Building Regulations approved document 'C'. *Site Preparation and Resistance to Contaminants and Moisture.* OPSI, Richmond, Surrey.

Office of Public Sector Information (OPSI) (2013c). Building Regulations approved document 'K'. *Protection from Falling, Collision and Impact.* OPSI, Richmond, Surrey.

Office of Public Sector Information (OPSI) (2016). Building Regulations approved document 'L1A'. *Conservation of Fuel and Power in New Dwellings.* OPSI, Richmond, Surrey.

Parnham, P. and Rispin, L. (2001). *Residential Property Appraisal.* E&F Spon, London.

Redland (2014). *The Redland Guide to BS 5534.* Monier Redland Ltd, BMI House 2 Pitfield, Kiln Farm, Milton Keynes MK11 3LW.

RICS (2019a). *RICS Valuation – Global Standards 2017 Jurisdiction Guide: UK.* RICS, London.

RICS (2019b). *RICS Professional Statement, Home Survey Standard*, 1st edition. (HSS) Royal Institution of Chartered Surveyors, Great George Street, London.

Rock, Ian (2006). *The Victorian House Manual.* Haynes Publishing, Sparkford, Yeovil, Somerset BA22 7JJ.

Smith v. *Bush* (1990). 1 AC 831.

Watt, David S. (2007). *Building Pathology – Principles and Practice,* 2nd edition. Blackwell Publishing Ltd, 9600 Garsington Road, Oxford OX4 2DQ.

Watt, David S. (2011). *Surveying Historic Buildings,* 2nd edition. Donhead Publishing, Lower Coombe, Donhead St Mary, Shaftesbury, Dorset SP7 9LY.

9 External joinery, other parts and decorations

Contents

9.1 Introduction

Most properties have some external timber components, which can include windows, doors, external cladding, fascia and bargeboards and so on. Depending on how well these have been constructed and maintained, there will inevitably be some minor shortcomings. As Melville and Gordon (1992, p. 177) point out, a few casements that are difficult to shut may be tolerated by most building owners, whereas when timber windows are rotten to the point of replacement, not only will the client be facing a large bill but they will be very angry with their surveyor if he or she did not point out the problem. Consequently, external joinery needs very close and careful attention. Nowadays, many of these components are made of PVC and metal, so this chapter also considers these materials.

9.2 Windows and doors

9.2.1 General defects to all types of windows and doors

Defects that are common to all types of windows and doors are described in this section.

Timber decay

According to the BRE (1997) windows manufactured between 1945 and 1970 were often of poor quality and decayed prematurely. In subsequent years, some of this problem was reduced by pre-treatment of frames and components. Because this has only been general practice since 1970, there are still many components that are vulnerable to deterioration. The BRE point out that the more effective use of preservatives should not be a substitute for good design (BRE 1985a). Several locations on windows and doors are particularly vulnerable. Where these are accessible, they should be probed to see if wet rot has developed. In many cases, the paint film may be intact but the timber below completely rotten. Windows and doors on the most exposed elevations (normally south- and west-facing) are particularly at risk and so should be inspected closely. The precise locations include (see Figure 9.1):

* The bottom rails and sections of any door and window frame. Most of the rainwater run-off will flow over these components;
* The junction of the side mullions or jamb sections and the timber sills;
* Around any rebates where the glazing putty has cracked or dropped away. Thin, dry bedded glazing beads can also present a problem;
* The bottom edge of sills, especially where they are bedded directly onto stone or tile sub-sills;
* Any open or gaping joints between timber components;
* Any obviously rotten areas of timber or parts that have been spliced, repaired or made good with a proprietary filler.

Once wood rot has begun, it's difficult to repair properly. However, if the amount and extent of rot are limited and the workmanship is good, the decay can sometimes be arrested by cutting out and splicing in new timber. The insertion of borax rods or special resins can also slow down the rate of timber decay. If the problems are more extensive, replacement of the whole window may be the only cost-effective option.

Where replacement is the chosen option, the building's heritage status will influence the choice. For example, if the property is listed or in a conservation area (especially one with an Article 4 designation), then the local planning authority may

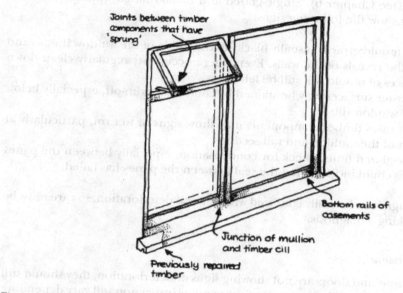

Joints between timber
components that have
'sprung'

Bottom rails of
casements

Junction of mullion
and timber cill

Previously repaired
timber

Figure 9.1 Vulnerable parts of a typical casement window.

require replacement of like with like. This should be reflected in the report to the client.

RAIN PENETRATION

Rain penetration can occur around the jambs, sills and heads of window and door frames if the DPCs that should be present in the wall reveals have not been installed properly or the sealants between the frame and the jambs are ineffective. The exposed elevations of dwellings are particularly vulnerable. Key indicators include:

- Damp staining to the internal reveals of window and door openings. In the worst cases, this could lead to a breakdown in the plaster and decorative finish;
- A lack of sealants at the junction of the frame and the jamb or defective sealants that are cracked, split or sagging (BRE 1985b);
- Lack of an overhanging sill with a throating that can shed water run-off from the face of the wall below.

Remedial work may range from re-application of an appropriate sealant through to removal of the frame and re-forming of the DPC where there has been a fundamental failure.

CONDENSATION

All dwellings will suffer from condensation from time to time. In some dwellings, condensation can be such a regular occurrence that it affects the durability of elements of the building (see Chapter 6). Single-glazed and metal-framed windows are most affected and can show the following signs:

- Evidence of mould growth (usually black) to the inside of the window frames and sometimes the reveals of the walls. Even where occupants regularly clean down surfaces, traces of mould will still be left;
- Adjacent plaster surfaces may be stained by condensate run-off, especially below the internal window sill;
- In the worst cases timber components may show signs of wet rot, particularly at the junction of the mullion and rail sections;
- With double-glazed units, check for condensation, especially between the panes of glass. This could indicate that the seal between the panes has failed.

Condensation signs can be easily confused with general deterioration, so care must be taken when making a diagnosis.

Functional performance

Even if the windows and doors are not showing signs of deterioration, they should still operate effectively. Although the extent of the window inspection will vary depending on the level of service agreed with the client (see Chapter 3 for more discussion), the surveyor should carry out the following checks:

- Windows and doors should be opened and closed. Comments should be made on whether they are easy and convenient to use. If keys are available, any security locks should be released and re-secured. If the keys are not available, this should be pointed out in the report;
- Are the doors and windows reasonably draught free? This may be difficult to assess during the summer without a wind. It can be estimated by seeing if there is any draughtproofing and how close the casements and doors fit to the frame;
- Do the windows let in enough daylight? Does the room seem dark? In some cases, replacement PVC windows with large mullions may have reduced light levels;
- If secondary glazing has been fitted, is it the right quality? Some fittings are insufficiently robust and distort when used. Is there any condensation between the two sets of glazing?
- Security is always a concern for clients. The BRE (1994) suggests that secure windows will have the following characteristics:

  - Glazed areas should be as large as possible so the noise of breaking glass can deter a burglar;
  - Look for small opening lights that may allow access to locks or catches on adjacent windows;
  - Louvre windows are very insecure, whereas trickle vents allow ventilation without having to open windows;
  - Externally fixed glazing beads can easily be levered off and internally fixed beads can be 'kicked in' if they are too small;
  - Window locks should need a key to open them. Because of the need to escape during a fire, a key should be kept close by the window (but concealed).

Many security recommendations contained in the official police initiative known as 'Secured by Design' (Police Crime Prevention Initiative 2019) were incorporated into the Building Regulations Part Q. Regulation Q1 applies to easily accessible doors and windows. The term 'easily accessible windows' is defined as:

- A window (or doorway), any part of which is within 2m vertically of an accessible level surface such as the ground or basement level, or an access balcony; or
- A window within 2m vertically of a flat or sloping roof (with a pitch of less than 30°) that is within 3.5m of ground level.

Where appropriate, clients should be warned when the windows represent a security threat.

Sash windows

Double-hung vertical sliding sash windows are probably one of the most common of all older windows in this country, 'introduced from Holland in the seventeenth century' (Muthesius 1982, p. 49). They consist of top and bottom glazed 'sashes' that are hung on sash cords, pulleys and weights within a boxed frame on either side (see Figure 9.2). In good condition, they can perform well, but if poorly maintained, they can have many problems. They can also be draughty and rattle in the wind, which some clients may find annoying. Some sash windows slide horizontally left and right, but there are few such units. Good repair of a defective sash window can be difficult and therefore expensive, and the client should be warned of this.

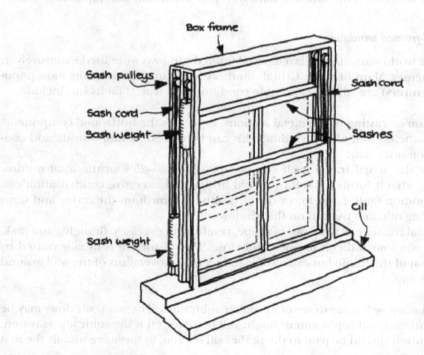

Figure 9.2 Sketch showing the construction of a typical vertical sliding sash window.

Key inspection indicators include:

- Do the sashes open? Are they painted shut? If they do open, are they easy to operate?
- Are the sashes well balanced? When unlocked and operated, do the sashes stay in one position or do they drift up or down on their own?
- Are the sash cords frayed or broken?
- Are the sashes well fitted or loose and rattling?
- Are there signs of wet rot? Look carefully at the junction of the timber and stone subsill;
- Look carefully internally at the parts of the box frame that contain the sash weights, since conditions in these parts are sometimes ideal to help dry rot thrive;
- Have the windows been fitted with secure and robust window locks? Sash windows can be easy to open from the outside unless properly secured.

Many older sash windows may add considerable character to the elevations of older dwellings. If replacement is recommended, advise the client that the new windows should at least complement the design of old-style sash windows. If the building is listed or is in a conservation area, replacement windows that exactly match the originals will be required.

In some properties, a modern version of the traditional sash window may be used. Called a spring balanced sash window, this eliminates the need for sash cords, box frames and weights and instead uses a spring-loaded device to counterbalance the sashes. Although this can reduce the overall size of the window, in our experience, these can be difficult to operate and hard to repair when they develop faults.

9.2.2 Metal-framed windows

Between the world wars, single-glazed metal-framed windows were used extensively in housing schemes. Many of these 'Crittal' windows (the name of one of the most popular manufacturers) are still in serviceable condition, but typical faults can include:

- Corrosion or rusting of the metal sections, especially the horizontal components towards the bottom of the window. This can be caused by rain outside and condensation internally;
- Because the actual frame itself can occupy up to 20–25% of the total window opening area it forms a significant cold bridge and excessive condensation can be a common fault. Look for evidence of mould growth on the frame and signs of pooling of condensation on the window sill;
- The metal frames can warp out of shape, resulting in excessive draughts and making the windows difficult to open and close. This distortion is usually caused by corrosion of the frame but can also be induced by movements of the wall around the opening.

Many metal window frames were set in timber subframes. The main windows may be in good condition, but replacement might still be required if the subframe is rotten. Special attention should be paid to the timber sill section, as these are usually the first to rot.

Modern metal-frame windows can include single-, double- and triple-glazed aluminium-framed windows and powder-coated steel windows.

Both types can include sliding or hinged casements and can be set directly against the openings or in timber (usually hardwood) subframes. Some types can have 'thermally broken' frames. This is where the inner and outer faces of the frame are isolated from one another by a solid resin or other insulating material. It is claimed that frames of this type are more thermally efficient. The criteria for evaluating these are similar to those identified for replacements in Section 8.2.3.

Key inspection indicators include:

- Are the windows stable, or do some parts flex and distort when operated?
- Do the frames suffer from surface condensation? Investigations by the authors have cast doubt over the real thermal value of 'thermally broken' frames, particularly earlier types;
- Sills to some windows are provided with holes so excessive condensation run-off can be drained to the outside. These can often become blocked, allowing condensation to overflow onto the internal sill board and wall finish below.

9.2.3 Installation of new windows and doors

Over the last 40 years or so, windows and doors of many properties have been replaced, usually in PVC, as the 'fashion' for replacement windows has grown. This type of improvement has been seen by many owners as the best way of reducing maintenance costs (costly redecoration), adding value to their property and improving thermal performance. Consequently, many perfectly sound and serviceable timber windows have been expensively replaced before their time.

In most cases, competent contractors have installed acceptable-quality components. However, occasionally, replacement windows result in more problems than they solve. Therefore, even if the windows are relatively new, they must still be assessed carefully. The following key questions will help in this process:

- Has the owner got any details of the installation work available? Was the contractor a member of an appropriate trade association and/or competent person scheme (see Section on legal issues below)? Did they install components that were constructed according to British standards?
- The openings may have been adjusted during the installation of the windows, for example sills and jambs built up, sub-sills removed, and so on. Look for new or altered brickwork around the opening. Has this been done properly? Are the units properly sealed, or is there any evidence of water penetration around new work?
- Older window frames often had an informal structural function, especially in bay windows. Once removed, the weight of the supported bay structure is transferred to new frames that may not have been designed for that purpose. Typical signs could include distortion of the construction above the window, including cracking of walls, sloping of upper floors and so on. The new window frames may also have distorted, making it difficult to open casements;
- Are the replacement windows secure against forced entry? Can the external beading be easily removed, allowing the glazing to be removed? Carefully check large lounge and patio doors. Because of possible deflection of the framing members,

locks and catches can easily be 'sprung' through the application of force to the element;

- Check for misting between the double glazing, indicating seals of double-glazed units have failed. It is good practice to mention this possibility in the report, even if there is no evidence of the problem, particularly in older units. Such condensation is often infrequently present, depending on the weather conditions;

- How energy efficient are the double-glazing units? Closely inspect the units and/or question the vendor and any documentation. Later versions are likely to be better, for example:

    - An old glazing unit with a 6 mm gap between glass sheets has a 'U' value of 3.1;
    - A 16 mm gap provides a 2.7 'U' value; but
    - That same gap if filled with argon gas gives 1.2 'U' value.

- Do the PVC windows and fittings need any maintenance? Such units often require regular adjustment, repair and replacement of items such as hinges, stays and ventilators – they do not last forever.

Replacement windows – legal issues

Since the first edition of this book, the statutory control of window replacement has changed:

- Since April 2002, replacement of windows should have been carried out in accordance with the Building Regulations following service of the appropriate notices, usually to the local authority building control department or approved inspector; or by a contractor registered with a competent person scheme approved by the government. Such firms can certify their own work complies with the Building Regulations and effectively issue their own certificates of compliance. The client should be advised that their legal adviser must check to ensure there is either a competent person's scheme certificate, for example 'FENSA' or 'CERTASS' (there are a total of six competent persons schemes), or approval from the local authority or approved inspector (DCLG 2018);

- Replacement of windows in a conservation area or a listed property usually requires the appropriate permission – the legal adviser needs to confirm any such replacement has approval;

- Most replacement work should be carried out with the benefit of a warranty or guarantee.

Replacement windows – standards

The Glass and Glazing Federation has comprehensive guidance available, especially their Glazing manual (GGF 2022). British standard publications you may find helpful in relation to windows and doors include:

- BS 5713 – this is sometimes seen stamped on the spacer bar in double-glazed units. It refers to the 1979 specification for hermetically sealed units. Anecdotal evidence suggests that any double-glazed units with this number on should be carefully checked for condensation since the standard of manufacture was not as robust as the more recent BS, in addition to which the unit is likely to be old in any event;

- BS EN 1279 – this standard dates from 2002, more than 20 years after the previous document. Again, the writers' experience is that units with a spacer bar so stamped are less likely to suffer condensation;
- BS 6206 – this 1981 standard confirms the glass is safety glass. The stamp is sometimes contained in a BSI heart-shaped motif;
- BS EN 12150 – the later replacement standard for safety glass in windows, doors and other vulnerable locations such as glazed shower screens. Other safety glass numbers include BS EN 14449 and BS EN 14179.

There is sometimes other helpful information stamped on the spacer bar between the glass sheets, including the manufacturer's name and a date on manufacture; thus, '(10 / 12)' confirms October 2012.

Glass in replacement windows may also threaten the safety of occupants. See Chapter 13 on hazards, in particular:

- Glass at low level can be dangerous if someone should fall against it. These should conform to appropriate safety standards;
- If the opening lights are small and, or positioned high up in the window, escape in the event of fire may be impeded if not prevented altogether.

9.2.4 Doors and door frames

Assessing the condition of doors and their frames is a similar process to that of windows. One of the main differences is that doorways are the main ways in and out of the dwelling. Therefore, they must be robust enough to withstand normal usage and prevent illegal entry. The following checklists may assist with the assessment.

Condition

Doors to domestic dwellings vary; most are made of timber, but many are now made of PVC or other materials. Panelled and partially glazed doors are particularly vulnerable to deterioration. *Key inspection indicators* include:

- Have the glazing putties or beads deteriorated?
- Is there any rot at the junction of mullions, rails and weatherboards, especially where there are open joints?
- Check the door frame where it meets the sill and make sure the frame is securely fixed back to the wall;
- Modern requirements in the Building Regulations are that properties should be reasonably accessible for people with disabilities. In practice, this means the main entrance door should be a 'level access' door. In addition, there should be a ramp outside the property with a 'reasonable' slope up to the door to provide direct access up to the door. Lack of such a door and ramp means it would be difficult for people with disabilities, for example, somebody in a wheelchair, to use the door. However, level access can sometimes cause moisture problems inside if the damp-proof course has not been designed and installed properly. Therefore, you should check for moisture related problems on the floor, wall and skirting board surfaces inside the dwelling.

Security

According to the BRE (1994), as many as 30–40% of residents in some inner-city areas live in fear of burglary. Yet the actual risks are much lower than this and statistics suggest the number of burglaries is steadily declining.

The Building Regulations Part Q of 2015 deals with secure door design. Any survey report should give owners useful advice and the entrance doors are one of the key components. Secure doors should have most of the following features.

MAIN ENTRANCE DOORS

- They should be robust and be able to resist being kicked or charged. Usually, they should be at least 44 mm thick;
- The frame should be securely fixed within the opening although this will be difficult to check during the survey;
- If 'external' door sets are installed in lightweight framed walls, most likely in flats or similar; walls should incorporate a 'resilient layer' to reduce the risk of anyone breaking through the wall and accessing the locking system. For example, the resilient layer could be:

  - Timber sheathing at least 9 mm thick; or
  - Expanded metal or similar;
  - Either of them being for the full height of the door and 600 mm either side of the doorset.

- The door should be hung on three 100 mm steel butt hinges;
- There should not be any glazed panels that could allow easy access to locks if the glass was smashed;
- The door should be fitted with the following (see Figure 9.3):

  - A mortice deadlock operable from either side with a key (minimum 5 lever);
  - A rim automatic deadlock;
  - A door chain or limiter;
  - A door viewer;
  - A letter box that is at least 400 mm away from any locks and with a flap to prevent insertion of a hand.

In flats, a quick release of the mortice lock in the case of fire is required.

OTHER EXTERNAL DOORS

Other external doors do not normally have to perform as escape doors and can generally be secured from the inside before the occupants leave the house or go to bed. They should be fitted with the following:

- Mortice sash lock;
- Two or more surface mounted or mortised bolts, preferably the type that are key operated.

These doors should conform to the general standards laid down for front entrance doors. This is because they tend to be in places that are not so observable, so they may be subject to prolonged and violent attacks, for example, with hammers, crow bars

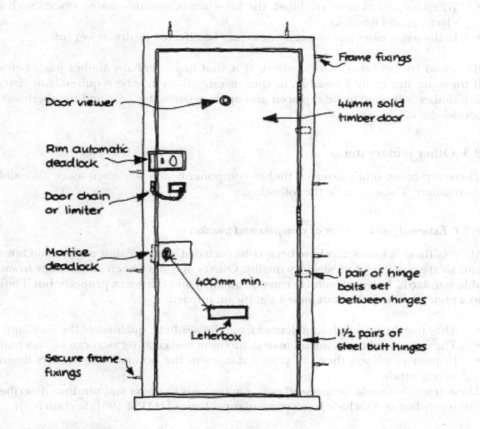

Figure 9.3 Security fittings to a main entrance door (reproduced courtesy of BRE).

and so on. Therefore, special reinforcing plates and other precautions are advisable. Interconnecting doors between attached spaces, such as a garage, are particularly vulnerable.

EXTERNAL DOOR THRESHOLDS

Where external entrance doors are exposed to the elements, rain can often penetrate across the threshold (lower part of the door frame) and around the frame (BRE 1985c). Key inspection indicators will include:

- Is there a water bar in the door threshold and in the top and sides of the door frame (usually in the form of a groove)? This helps to prevent water penetration between the door leaf (the opening part) and the door frame, particularly in conditions of driving rain;
- Staining to the timber threshold internally;
- Discolouration and high moisture meter readings to the floor finish just inside the entrance door (for example, stained carpet, hardboard, floorboards, and so on);

- When carpets or mats are lifted, the presence of moisture-loving insects such as silverfish and woodlice;
- In the worst cases any adjacent areas may be affected by dry or wet rot.

If a wood rot outbreak has occurred, it is vital to inspect any timber joists below. If these are not easily accessible, further investigations may be required. Any defective timber will need to be replaced and appropriate water bars and weatherboards provided.

9.3 Other joinery items

There are many other external timber components and some are more accessible than others. These include the following.

9.3.1 External features such as canopies and porches

Many of these features may have been constructed at the same time as the main house and so are possibly of satisfactory quality. Others may have been added later to variable standards. External features can add amenity and value to a property, but if built to a poor standard, they can have a significant impact:

- They may be unsightly and detract from the aesthetic qualities of the dwelling;
- The expense of repairing them or demolition and clearing away can be very high;
- In poor condition they can pose a danger to the occupants, especially during strong winds.

These features should be assessed on a similar basis to doors and windows described previously but also include some particular problems (HAPM 1991, Section 5.3):

- Foundations and supporting structure are sometimes less effective than the main walls. Timber columns in particular should be well supported in appropriate shoes at ground level;
- They can have poor water-shedding properties, which can lead to increased rates of deterioration of the structure below. Formal rainwater drainage systems should be provided for all features, apart from those with very small roof areas;
- Timber gallows bracket supports to entrance canopies or bays can suffer from rot where not isolated from damp masonry. Checks should be made for signs of movement;
- In a similar manner to dormer windows, small pitched roofs over porches and bay windows, etc. can be difficult to form using the same tiles as on the main roof. Smaller plain tiles are often more effective. Standard ridge and hip tiles may also be inappropriate.

For more information on how to tackle porches, see Section 9.3.5.

9.3.2 Fascias, eaves and bargeboards

Most roofs will have several timber components at or above gutter level. This is especially true around dormer roofs, where it might be difficult to see fascia boards, never mind inspect them. On some classically designed older houses, bargeboards and fascias might be large and very ornate. The cost of repairing these features could be very high. Where the building is in a conservation area or listed, the features might have to be replaced exactly.

Inspecting these features can often be difficult. *Key inspection indicators* include:

- Where the features are less than 3 m off the ground they should be probed at regular intervals (say 1 m) and at vulnerable positions (joints for example);
- At higher levels, the features should be carefully assessed with selfie-sticks, camera poles, drones or binoculars. Key features would be peeling and missing paint, obvious signs of rot or a 'waviness' to painted surfaces and any open or sprung joints between components;
- In many older properties, for example those built in the 1960s, or ex-local authority properties; soffits were often made of asbestos-cement materials. Clients should be warned of the safety implications, especially if they need to be decorated;
- Surveyors should be especially careful when they see new plastic fascias because they are sometimes fixed directly over original fascias that may contain asbestos, or rotten timber fascias. Look at similar properties in the road for an indication of the likely original material.

9.3.3 Cladding and boarding on low-rise properties

Many dwellings, especially the large number of timber-framed dwellings built between 1960 and the early 1980s, have external horizontal or vertical timber boarding. Assessment of these claddings is similar to any other timber component. Special attention should be given at vulnerable points, such as:

- Along the bottom edge of the cladding where it terminates against a sill or some other form of construction;
- At any external corner junction where one board meets another;
- Around window and door openings where the timber is relied upon to seal the junction with the window frame and sill;
- In any areas where excessive rainwater run-off is evident, for example, beneath a leaking gutter or a defective overflow. The risk of rot in these locations is high and so should be inspected carefully.

Many areas of timber cladding have subsequently been replaced with lookalike plastic panels. Early types of such cladding sometimes suffer from significant discolouration, and they can become brittle due to exposure to ultraviolet light.

9.3.4 Cladding and timber components on flats and houses in multiple occupations

Surveyors must be especially careful with components such as cladding and timber balcony structures on flats since the increased awareness following the tragedy at Grenfell Tower in 2017 and other fires relating to timber components such as balcony guard rails. They must be very aware of the duty of care they owe to owner-occupiers, landlords and tenants. In addition, local authorities have powers in the Housing Act 2004 to take enforcement action against landlords in cases of serious hazards – if a surveyor fails to inspect and report appropriately, the occupiers could be at risk and their client could be seriously disadvantaged financially, including being fined.

The Grenfell fire occurred due to the presence of highly combustible cladding materials on the external surfaces of a high-rise block. This was despite the fact that the materials had been provided and fixed with Building Regulations approval (Grenfell Tower Inquiry 2022). It is alleged that the fire spread, in part, due to lack of cavity fire barriers. The problem for the surveyor is that it is not possible to inspect such barriers once the cladding is fixed in position. Initial concern was expressed for buildings over 18 m, but such issues can apply to lower-rise buildings.

Desktop study and knowledge of locality are very important in relation to this issue. The surveyor must enquire of the managing agent and owner regarding whether any cladding material has been inspected and certified by a suitably qualified person. It is difficult to be more precise because this is an area where statutes, regulatory control and professional guidance are changing so quickly.

It is important you keep yourself updated with developments in this area and we suggest the following sources of information as starting points:

- Information relating to the fire at Grenfell Tower – www.gov.uk/government/collections/grenfell-tower
- UK finance on cladding – www.ukfinance.org.uk/area-of-expertise/mortgages/consumer-q%26a-cladding
- RICS External wall systems FAQs – www.rics.org/uk/news-insight/latest-news/fire-safety/cladding-qa/

Any certificate older than 5 years should be treated with caution. There is a list of approved bodies whose members may be able to carry out assessments of cladding available from RICS (RICS 2020).

Key inspection indicators include:

- Inspection should be made from available public spaces and from within the boundaries of the development, visually and with binoculars, in particular:

  - From the windows of the subject flat;
  - On any private balconies; and
  - From public access ways;

- The surveyor must make careful note of any definite, or possible, combustible materials to the outside of the flat;
- The surveyor's consideration of this matter should be made together with other relevant issues including:

  - The means of giving warning in the event of a fire in the premises [that is, the smoke and heat detector system(s)];
  - The availability of a satisfactory means of escape from the building in the event of a fire (see Chapter 13 for more information about relevant Building Regulations requirements), and

- 'Carbon monoxide alarms ... gas appliance checks ... electrical compliance ... (and) ... furniture compliance' (Strong et al. 2019, p. 14).

All relevant matters must be flagged for consideration by the client's legal adviser. The surveyor should be prepared to discuss this matter further with the legal adviser and client.

At the time of writing, the formal public inquiry was continuing, with reports of potential legal action to be taken against designers, contractors and others implicated in the deaths. Fire safety will continue to be a matter of serious and significant concern for all surveyors.

9.3.5 Conservatories

This part of a home has become more important over the years as thermal performance and comfort levels (provision of heating and cooling systems, etc.) have improved. Homeowners spend a lot of time in them. In response to increasing numbers of negligence claims, because many surveyors treated conservatories as 'outbuildings' with a reduced inspection regime, RICS made a conservatory part of the main building in their survey products. The same level of inspection, recording and reporting is therefore required for a conservatory as for the main building.

Key inspection indicators include:

- Many conservatories are built without Building Regulations approvals, with shallow foundations of inadequate width – use the spirit level to check this, and inspect the junction with the main dwelling;
- Conservatories have lots of glazed areas – look for presence of safety glass and failure of sealed units causing condensation;
- Modern glazed or even polycarbonate roofs can have a significant weight and homeowners like bright, airy, conservatories, with no lateral ties; thus, the roof structures on many conservatories are untriangulated – check with spirit level and by eye for distortion and bulging in the frame and bowing of gutters;
- A translucent roof can cause overheating of the space beneath if there are no blinds or similar shading devices fitted;
- Any gutter between the main property and the conservatory roof may be difficult to keep clear of leaves and moss on a regular basis because access is difficult, causing water penetration;
- Where the roof meets any wall of the main property there would nowadays be a cavity tray. It is difficult to install a cavity tray in an existing wall in this position when a conservatory is added to an existing property – check for damp inside, especially in areas where wind-driven rain is more likely;
- Presence of a conservatory means access for maintenance purposes to the different parts of the main building above the conservatory is restricted, meaning maintenance and repairs will be more difficult and therefore possibly more expensive because of a requirement for special access equipment – warn the client;
- Most modern conservatories are made of PVC; older structures may be of timber and, unless well built in hardwood, are more likely to be suffering wood rot – check carefully in locations suggested previously and consider reporting it as a temporary structure.

9.4 External decorations

9.4.1 Assessment

If external decorations are not regularly renewed, the durability of windows, doors and other components can be reduced. Consequently, it is important that the state of the external decorations is assessed so an accurate estimate of future redecoration can be given.

9.4.2 Mechanisms of deterioration

Paint failures are usually associated with inadequacies in undercoat or primer (Richardson 1991, p. 126) and most failures occur at the junction between components. There are many ways a paint film can fail, the main ones including:

- Blistering – Blisters in the paint film are usually caused by moisture trapped underneath or within the paint film;
- Brittleness or flaking – Flaking and cracking of the paint film is usually caused by internal stresses when drying. Another cause could be dimensional changes in the substrate (i.e., moisture movement in timber);
- Chalking – Where the paint loses its gloss, leaving white or slightly tinted powder on the surface, caused by photochemical breakdown of the surface layer releasing the pigment.

A more complete description has been given by Cook and Hinks (1997, p. 417).

If any of these features are noted, the affected area of paint should be removed back down to the substrate and the decorations renewed.

9.4.3 Other decorating issues

External wall surfaces of many dwellings may have been painted in the past to either enhance the appearance of the building or offer added protection against the elements. In some cases, these coatings might have been proprietary applications containing additives such as mineral substances or silicones. They are often thicker than normal paints and if the surface was not properly prepared, these coatings can soon blister and flake. In the worst cases, they can hold moisture in the wall, preventing it from drying.

One of the drawbacks of painted walls is that the owner must regularly redecorate those large external areas. This can be expensive, and clients should be made aware of this additional liability.

Black or any other dark colour on outside timbers can cause over – heating in sunlight and therefore splitting of the timber. Any splits can allow water penetration, causing wood rot. A lighter colour tends to perform better in practice, since the sunlight, and therefore the heat, is reflected away from the surface.

Gloss paint containing lead was used in outside decorations in the past, probably until the mid-1970s. This is usual for most properties of a similar age and type. Lead is a potential hazard and clients should be warned if the surveyor believes it may be present – see Chapter 13.

Outside redecoration, or altering the style of decoration, of a listed property or property in a conservation area usually requires local authority planning approval. This has legal implications.

For flats at high level, clients should be warned of likely higher costs of maintenance of external decorations when compared with low-rise properties. This is because, while such costs are likely to be met through a common repairing fund, the costs will be met by individual flat owners on a pro-rata basis.

References

BRE (1985a). 'Preventing decay in external joinery'. *Digest 304*. Building Research Establishment, Garston, Watford.

BRE (1985b). 'External walls: Joints with windows and doors – applications of sealants'. *Defect Action Sheet 69*. Building Research Establishment, Garston, Watford.

BRE (1985c). 'Inward-opening external doors: Resistance to rain penetration'. *Defect Action Sheet 67*. Building Research Establishment, Garston, Watford.

BRE (1994). *BRE Housing Design Handbook: Energy and Internal Layout*. Building Research Establishment, Garston, Watford.

BRE (1997). 'Repairing timber windows: Investigating defects and dealing with water leakage'. *Good Repair Guide 10, Part 1*. Building Research Establishment, Garston, Watford.

Cook, G.K. and Hinks, A.J. (1997). *Appraising Building Defects*. Longman Scientific and Technical, Harlow, Essex.

DCLG (2018). *Competent Person Scheme – Current Schemes and How Schemes Are Authorised*. Department of Communities and local Government. Available at www.gov.uk/guidance/competent-person-scheme-current-schemes-and-how-schemes-are-authorised#types-of-building-work. Accessed 8 January 2022.

GGF (2022). 'Glass and Glazing Manual, GGF Technical Hub'. Available at www.ggf.org.uk/technical-hub/. Accessed 8 January 2022.

Grenfell Tower Inquiry (2022). Available at www.grenfelltowerinquiry.org.uk/. Accessed 8 January 2022.

HAPM (1991). *Defect Avoidance Manual – New Build*. Housing Association Property Mutual Limited, London.

Melville, I.A. and Gordon, I.A. (1992). *Structural Surveys of Dwelling Houses*. Estates Gazette, London.

Muthesius, Stefan (1982). *The English Terraced House*. Book Club Associates, London.

Police Crime Prevention Initiative (2019). *Secured by Design Homes 2019 version 2*. Secured by Design, London.

Richardson, B.A. (1991). *Defects and Deterioration in Buildings*. E&FN Spon, London.

Royal Institution of Chartered Surveyors (RICS) (2020). *EWS Form – List of Relevant Professional Bodies*. RICS, Great George Street, London.

Strong, Gary, et al. (2019). *Royal Institution of Chartered Surveyors (RICS) A Clear, Impartial Guide to Fire Safety, Version 3*. RICS, Great George Street, London.

10 Internal matters

Contents

DOI: 10.1201/9781003253105-10

10.1 Introduction

Defects to the internal elements of a dwelling are often overshadowed by the problems associated with the major elements of a building, such as the external walls and the roof. However, internal defects are important for two reasons:

- They may give an indication or confirm possible defects that exist in the main superstructure of the dwelling; and
- They may be significant defects in their own right for which the remedial works would be very costly.

This chapter will concentrate on defects that are closely associated with internal features. Signs associated with defects, such as subsidence and moisture-related problems, are mostly included in the appropriate chapters.

10.2 Walls and partitions

10.2.1 Types of partitions

The main function of internal walls within a property is to divide space. Over the years, a wide range of materials have been used:

- Solid brick walls – these can vary between 240–265 mm (one brick thick) down to 130–150 mm (half a brick thick). These measurements include plaster to both sides. The walls will give out a solid sound when tapped. Some older walls may measure even less, at around 90 mm. This may indicate that the original builder laid bricks 'edge down', which results in a slightly weaker wall;
- Timber walls with brick infill – a traditional variation in some areas. This type of wall usually consists of vertical timber studs and horizontal noggins with brick-on-edge infill panels plastered over both sides. Usual thickness is about 140 mm, depending on the size of the bricks;
- Timber stud partitions – in older properties two types may be encountered:

  - Trussed partitions – these can consist of a series of struts, braces and ties. They may have been designed to take a considerable load from floors and/or internal partitions above and may be a very important part of the building's structure. If any part of this is cut through during alteration work, the building's structure could be affected. Thicknesses vary, sometimes regionally, but are typically around 140–150 mm;
  - Conventional stud partitions – still used widely and comprising vertical timber studs with horizontal noggins clad with either plasterboard or lath and plaster finish. Thickness is usually based around timbers 75 mm wide, so with plaster added, a measurement of around 100–110 mm is typical, but surveyors will find local variations based on their knowledge of locality;
  - Both timber partition types will sound hollow when rapped with the surveyor's knuckles.

- Blockwork partitions – a variety of blocks have been used since the 1920s including 'breeze' blocks (made from coke 'breeze', or ash, from power stations; mixed with cement) and hollow clay versions. Modern blockwork is made from concrete with a wide variety of aggregates, from a lightweight non-load-bearing block 75 mm thick to more substantial dense load-bearing versions 215 mm thick, plus the thickness of two plaster finishes (in the region of 30 mm).

10.2.2 Load-bearing partitions

One of the most critical judgements to make is whether the partition is load-bearing (see Figure 10.1). For example, an internal partition may take the loads from:

- Struts and other components from a traditional 'cut' or purlin roof. This includes the loads of the roof structure, coverings and intermittent live loads such as wind and snow;
- Water tanks in the roof space;
- The self-weight and imposed loadings from intermediate floors and staircases;
- The weight of the partitions directly above.

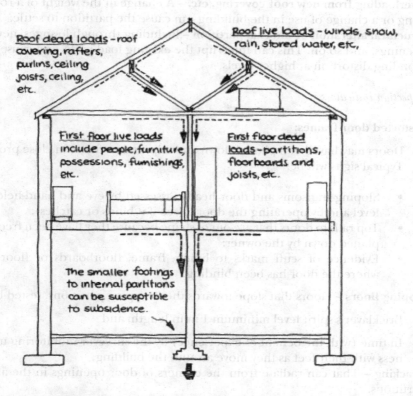

Figure 10.1 Load paths through an internal partition in a typical domestic property.

In some circumstances, they may also provide buttressing, bracing and lateral stability to the whole structure. As a general guide:

- Older properties with traditional roofs often have load-bearing partitions on all floors;
- Newer dwellings with trussed rafter roofs may have load-bearing partitions at ground floor only, as most loads from truss roofs are sustained by external walls. Where properties are three storeys or more, load-bearing partitions may extend higher than the ground floor.

10.2.3 Structural failures in load-bearing partitions

There are several reasons for a partition to develop structural defects resulting in distortion. They include:

- **Poor foundations** – In some cases the loads carried by internal walls can exceed those supported by some external elements. Yet they can have minimal foundations much shallower than the footings of external walls. This is often the case in older properties. This can result in differential settlement between the external envelope and the internal structure;

- **Overloading from new roof covering, etc.** – A change in the weight of a roof covering or a change of use in the building can cause the partition to settle;
- **Structural alteration of internal partition** – including through lounges, new door openings, and so on. This can interrupt the existing load paths and cause corresponding distortion at higher levels.

Key inspection indicators

- **Distorted door frames:**
  - Doors may have been planed down and adjusted so they can close properly;
  - Typical signs will be:
    - Sloping transoms and door heads (assessed by eye and hand-held spirit level and by operating the door to see if it binds or catches);
    - Top rails to doors that are out of shape because they have been frequently planed down by the owner;
    - Evidence of scuff marks to lining, frame, floorboards or floor finish where the door has been binding.
- **Sloping floors** – Floors that slope towards the dropping partitions, tested by:
  - Bricklayer's spirit level minimum 1 m in length; and
  - In time (with the benefit of experience) by the surveyor monitoring unevenness with their feet as they move around the building;
- **Cracking** – That can radiate from the corners of door openings in the affected partitions.

In many older properties, these distortions might have occurred many years ago. Buildings often reach an equilibrium, albeit a little out of true. The key factor here is how recent these movements have been. The definition of 'recent' is any indication of movement having occurred in the last 5 years (see Chapter 5 for more discussion of this). The following signs can indicate whether the movement is contemporary or not:

- If the door shows clear signs it has been recently adjusted and planed. This could include freshly exposed timber, edges of the door recently touched up with paint, and so on;
- Cracking that extends through decorations that are relatively new; (for example, less than 5 years old);
- Evidence of building alterations that match the age of the movement.

If these triggers are present, the principle of 'follow the trail' applies where you should carefully consider the matter and/or recommend further investigation where appropriate.

10.2.4 Non-load-bearing partitions

This type of partition mainly divides spaces and will rarely carry any formal loads apart from its own self-weight. Key issues include the following.

Structural stability

Although non-load-bearing partitions may support nothing but their own self-weight, they must be properly supported by the structure beneath, especially on upper floors. A common problem is associated with timber intermediate floors that support block partitions. Ideally, where these partitions run parallel to the floor joists, they should be positioned directly above a larger single, double or treble joist or, in some cases, a steel beam. It may be necessary to calculate the adequacy of the support just to make sure. If not, deflection of the floor can occur. Where partitions run at right angles to the joists, a line of solid noggins should be provided. In this case, it is vital that the block partitions are not built directly off the flooring (HAPM 1991, Section 3.2).

Plaster surfaces to block partitions can also develop defects, especially if the blocks are too dry when plastered. In this case, water can be drawn from the plaster, resulting in poor adhesion characterised by a usually fine pattern of surface cracking.

Moisture problems to the base of internal walls

Moisture problems were discussed in some detail in Chapter 6. Because the subsoil adjacent to external walls is generally wetter than that beneath the building itself, moisture problems to internal partitions are not a widespread problem. Even so, to ensure moisture does not affect internal surfaces, partitions must still have effective damp-proof courses properly linked to adjacent damp-proof membranes. If not, moisture can affect the wall. Failure to link DPCs to DPMs is a relatively common fault in the writers' experience.

A particular problem that can occur with ground-floor timber-framed partitions, especially in older properties, is wood rot in the sole plate. If this is not protected by a DPC, or rubble has breached the existing one, conditions for the development of wet or dry rot may exist. The position where a suspended timber floor changes to a solid floor is particularly vulnerable (see Figure 10.2). In these locations, special care

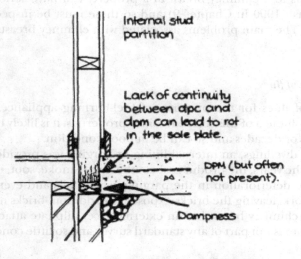

Figure 10.2 Sketch section showing the junction of a solid and suspended floor with an internal partition.

must be taken to test the lower parts of walls, skirtings, door linings and frames and other timber components such as cupboards in contact with those walls for moisture and visual signs of wood rot.

Surveyors should be careful during their inspections of old buildings, where dry linings may have been fixed to internal walls to hide the effects of moisture in all its forms. This boarding is usually fixed with 'dabs' (blobs) of mortar/adhesive or timber battens. Mortar dabs can sometimes transmit moisture or salts from any old, wet wall to the dry-lining and this can cause wood rot and wood-boring insect attacks where timber boarding has been used. Alternatively, if dry-linings are fixed with timber battens, the battens can suffer similar problems if they are not correctly isolated.

Surveyors should treat internal walls that have a decorative textured coating finish with similar caution. While some homeowners may have had such a finish applied in the past for fashion reasons, textured coatings are sometimes applied to mask moisture (or hide cracks) since they are more resistant to moisture than normal decorations and flexible as they are plastic by nature.

10.3 Fireplaces and chimney breasts

Most properties will have at least one chimney breast. In older dwellings, there may be a considerable number, with flues serving every room in the house. The primary function of the chimney breast is to provide a passage for the products of combustion, but many often have an informal structural role. When they are properly bonded into the main walls, chimney breasts can have two opposite effects:

- They can provide a measure of stability and restraint to the main structure, acting as a buttress – Building Regulations Approved Document A has more details;
- Because of their mass and self-weight, they can result in subsidence if the foundations are not satisfactory.

Because of their structural influence and the significant loads associated with them, any modifications to a chimney breast in a property can have serious ramifications (see Smith v. Bush 1990 in Chapter 9) and so these must be inspected very closely during a survey. The main problems associated with chimney breasts are outlined as follows.

10.3.1 Adequacy of flues

The adequacy of flues for the purposes of fuel-burning appliances is discussed in more detail in Volume 3 of this series. In older properties, it is likely the chimneys will have been used for decades and so can be in poor condition.

In most older dwellings, an internal lining or parging was provided in a lime and sand mix. Over the years, the products of combustion (smoke, soot, tar, salts and condensation) cause deterioration in the parging, which fails and eventually falls away from the brickwork, leaving the bricks exposed. In addition, bricks may have deteriorated and if the chimney breast has an external face, sulphate attack may be active. The testing of flues is not part of any standard survey and so little conclusive comment can be offered, although most stacks built after 1966 should have a flue lining, when this became a requirement of national Building Regulations. However, there are several signs indicating possible significant problems:

- Where it is appropriate and safe to do so, the surveyor should use a torch to look up the chimney flue to see the condition of the original flue or if another lining has been installed;
- Soot falls in the fireplace area, possibly mixed with old parging;
- Brown watery stains to the surface of the chimney breast within the room. This usually indicates soot and salts have migrated through mortar joints (exposed by failure in the parging) and into plaster;
- Testing at regular intervals with a moisture meter is necessary for two reasons:

  - To see if there is any water penetration from the chimney above; and
  - For any hygroscopic salts (moisture-attracting) in the plaster. High readings may suggest failure in flue linings;

- Deteriorated mortar joints and brickwork to the chimney stack in the loft space. In extreme instances, the surveyor will sometimes be able to see into the flue – this is a safety hazard due to the possibility of escape of sparks and fumes if the flue is still used. White salt crystals may also suggest moisture-related chemical defects;
- Old chimney pots and flaunching in poor condition.

Even where these signs do not exist, it cannot be assumed that the flues are adequate for use. Clients should always be advised that flues should be tested and regularly cleaned if they are to be used for an open fire or other combustion appliance.

10.3.2 Blocked fireplaces

Fireplaces may have long since been removed and the chimney breast blocked. This is often done in brick or blockwork and plastered over, but sometimes a less satisfactory method of sheet boarding on a timber framework is used. Surveyors should note old boarding often included asbestos-containing materials and clients should be warned appropriately.

If done incorrectly, there could be a moisture build-up in the redundant flue, causing moisture staining on internal wall surfaces. The BRE (1987b) have identified a number of different situations (see Figure 10.3):

- Chimney on internal wall – the chimney stack should be taken down below roof level and capped. The flue should not be ventilated;
- Chimney on external walls (eaves elevation) – the chimney should be taken down to below roof level and capped. Ventilation to external air is provided at the head and foot of the stack;
- Chimney on external wall (gable elevation) – the flue should be ventilated at the head and foot to outside air. The top of the stack can be vented by providing a proprietary venting hood or by capping the stack off and venting via air bricks;

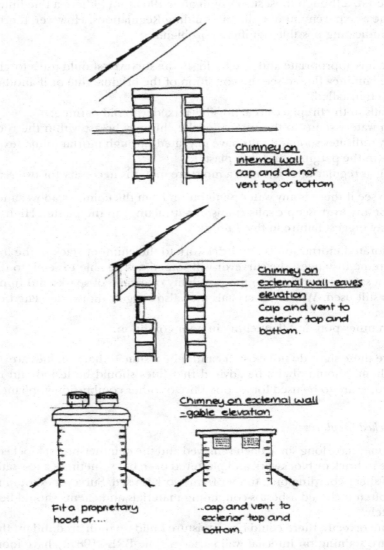

Chimney on
internal wall
Cap and do not
vent top or bottom

Chimney on
external wall-eaves
elevation
Cap and vent to
exterior top and
bottom.

Chimney on external wall
-gable elevation

Fit a propnetary
hood or....

..cap and vent to
exterior top and
bottom.

Figure 10.3 Sketch of different ventilation arrangements for chimneys when they are capped off (reproduced courtesy of BRE).

- Disused chimney stack that has been left in place above the roof line – the following advice should be given:

    - The existing flue should be vented at the top with a proprietary venting hood that excludes moisture;
    - Ideally the chimney should be swept and cleaned prior to this work being completed.

Other problems to be alert to:

- Make sure a flue ventilated to an internal room does not conduct water vapour into the roof space. This will cause condensation in the roof void. Check the chimney flue in the roof space to make sure there are no gaps/openings;
- A disused flue that is part of a larger stack containing others still in service should not be capped. There is a risk that gases may leak into the disused flue and spill into the living accommodation at lower levels.

10.3.3 Structural alterations

Many owners remove existing chimney breasts to create extra living space and sometimes because the fireplaces have become redundant following installation of central heating. The structural implications can be serious. As well as the Smith v. Bush (1990) case, some interesting discussion about the removal of chimney breasts occurred in Sneesby and another v Goldings (1995). In this case, a surveyor carrying out a mortgage valuation failed to notice that where a chimney breast had been removed, the structure above had been inadequately supported. The surveyor had realised the chimney breast was missing and even checked in the room above for signs of structural distress but found none. The inadequate structural support had been boxed in by new kitchen cupboards. If the cupboard doors had been opened, inadequate and shoddy work could have been seen. Soon after the plaintiffs had moved in, large sections of ceiling plaster began to fall down. The chimney breast had to be rebuilt to prevent further structural damage. The court held that the surveyor was liable despite the checks he had made. This was because the removal of the chimney breast had created a 'trail of suspicion that should have been followed and resulted in a more complete inspection at the site of the chimney breast itself to see what was done there'. The doors of the kitchen cupboard could have been quickly opened, revealing signs of the defective work. There are many lessons from this case. When inspecting a property:

- All chimney stacks should be located. This can be done externally, but care must be taken where chimneys have been removed to below the level of the roof covering. Always check for a chimney from an old kitchen or 'scullery'. Smaller chimneys may have served an old 'copper' or water boiler long since removed;
- Track each of the chimney stacks through the property to identify any missing or altered chimney breasts;
- Where chimney breasts have been removed:
  - Check in the room or space above for any signs of distress or structural movement;
  - Open any cupboards, wardrobes, go into accessible roof spaces to see if there is any evidence of completed support work. Going beyond the usual remit of a standard inspection may be necessary in these cases;
  - Ask the vendor if they have any details of the work that was carried out including drawings, copies of guarantees, Building Regulations approvals and so on. Most building control authorities would classify such an alteration as 'notifiable work' requiring approval.

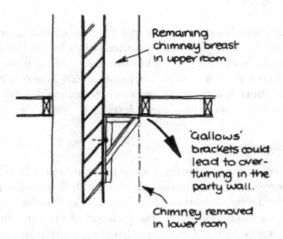

Remaining chimney breast in upper room

'Gallows' brackets could lead to over-turning in the party wall.

Chimney removed in lower room

Figure 10.4 Sketch of a typical 'gallows' bracket that was often used to support the upper portion of a chimney breast.

Where a chimney breast has been removed and the supporting work can be inspected, a substantial construction should be expected: for example, a substantial concrete or steel lintel or beam properly pinned up to the remaining chimney breast, spanning between adequate supports, such as load-bearing internal and external walls.

In properties where chimney breasts have been removed some time ago, the remaining construction may be supported by a set of 'gallows' brackets welded together and bolted back to the main wall. Although this was a common method, most local authorities no longer allow this form of support, as it can impose additional stress on neighbouring properties. Consequently, clients should be warned of this (see Figure 10.4).

It is not uncommon to come across some surveyors who will judge a removed chimney breast as satisfactory if it has been removed for more than five years and there is no evidence of adverse structural movement; it may be safe to assume the work is adequate. Although this may appear reasonable, the structural equilibrium of the removed chimney breast could be disturbed if the neighbouring owner carries out similar alterations to their property. Consequently, the work should be assessed as described previously, even when the work was carried out some time ago.

10.3.4 Moisture at the base of chimney breasts

Any chimney breast is a continuation of the main wall of the property and so should be properly isolated with a damp-proof course. Below ground floor level, brickwork was usually built up and backfilled with rubble to form a base for the hearth. In these areas, poor constructional techniques often resulted in the DPC being breached. Some builders omitted a DPC here, considering it unnecessary, since the chimney would be warm from use of the fire for most of the year. Additionally, the combustion of solid fuels for many decades may have contaminated the brickwork and some of the contaminants could be hygroscopic, which could result in moisture-related problems. Therefore, it is important to check the cheeks and face of the chimney breast at a low level for moisture.

Any timber joists and floorboards bearing on the hearth may be in contact with damp materials and so these should be checked as well.

10.3.5 Hearths

The primary function of a hearth is to provide a non-combustible area should any spark or burning fuel fall from the fire. Many hearths are not big enough. For example, for solid fuel appliances including open fires, the current Building Regulation Approved Document J requires constructional hearths to extend 500 mm beyond the face of the chimney breast and at least 150 mm beyond the sides of the fireplace opening (see Figure 10.5). Where dimensions are less than these and the fireplace is likely to be used for an open fire, the client should be warned of the dangers of such a feature.

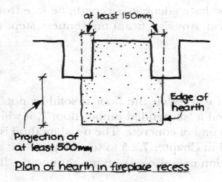

Plan of hearth in fireplace recess

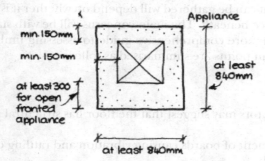

Plan of free standing hearth and open fronted appliance

Figure 10.5 Examples of minimum sizes of different hearth constructions (Crown copyright. Reproduced with the permission of the Controller of Her Majesty's Stationery Office).

10.4 Floors

10.4.1 Introduction

Floors have been mentioned in two previous sections:

- Chapter 3.4.5 where the lifting of floor coverings was discussed; and
- Section 7.2.5 which identified the key indicators that could help assess the condition of a timber ground floor, even though the space beneath could not be inspected.

These aspects will not be repeated here. This section will focus on those defects not previously covered.

The main purpose of floors is to:

- Provide localised lateral support to external walls – see Chapter 5; and
- Safely and adequately support people, furniture and other loads.

To do this, any floor must have adequate strength, be free from excessive moisture, and be even, level and firm. Any significant unevenness, steps and cracks could be a trip hazard.

10.4.2 Lower floors

The first fact to establish is whether the floor is solid or not. If the floor is timber with a gap beneath (called a 'suspended timber floor'), it will require a very different inspection routine to one of concrete. The usual method is called the 'heel drop test' and this is described in Chapter 7.2.5 to see if it gives out a hollow sound. If it is timber, the level of vibration may also give an insight into the floor's stability.

Timber lower floors

The information that can be gathered will depend on whether it is possible to visually inspect the floor space beneath. The main concerns will be with stability, rot and beetle infestation. For a more comprehensive guide to assessing timber floors, see BRE (1997a), but the main points are summarised as follows.

TYPICAL DEFECTS

The following indicators may suggest that the floor has structural problems:

- Excessive movement of boards causing vibration and rattling of ornaments when heel drop tested;
- Perceptible slope detectable by 'feeling' with the surveyor's feet and/or with a spirit level.

There could be several causes:

- **Inadequate size of floor timbers** – Undersized timbers may have been used in the floor resulting in excessive deflection. A method of getting a rough assessment of the floor's adequacy is to compare the floor joists with the tables published by TRADA (the Timber Research and Development Association) or in the current issues of the National House Building Council Standards. It is important to remember that the span is measured between effective supports. For example, the outer walls, any internal walls and any honeycombed sleeper walls. It is not the total length of the joist;

- Poorly constructed floors – much of the mass-produced Victorian housing suffered from variable standards of design and workmanship (Douglas 1997). Suspended floors could also have been altered over the years, resulting in joists that are poorly seated on wall plates. Sleeper walls have often been removed for access purposes (especially by electricians and plumbers) and not properly rebuilt;
- Floor timbers weakened by rot and woodworm – older timber floors are very vulnerable to outbreaks of wet and dry rot and attacks of wood-boring beetles. Joist ends were often built into supporting walls without adequate DPCs and debris may have built up over the years to partially bury timber components. Until the early parts of the 20th century, the underfloor surface (called the 'oversite') was rarely sealed with concrete but often left as bare earth. This can result in high moisture levels, encouraging rot and beetle attack. This is often compounded by poor ventilation to the subfloor area.

Assessing the adequacy of ventilation to subfloor areas

When deciding whether there is enough subfloor ventilation, there are three separate issues to consider:

- Are there enough air bricks? Building regulations Approved Document 'C' suggests on technical solution might be suspended floors should be vented as follows:

  - Two opposing external walls should have ventilation openings so that air can ventilate all sides and all parts, 1500 mm² for each metre run of wall, or 500 mm²/m² of floor area (whichever is the greater);
  - The space between the ground under the floor ('oversite') and the underside of the joists should be at least 150 mm (OPSI 2013b, p. 26).

  This would equate to approximately one 225 × 150 mm air brick at every 1.5 m centre around all accessible sides of the building. However, these requirements are based on modern construction. For older properties, increasing the level ventilation to the subfloor voids would be sensible.
- Are these air bricks properly distributed to prevent stagnant air pockets where moisture might build up? Even if there are enough airbricks, they may not be in the right position (Ridout 2000, p. 204). Melville and Gordon (1992, p. 195) identify the typical problem of a terraced house with a back addition or 'offshot' with a solid floor (see Figure 8.22). Although there are adequate airbricks in front and back, there is a whole area of stagnant sub-floor void. These are the areas at risk from rot and beetle attack;
- Are the sleeper walls adequately perforated to allow good cross-ventilation? If not, bricks may need to be removed (BRE 1986b).

A more detailed review of the moisture that collects beneath suspended timber floors has been produced by Harris (1995, p. 11).

Solid lower floors

This type of floor, called a ground-bearing or supported slab, is usually of concrete and is often regarded as being a very durable material. The material has good compressive strength (when subjected to a load and supported beneath) but is weak in tension (under the same load, but unsupported beneath in whole or in part). So it is strong but brittle (Watt 2007, p. 64). Concrete floors are not defect free and many older solid floors contain little concrete. For example, solid floors in older rear kitchens and sculleries may be no more than a thin screed laid over ash or clinker hardcore to provide a level bed for a quarry tile surface (BRE 1997a, p. 1). The main defects include:

- Settlement of the slab – where support is removed from beneath, causing the concrete bed to settle. Because most concrete floors are unreinforced and the material is weak in tension, they crack haphazardly as they settle down to their new level. This might be caused by:

  - Consolidation of the hardcore and subsoil following construction, although such movement usually ceases after around 10 years. This can be a problem, especially around the perimeter of the building where the floor slab can become a cantilever as the hardcore in the foundation trench of the main wall settles, since the hardcore is much deeper than it is under the slab;
  - Subsidence caused by external factors such as drains or trees affecting shrinkable clays, mining activities and so on.

Key inspection indicators:

- Gaps below skirtings (see Figure 10.6; BRE 1990);
- Cracks or unevenness in floor finishes that suggest cracks in the slab below;
- Sloping floors (detectable by eye and/or with a large spirit level) and dished or isolated depressed areas of floor surface;
- A hollow sound when the floor is robustly tapped, indicating hardcore has settled away from the slab. In this case, the slab is no longer fully supported and can 'snap';

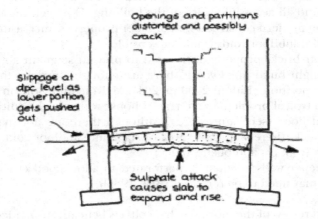

Figure 10.6 Sketch showing effects of sulphate attack on a solid floor.

- Ill-fitting doors in partitions built off the defective slab (very common in 1950s – 1960s buildings).
- **Swelling of the slab** – in some cases the concrete floor may swell upwards. This will cause the concrete slab to crack in a similar manner to that described previously, but for opposite reasons. This may be caused by:

  - Sulphate attack to the underside of the concrete slab – hardcore or subsoils rich in sulphates can combine with moisture and cause chemical change in the concrete. The main effect is to cause the concrete to expand and the floor to swell upwards. For more information on this aspect, see BRE 1996;
  - Swelling of the ground and hardcore – where a tree has been removed, clay subsoil may swell as moisture returns. A similar effect can occur where moisture affects some hardcores, such as steel slags and colliery wastes, although this is rare.

Key inspection indicators:

- A clear rise in floor level, especially (though not always) towards the centre of the slab;
- Concrete slab or floor finish showing random cracking;
- Binding of doors that open into the room;
- Distortion of internal partitions supported on the slab;
- External walls pushed outwards by the force of the expanding slab. The wall tends to slip at DPC level (see Figure 10.6);
- Is it known locally that the subsoil or groundwater is high in sulphates? Or did local building techniques often incorporate unstable hardcores below ground-floor slabs?
- **Failure of the damp-proof membrane (DPM)** – problems associated with moisture have been outlined in general terms in Section 6.4.3. This section looks at those problems closely related to solid floors. The floors to many older houses simply do not have a DPM, and those built between 1950 and 1966 may rely on thermoplastic tiles stuck down with a bitumen adhesive, an asphalt topping, or other method (BRE 1997b; Rock 2006, p. 153).

  Therefore, where the solid floors to older properties have not been replaced, they must be viewed with caution. For newer construction, Building Regulation Approved Document 'C requires a DPM must be provided within the solid floor and this may typically include a plastic sheet with sealed joints, lapped onto the damp-proof courses in the walls (OPSI 2013b, p. 26). This may be above or below the concrete floorbed. Because well-laid, good-quality concrete is virtually waterproof, it can cope with a few minor problems with the damp-proof membrane. These typically occur from sharp stones, work boots, around services and drainpipes and in corners (where the DPM is often cut to make it fit). Where the defect is significant, high levels of moisture can enter the floor slab and cause serious deterioration, especially where the local water table is high. Moisture can affect a floor by several routes:

  - Missing, poorly lapped (at the joins between plastic sheets) or partially missing DPM;
  - Badly punctured DPM;
  - Poor link between DPM of the floor and DPC of the wall.

In each case, the slab and any overlying screed can act as a huge reservoir for moisture. This may take many months to dry out, even when the defect has been rectified.

Key inspection indicators:

- Deterioration or staining of the underside of the floor coverings and finishes;
- High moisture meter readings to the surface of the floor;
- In some cases where the wall plaster is in contact with the floor surface, high moisture meter readings to the base of the plaster with characteristic tide-mark staining;
- High moisture meter readings to any skirting boards.

The main problem with this type of defect is that it can easily be confused with others. For example:

- Low levels of moisture problems at a low level in the wall. This can result in lateral (sideways) movement of moisture in the slab of up to 1 m;
- Cold bridging in uninsulated slabs causing condensation moisture on the floor surface;
- Hot and cold-water pipes leaking into the screed (BRE 1988);
- Condensation beneath a sheet vinyl floor covering; and
- Many other household practices such as regular washing of bathroom and kitchen floor coverings, spillage from baths and canine and feline toilet habits.

Other solid floors can incorporate timber boards or blocks (such as hardwood parquet tiles) fixed directly onto and/or into the concrete or other material beneath. Deterioration in the DPM or the fact that the nails fixing the boards sometimes punctured the DPM can cause damp ingress into the floor (Rock 2006, p. 151). This can lead to wood rot.

Recognising that there is a problem is straightforward, but finding the true cause and organising the correct remedial work is far more complicated. Indeed, some commentators believe that diagnosis of this problem is possibly one of the most difficult faced by any surveyor, requiring very careful consideration. Burkinshaw for example stated '. . . inspecting a concrete floor poses a much greater challenge to a surveyor than a timber floor' (Burkinshaw & Parrett 2003, p. 64). The duty of the surveyor in the first instance is to identify a problem exists.

'Historic' solid ground floors

Surveyors will sometimes be asked to inspect solid ground floors in older, 'traditional' properties. Such floors were sometimes 'bedded directly on the earth or on permeable fills or mortars with no damp-proof membrane' (Historic England 2016b). This type of floor is based on moisture evaporating through the floor and then dissipating throughout the property because of the higher levels of natural ventilation that often characterise older properties (see Figure 10.7). Figure 10.8 illustrates the consequences of replacing such an older floor with a 'modern' version – moisture can be driven into what were 'dry' walls.

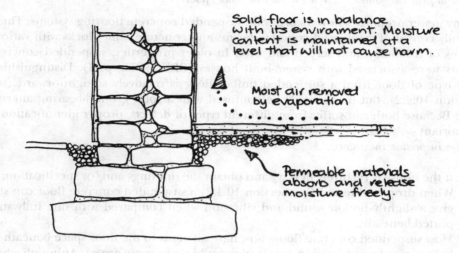

Solid floor is in balance with its environment. Moisture content is maintained at a level that will not cause harm.

Moist air removed by evaporation

Permeable materials absorb and release moisture freely.

Figure 10.7 Example of a breathable 'traditional' floor (reproduced by kind permission of Historic England).

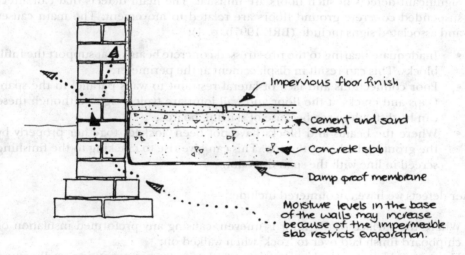

Impervious floor finish

Cement and sand screed

Concrete slab

Damp proof membrane

Moisture levels in the base of the walls may increase because of the impermeable slab restricts evaporation.

Figure 10.8 Implications of replacing a breathable 'traditional' floor, with a concrete floor including a DPM (reproduced by kind permission of Historic England).

Surveyors will need to explain the particular nature of floor construction in older buildings and the long-term implications. Floors of this type are common in listed buildings and their replacement would be an issue for the legal adviser.

Ground floor suspended concrete ('beam and block') floors

Many newer properties now incorporate suspended concrete flooring systems. These usually consist of reinforced concrete beams and concrete infill blocks with various forms of insulation and screed laid over. In older properties, suspended concrete floors were associated with system-built houses (BRE 1997b, p. 4). Distinguishing this type of floor from a suspended timber floor is relatively straightforward (see Section 10.4.2), but it can easily be confused with a solid ground-bearing alternative. Because both can suffer from different types of defects, proper identification is important.

  Key inspection indicators:

- If the house is relatively new try and obtain the drawings and/or specification;
- When drop-heel tested (see Section 10.4.2) a suspended concrete floor can still give a slightly hollow sound and vibration when compared with one fully supported beneath;
- Most suspended concrete floors now have air vents to the floor space beneath to help prevent condensation (and radon build-up in some areas). Although when this type of floor first came into use (mid-1980s), many had no ventilation, Building Regulation Approved Document 'C' requires underfloor ventilation to the same standard as a timber suspended floor;
- In areas where 'flooding is likely', consideration should be given to means of inspecting and clearing out the voids under the floors (OPSI 2013b, p. 27);
- Significant defects in such floors are unusual. The main defects that can affect suspended concrete ground floors are related to movement. The main causes and associated signs include (BRE 1997b, p. 4):

  - Inadequate bearing to the pre-stressed concrete beams that support the infill blocks. This can result in displacement at the perimeter;
  - Poor connections and lack of lateral restraint to walls parallel to the span. Gaps and cracks at the floor and wall joint are typical signs, although these can be difficult to see because of skirting boards;
  - Where the beams and blocks have not been 'locked' together properly by the grouting laid over the top. This can give rise to cracking to the finishing screed in line with the span direction.

Other defects we have encountered include:

- Where the surface of the floor is uneven, causing any preformed insulation or chipboard finish laid over to 'rock' when walked on;
- Moisture/water from leaking heating or other water pipes, causing swelling and distortion in chipboard or similar timber flooring. In extreme cases, wood rot can result.

Thermal insulation of ground floors

Since 1995, the Building Regulations have generally required provision of insulation to ground floors in new construction, including extensions and alterations. This helps to reduce energy costs, energy loss and greenhouse gas emissions.

Concerns about global warming have resulted in many older properties being ret-rofitted with insulation and this trend is set to increase in the future. More detailed information on how older floors can be insulated is available from Historic England (Historic England 2016a, 2016b) but the following principles may be useful:

- Timber floors: a number of different methods and materials can be used rang-ing from rigid foam slabs wedged between joists to mineral fibre hung in nets. Whatever the approach, the installers should have made sure the existing floor structure was in a satisfactory condition and adequately ventilated. If not, the alteration of the thermal and moisture balance can increase moisture content and make them more prone to wood rot and wood-boring insect attacks. The joist ends should be checked if they are accessible.
- Retrofitting insulation to a solid or suspended reinforced beam and block floor is more difficult. Unless the floor is being replaced for other reasons (for exam-ple, sulphate attack), the only feasible option is to overlay the existing floor with rigid insulation boards and a new floor finish. However, there are considerable drawbacks:

  - The insulation could make an existing moisture problem worse if the exist-ing floor is not properly protected by a DPM/DPC (see Section 6.4.3);
  - Permanent fittings and fixtures will have to be removed, set aside and refixed once the insulation has been laid. A typical example would be kitchen units. If not, then large areas of the floor beneath the units will form a 'cold bridge' and this could result in condensation and mould growth.
  - Adding an additional thickness to the existing floor will mean all doors will have to be altered and if there is a staircase, the height of the first riser will be reduced. Varying riser heights is a common cause of trips and falls and so is a potential safety hazard.

In all cases, the existing construction has to be properly assessed, appropriate remedial works carried out and new insulation methods and processes designed and installed to match the characteristics of the dwelling. If not, then the so-called improvements could be the cause of problems.

Spray foam insulation to the underside of timber floors

In Chapter 8, we warned about the problems caused by spray foam insulation that is applied to the underside of existing roof structures. Sadly, some of the same installers are insulating the underside of existing timber floors with the same material. In some cases, remote-controlled mechanised robotic devices are used to spray the foam.

In our opinion, the use of spray foam insulation on the underside of timber floors is more problematic than for roofs because:

- Access is normally restricted so the quality and accuracy of the work is likely to be worse;
- Ventilation rates are likely to be lower and moisture levels are likely to be higher than in roof spaces. These factors will increase the likelihood of wood rot and wood boring insects.

If you come across this method of insulating a floor, you should warn your client of the implications and recommend the foam should be removed.

10.4.3 *Upper floors*

In most cases, the upper floors of residential properties are timber. The exceptions are purpose-built flats and maisonettes, which may have solid concrete party floors.

Timber upper floors

In many ways, intermediate timber floors are constructed in a similar manner to the ground floor equivalents and so the same rules will apply.

These are summarised as follows.

- Deflection check – the standard heel drop test is appropriate for upper floors. Any excessive deflection may also be evidenced by:

  - Cracking to the ceiling and ceiling/wall junction in the room below;
  - Distortion of partitions supported by the defective floor.

The causes of this could be:

- Inadequate joist size for the span of the floor – this can be checked by using the same approach as described in Section 10.4.2. The depth of the floor joists can be estimated by measuring the depth of the whole floor where visible in the stairwell. Deductions are then made for floorboarding (say 13–19 mm) and the ceiling (allow 20 mm maximum). If a carpet can be lifted and the floorboarding exposed, the spacings of the joists can be estimated by measuring the distance between the nails in the floorboarding. This is not a certain way of assessing the adequacy of the floor. Hidden notching or other constructional faults may still cause problems. Despite this, it can help a surveyor add value to the assessment of condition;
- Lack of strutting at mid-span – most suspended timber floors require some form of strutting at mid-span to stop the joists twisting and to stiffen the floor up. Without lifting several floorboards, it is impossible to determine whether this is present;
- Inadequate end support – joists in older properties were built into solid walls without any sort of restraint or positive fixing. If the wall moves out of plumb (see Section 5.6.1), the joists can be left with little end-bearing. In new properties, where joists have been hung on joist hangers, the end bearing can be badly made, offering little effective support to the joist or wall (BRE 1984);
- Inadequate support around openings in the floor – where staircases and chimneys pass through the floor, a series of trimming joists are often used. They normally consist of wider joists or two standard joists bolted together. These will then connect onto two other trimmers that collectively define the opening. Because the trimmer joists are carrying much higher loads, their design, sizing and jointing is far more critical than for other joists. Landing areas should be closely inspected for signs of deflection to the floor;

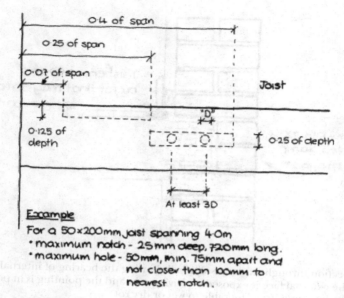

Figure 10.9 Recommended notching sizes and spacings.

- Deflection due to excess notching – many older properties may well be on their second or third 'replumb' and 'rewire'. Each time, contractors may have notched or drilled the joists inappropriately (see Figure 10.9). In some cases, this might lead to deflection, especially where many pipes pass through the floor in a small area (BRE 1987a). Special checks should be made around hot and cold-water tanks, where a high number of pipes, cables and associated notches can be anticipated. Plumbers often leave conveniently loose floorboards in airing cupboards after repair work. Surveyors can lift the boards to get a better appreciation of dimensions and also check timbers around and under these cylinders – a favourite place for wood-rot due to slow leaks from pipe and other joints.

There are numerous other defects that could affect intermediate floors. These are related to problems with other parts of a building, emphasising that any deficiency should be looked at holistically rather than just in isolation:

- **Wood rot to joist ends** – joists built into older solid walls may be as little as 100 mm away from the external face (see Figure 10.10). If the bricks are porous or the pointing is in poor condition, dampness can affect the timber and cause wood rot. Leaks from WCs, baths and washbasins can also damage floors. Key inspection indicators include:

  - Poor pointing and deteriorating masonry externally;
  - Evidence of wood rot to skirtings along with high moisture meter readings;
  - Water stains and/or mould growth to adjacent walls and ceiling surfaces.

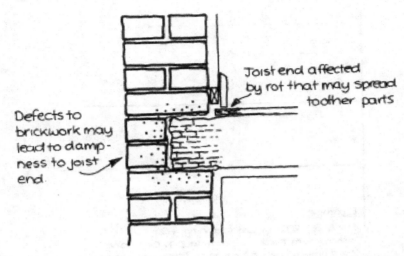

Joist end affected
by rot that may spread
to other parts

Defects to
brickwork may
lead to damp-
ness to joist
end.

Figure 10.10 Section through external solid wall showing the bearing of internal floor joists. If the external face is exposed to driving rain and the pointing is in poor condition, the joist ends are vulnerable to wet or dry rot.

- **Beetle attack** – The effects of wood-boring beetle are discussed in detail in Section 10.4.2. Upper floors can be affected, so floorboarding should be inspected wherever floor coverings, furniture and possessions allow;
- **Lack of lateral restraint** – It is important the floor and the external walls of a building are structurally connected to ensure the wall is laterally stable (see Chapter 5 including Sections 5.3.1 and 5.7). Many older properties will not have this, and the wall may bulge outwards. The presence of strapping cannot be deduced without taking up floorboards, but several key inspection indicators may suggest that lateral instability can be a problem:
  - Gaps between the skirting and floorboarding often masked by timber beading;
  - Cracks at the junction of the affected external wall and the ceiling of the upper floor;
  - presence of tie bars externally can indicate remedial work has been carried out to make a better connection between the external wall and floor, thereby supporting the wall.

- **Defects with panel flooring** – chipboard flooring panels are commonly used in new housing for the floor deck of intermediate floors and can be very satisfactory if the correct type is used. But investigations have shown that this is not always the case (BRE 1983). Using the wrong type in areas of high humidity, not allowing space around the edge for expansion and inadequate nailing, gluing and edge support (noggins) can all lead to problems. Typical indicators can include:
  - Sagging and buckling panels – possibly causing a trip hazard;
  - Loose and squeaking areas of flooring – many clients find this annoying and rectification costs can be high. Some builders now secure boards using

polyurethane foam, which is effective, but render boards very difficult to lift for maintenance or repairs;
- Dramatic loss of strength in wet areas – resulting in collapse in extreme cases.

- **Incorrect thickness of timber boarding for the spacing of the joists** – Although this is not such a serious defect it can give rise to excessive vibrations. The thickness will be difficult to determine without lifting a floorboard or panel, but the BRE (1983) suggest:

  - 16 mm boarding for 450 mm joist centres;
  - 19 mm boarding for up to 600 mm joist centres.

- **Deflection under load from heavy internal partitions** – Upper floors are sometimes uneven where they have bent (deflected) due to the weight of the solid or heavy timber partitions or walls on the upper floor(s) that are not situated over walls beneath. This is typically around airing cupboards or bathrooms where loads are imposed by the mass of water in hot water cylinders and baths (1 m^3 of water = 1 tonne). Nowadays, in modern properties, there are thicker timber, steel or concrete lintels in such locations to prevent such distortion.

The preceding section describes how conventional timber floors can be assessed. However, in recent years, technological advances have led to the introduction of modern methods of construction. In relation to floors, these include:

- 'Webbed' joists, comprising steel lattice bracing supporting longitudinal timbers (flanges) top and bottom, thereby providing 'easy access for the installation and maintenance of the services in the floor zone' (MiTek 2020); and
- 'I-joists' with an oriented strand board web (or OSB for short) and timber flanges top and bottom, all glued/fixed together.

The modern webbed joists with steel lattice bracing should suffer less from carpenters and electricians retro-fitting pipes and cables since the spaces for these services are built-in. However, the modern 'I beam' floor joist can be more vulnerable. This type of joist is precisely engineered and although they should support their design loads, their 'reserves of strength' may be minimal. For example, the notching of the softwood top or bottom flange would be far more serious than with traditional timber joists. Consequently, where more modern properties have used some of these techniques, you should critically evaluate any recent alterations. For example, where a new ensuite shower room has been added, check the stability of the adjacent floor by the heel-drop test and spirit level.

Concrete upper floors – usually in flats

A number of purpose-built residential blocks have concrete 'party' and internal floors constructed using a wide variety of materials and techniques. These may include:

- Reinforced concrete that may incorporate secondary steel beams and/or a variety of hollow clay or concrete pots;
- Pre-cast concrete beams with concrete blocks or panels laid between;
- Pre-stressed concrete structures where the reinforcing wires are put under tension and sealed within the concrete. (Marshall et al. 1998, p. 116).

Because the concrete is usually not exposed to external influence, the likelihood of serious defects is much less than for external concrete (see Section 4.7). Despite this, several problems can occur:

- Stability problems at the end bearing – this is where the floor bears onto the wall. Examples could include:

  - Lateral instability in the wall that reduces the end bearing of the edge of the floor;
  - Differences between the expansion rates of the concrete slab and the wall construction can lead to cracking and spalling of adjacent brickwork.

- Deterioration in the concrete where it extends to the external face of the building (see Section 5.7.9 for more information);
- Problems where the floor extends to form a balcony. Poor design, detailing or material breakdown can result in the failure of the cantilevered balcony. The signs can include cracking to the top side of the balcony adjacent to the external wall;
- Although moisture-related defects are not common, it can affect intermediate floors in several ways including:

  - Through faulty damp-proof course and cavity tray detailing at the junction of the floor, balcony and wall (see Figure 8.25). Water staining on internal surfaces may be an indicator. This problem can be accentuated by condensation in older blocks where a lack of thermal insulation can result in cold bridging;
  - Leaking water and heating pipes, faulty sanitary fittings, and so on that may result in large amounts of water entering the screed above the main slab. This can give rise to staining to the ceilings below. In some cases, 'tide marks' to internal partition walls can occur and resemble typical 'rising dampness' signs. The authors know of defects like this that have led to the injection of a chemical damp-proof course into walls of a 3rd floor flat.

- General structural problems with the floor as a whole that may result in distortion. Although not common, some pre-1975 buildings may have been constructed with high alumina cement concrete (HACC). This was used to develop a high early strength in the concrete, but it often resulted in a dramatic loss of strength shortly afterwards. Stability problems can occur, especially in warm, humid environments where the strength of the concrete can be further reduced because of chemical change. Dramatic collapses in the 1970s (one famous incident involved the roof of a swimming pool) highlighted the danger posed by these types of structures. Based on a visual survey, it is virtually impossible to identify whether HACC has been used. However, the following points could be useful:

  - It was often used in concrete buildings constructed before 1975 so buildings later than this time will not usually be affected;
  - Wide-span pre-cast concrete sections are the most at risk;
  - Warm, humid environments can accelerate the deterioration of the concrete. Therefore, high-risk uses include swimming pools, sports facilities or even where a concrete element regularly gets wet through other defects (e.g., leaking flat roof, faulty water pipes);
  - Deflection in the slab can suggest problems. Signs could include:

- Sloping floors;
- Gaps between the floor and skirtings; and
- Distorted internal door openings and binding doors.

Where these factors are noted, the surveyor should consider advising further investigations. A good 'knowledge of locality' and building types, history and practice will help the surveyor provide the necessary advice.

Thermal insulation in upper floors

You will sometimes encounter upper (usually first) floors of properties exposed beneath to the outside. Typical examples include:

- Pedestrian passageways at ground floor level through old terraces, giving access to rear yards and gardens, with shared bedroom or similar accommodation above; and
- Vehicle accesses on modern developments into parking areas, with similar accommodation over the archway.

Modern Building Regulations now require such a floor should include provision for thermal insulation to help reduce heat loss (and gain). In our experience, properties built before 1966 are unlikely to be insulated and will be vulnerable to condensation. Further information is available in Building Regulation Approved document 'L' (OPSI 2016, 2018).

10.4.4 Ceilings

Ceilings are included here because their condition is linked with that of the floor to which they are attached. Older ceilings can pose many serious problems, as they can collapse suddenly without warning, causing injury to occupants and damage to possessions. The two main types of ceilings are:

- Lath and plaster ceilings – Three coats of lime plaster were applied to a base key of thin strips of softwood or laths nailed to the joists above. The plaster was squeezed between the laths to form a key (see Figure 10.11);
- Plasterboard ceilings – The modern version used on most dwellings built since the Second World War.

Failure of plasterboard ceilings is rare; however, the same cannot be said for lath and plaster. Years of continual vibration and changes in moisture content, temperature, etc., can lead to failure of the plaster key of the ceiling. Whole sections of the ceiling can bulge downwards, sometimes only kept in place by the decorative lining paper. When a ceiling has reached its limit of stability, one more slammed door, child running up the stairs or negative pressure from wind blowing up a chimney can cause dramatic collapse. Failures can be assisted by:

- Water leaks from roofs and sanitary fittings above;
- Pugging or 'deafening' (in Scotland) material placed between joists on top of the ceiling for sound insulation;
- Heavy ornamented centre roses and cornices that add to the weight of the ceiling.

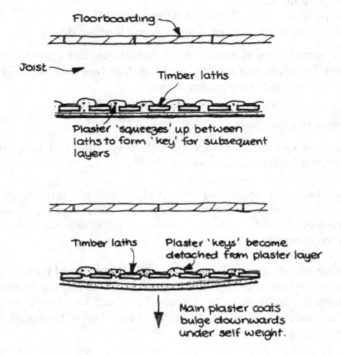

Figure 10.11 Sketch section through a lath and plaster ceiling. The lower drawing shows what can happen when there is a failure of the plaster 'key' to the ceiling.

Key inspection indicators include:

- Large cracks and bulges to the ceiling, sometimes showing through any lining paper (lightly tapping with a long spirit level is very helpful here);
- Water staining to the ceiling;
- Evidence of patch repairs to the ceiling, suggesting problems in the past.

Where these indicators are present, the client should be warned of the danger the ceiling may present and the high cost of replacing it. It is rarely possible to success-fully patch repair a lath and plaster ceiling. If the dwelling is a listed building, the local planning authority might require the owner to apply for listed building consent to replace it.

Plasterboard ceilings are generally more reliable. Some of the common defects include:

- Uneven ceilings, cracking at the plasterboard joints and at junctions with the walls. One of the main causes is where the appropriate noggins (timber bearers to the free edges of the board) have not been included or the plasterboard is too thin for the spacings of its supports (BRE 1986a). If this occurs in the ceilings of the uppermost floor, it might be possible to inspect the support arrangement from above in the loft space;
- Nail heads have popped out where they have not been properly driven home.

These faults may be minor defects, but occasionally, major building movement may also result in similar signs. The surveyor should always look broadly at problems like this.

Other ceiling types

You may also encounter other ceiling types and finishes, including:

- 'Textured coatings' sometimes contains small amounts of asbestos;
- Polystyrene tiles and covings – a fire risk and they also give off poisonous fumes when alight;
- In older buildings, 'reed and plaster' ceilings. Generally, they perform similarly to lath and plaster;
- Metal lath and plaster – used after the second world war and similar to lath and plaster;
- Sheets of mineral fibre-board, sometimes attached under failing lath and plaster ceilings and much used for repair of bomb damage. These are fire hazards and sometimes contain asbestos;
- Sloping ceilings at low level – a hazard for heads; and
- Sheets of asbestos – based boarding. Common in integral garages.

10.4.4 Sound insulation

Building Regulations have required reasonable sound insulation to be incorporated into new dwellings since 1985, with significant improvements to the rules in 1992 and 2004. For example, in 1992, the requirements were extended to include material alterations in dwellings such as flat conversions. Building Regulation Approved Document 'E' includes requirements that separating elements should have reasonable resistance against impact and airborne sound (OPSI 2015a, p. 12). Thus, internal walls and floors within properties (for example, between adjacent rooms vertically and horizontally) and to the internal walls and floors between adjoining properties (for example, between flats and adjoining flats and between flats and common stairwells and hallways) should achieve those requirements.

Sound insulation is not usually a problem in a single-family dwelling. The issues become more critical in flats, either in a purpose-built block or in a conversion scheme or between attached homes. Inadequate sound insulation can be very unpleasant for all concerned and the source of many neighbour disputes.

In refurbishment or alteration schemes, there are several different ways of achieving this. An example of one common method is shown in Figure 10.12 and is used in many conversion schemes. Figure 10.13 shows typical soundproofing to a staircase where it separates two dwellings. Establishing whether any floor has an adequate level of sound insulation is beyond the scope of the standard survey unless floorboards can be lifted, but a few key indicators can help give an insight.

Key inspection indicators

- If the flat has been recently converted, was it granted appropriate building regulations approval?
- At the entrance to the upper flat, is there a timber threshold piece that marks a change in level between the finished floor level of the flat and the common area?

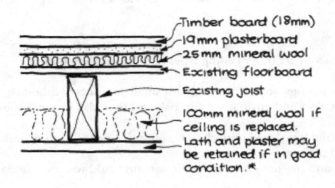

Figure 10.12 Section through an existing timber floor showing retro-fitted sound insulation.

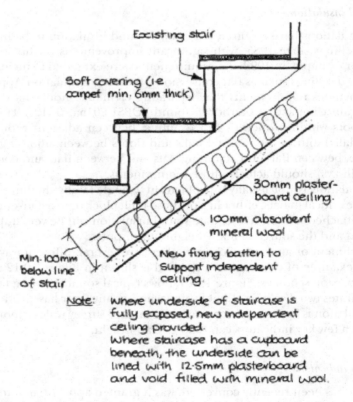

Figure 10.13 Typical soundproofing to the existing staircase between dwellings.

This could suggest that some form of sound insulation has been laid over the floor;

- If a carpet corner can be lifted, can floorboards be seen? If yes, it is unlikely that appropriate sound insulation has been provided;
- During the survey, can sounds be clearly heard from the property above or below? This will depend on whether the dwellings are occupied at the time and on the surveyor's judgement.

Clients should be warned whether it is likely noise insulation is present in the property and that even if it is present, there is a possibility of some nuisance from noise and sound transmission, especially for older conversions.

Sound insulation and historic buildings

In the building regulations, there is a general exemption to these rules for historic buildings. For example, for listed buildings that are refurbished or going through a change of use, the works should not 'prejudice the character of a listed building' and certain sound insulation requirements need not therefore be met (OPSI 2015a, p. 13). The description 'historic buildings' is wider than simply listed buildings and could include those in conservation areas and vernacular buildings of traditional form and construction.

Where nuisance from noise is more likely to be an issue or you have any other concerns (for example, in flat conversions), you may wish to consider recommending the legal adviser seeks positive confirmation that the requirements of Part 'E' have been complied with.

10.4.5 Fire safety and resistance

General comment

The ability of the internal walls, floors and ceilings to resist the passage of fire and smoke may mean the difference between life or death for the occupants. Most standard surveys do not include an assessment of the adequacy of the fire resistance of the property. However, public awareness of fire safety issues has significantly increased, especially following the catastrophic fire at Grenfell Tower in 2017. It was reported in 2006 that 'more than 400 people die each year as a result of accidental fires and more than 11,000 . . . injured' (DCLG 2006). You must therefore make appropriate note of all aspects of construction and safety relating to fire and report accordingly, particularly with buy-to-let property. This is because the client will become a landlord and owe a duty of care to the tenants. See the final part of this section titled 'fire risk assessments and tenanted properties' for further discussion.

Low-rise properties

When looking at this type of accommodation, there are several key features that can help you identify any clear deficiencies. These indicators are a summary of what can be seen and do not necessarily represent a breach of the regulations. A more detailed account of the requirements is contained in Part B of the Building Regulations, with which surveyors should be familiar (OPSI 2019). Requirements can be summarised as requiring:

- There should be a means of detecting and giving warning in the event of a fire in the dwelling. Specifically, this usually means a minimum network of interconnected, mains-powered smoke alarms in all circulation areas that form parts of the escape routes. This system should have additional integral battery backup power, plus a heat detector in the kitchen area (in most cases); and
- All dwellings should have a satisfactory means of escape from and through the dwelling to a safe location outside.

More specifically and in addition, for single-family dwellings:

- In basements, on the ground floor of houses and in bungalows, all 'habitable rooms' (bedrooms and reception rooms) should have:
  - A door opening into a hall with a 'final exit'; or
  - An emergency escape window or door with an unobstructed opening of at least 450 mm x 450 mm, minimum 0.33 m² overall size (for example, 450 mm x 750 mm approx.), the lowest part of the opening maximum 1100 mm above floor level; giving egress to
  - A place free from danger from fire (courtyards or inaccessible back gardens should be deep enough so occupants can avoid heat and flames).
- For houses with a storey less than 4.5 m above ground level (for example, a two-storey dwelling), escape from habitable rooms should be either:
  - Through a door or window as described above; or
  - Down a 'protected stairway' (see section on flats and maisonettes below for definition);
- Houses with a storey more than 4.5 m above ground level (typically a three- or four-storey house) should have a protected stairway down to the final exit, or in the case of a third floor or above, a sprinkler system;
- Loft conversions should be as described in Section 7.6.6.

For most properties, therefore (basements, bungalows or homes with two floors), key inspection indicators will include:

- Is there a compliant fire detection and warning system?
- In habitable rooms, do opening windows or doors comply with size and height described above;
- In the case of two or more storey dwellings:
  - Is there a protected stairway?
  - Is there access to a safe place outside?
- If there is an integral or attached garage, current Building Regulations require there should be 30 minutes fire-resistance to a floor, ceiling, doorway and so on between the garage and living accommodation such as a bedroom above.

Flats and maisonettes

Although the regulations are very complex, the main features should include:

- The principle of 'compartmentation':
  - Each flat should incorporate the required fire resistance (usually min. 30 minutes, but increasing to 120 minutes adjacent to firefighting shafts) in dividing floors and walls (and component parts such as doors) to prevent the spread of fire; and
  - Fire-stopping (examples include mortar, intumescent mastic and glass fibre) is particularly important where service pipes and cables pass between compartments, to prevent gaps fire can pass through;
- Blocks of flats with a floor more than 30 m above ground level should have a sprinkler system, although not in 'fire-sterile' common areas;
- Individual flats should have the same automatic smoke detection and alarm systems required in all other dwellings;
- Emergency lighting on common escape routes – usually corridors and stairwells;
- A building with a storey more than 18 m above access level must have at least one firefighting shaft including a firefighting lift;
- Escape routes should have a minimum clear headroom of 2 m;
- Fire doors with self-closing devices to all internal flat doors and the flat entrance door itself. There should always be two fire doors between an individual room and the common staircase or landing. The doors to any large cupboards or storage areas should also be half-hour fire resisting;
- A protected stairway from the flat door to the main entrance door of the whole building. A 'protected stairway' is defined as a 'fire sterile' area which leads to a place of safety away from the building. The enclosing walls, floors and staircases should be able to resist the passage of fire and smoke for at least 30 minutes. Any cupboards or storage areas off the main staircase should also have a 30-minute fire rating;
- A block of flats can be served by a single staircase providing that the top floor of the building is no more than 11 m above ground level and there are no more than three storeys above the ground storey. Above this level, the regulations become very detailed and specific but usually involve an alternative means of escape;
- Generally increased levels of fire resistance are required where a flat is situated over a shop – at least 60 minutes.

Fire risk assessments and tenanted properties

The Regulatory Reform (Fire Safety) Order 2005 confirms that the 'responsible person' (usually the owner or managing agent) for multi-occupied residential buildings must prepare a written fire assessment in communal areas (DCLG 2005). Such an assessment is therefore required in blocks of flats and houses in multiple occupations. For example, common hallways, stairwells, corridors and shared kitchens. The assessment considers matters such as the likelihood of a fire starting, the necessary precautions to prevent this occurring, fire-fighting equipment and how to ensure the occupants are warned of any fire and can safely escape. Recommendations on how often assessments should be reviewed vary, depending on the risk, but sometimes as often as every year and certainly if material alterations have been carried out.

In single property 'buy-to-let' properties, a responsible person must carry out a visual fire risk assessment (a written assessment is not necessary) and minimum requirements include:

- A smoke alarm on every floor;
- Carbon monoxide alarm in every room with a solid fuel burning appliance;
- A satisfactory means of escape;
- Test certificates for gas (updated annually) and electricity (every 5 years) installations.

Surveyors should flag up any non-compliances or other risks in their report and should, furthermore, ensure the legal adviser asks the vendor to confirm compliance with legislation, including a copy of the assessment where appropriate.

Fire safety law is a continuously changing area of surveying. Alterations in requirements arise from the regular updating of legislation and due to legal precedent. For those same reasons, you must maintain your knowledge of current best practices through your lifelong learning strategy.

It must be stressed that these indicators are a very broad way of giving an initial assessment of fire safety in a building. If the inspection reveals significant hazards or the surveyor is unsure regarding this vitally important issue, the Building Regulations should be consulted in detail or the matter referred to an appropriate professional for further investigation. LACORS (Local Authorities Coordinators of Regulatory Services) has published guidance (2008), which is particularly helpful for surveyors.

10.5 Internal joinery

10.5.1 Staircases

ROSPA reports that 'every year there are approximately 6000 deaths as the result of a home accident . . . falls are the most common accident' in the UK (ROSPA 2020). Many of those deaths occur from falls on steps or staircases.

The staircase in a dwelling should provide a safe and serviceable means of access from one floor to another. Defects are not the only cause of the danger, as the original design can create several other hazards. This is a problem for the surveyor. Many staircases in older dwellings will not conform to the current building regulations, yet they might be safe to use. On the other hand, winding and narrow stairs that lead up to attic bedrooms may be very difficult to use safely. This is an area where you must use sensible and well-balanced judgement.

Design of the staircase

A safe staircase should include the following:

- For a private stair in a dwelling the maximum pitch should be 42 degrees with a maximum rise of 220 mm and a minimum going of 220 mm;

- For common access areas in buildings that contain flats, nosings should be visually contrasting, with no open treads;
- For a staircase in a common hallway of a block of flats the maximum rise is 190 mm and the minimum going is 250 mm;
- There should be at least 2 m headroom above the pitch line of the staircase, except for a loft conversion where an average of 1.9 m is allowable, absolute minimum 1.8 m;
- For a staircase in a single dwelling there should ideally be a minimum width of 900 mm.
- A common stair in a block of flats should be at least 1 m wide;
- If stairs have more than 36 risers in consecutive flights, there should be at least one minimum 30-degree change of direction (for example, a landing). Open treads should overlap at least 16 mm and gaps to the open parts should not exceed 100 mm;
- There should be a landing at the top and bottom of each flight with at least 400 mm of clear space (see Figure 10.14);
- There should be at least one handrail to a flight. For flats without lifts, one on each side of the flights and landings;
- All staircases, landings and balconies should be protected by appropriate guarding, where there is a drop of more than 600 mm (roughly 3 steps). This should be between 900–1000 mm, depending on its location. To protect young children, a 100-mm ball should not be able to pass between the balustrades. This guarding should be robust and able to resist lateral pushing pressure;
- Any tapered treads should conform to Figure 10.15;
- Rails parallel to the landing or stair pitch encourage children to climb, so are not recommended;
- There should be no unevenness in the flight. For example, nosings on treads should be in a continuous plane;
- Steeper 'alternating tread stairs' can be used to provide access to a loft room in certain circumstances but should be treated cautiously.

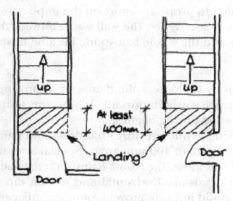

Figure 10.14 Safe arrangements for landings and doors on staircases (Crown copyright. Reproduced with the permission of the Controller of Her Majesty's Stationery Office).

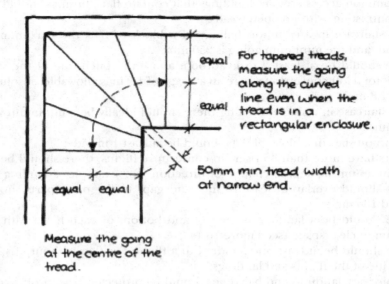

For tapered treads, measure the going along the curved line even when the tread is in a rectangular enclosure.

equal
equal

50mm min tread width at narrow end.

equal   equal

Measure the going at the centre of the tread.

Figure 10.15 Guidelines for assessing suitability of tapered treads on a staircase. It is important to note that these measurements apply to new work. Many existing staircases will not conform to these strict rules. It will be a matter for the surveyor's judgement as to whether the staircase is suitable (Crown copyright. Reproduced with the permission of the Controller of Her Majesty's Stationery Office).

For further information, see Approved document 'K' of the building regulations (OPSI 2013c).

Condition

Many of the defects to a staircase will originate in faults to other parts of the adjacent construction, including the walls and floors. Failure to comment on the condition of staircases has been frowned upon in the courts. In Sneesby and another v. Golding (1995), the surveyor failed to properly report on the implications of removing a supporting partition to a staircase. When the wall was removed, the newel post was left standing independently and the whole balustrade became unstable.

Key inspection indicators:

- Check treads and risers for loose, split, damaged or unsupported timber. Condition of nosings is particularly important. Where the timber is exposed (often beneath), look for wood-boring beetles;
- Carefully test (by using gentle pressure) handrails and balusters, noting missing balusters or those more than 100 mm apart. Special attention should be paid to the newel post. If this is loose, the whole balustrade may be unstable;
- Check whether the treads are horizontal and do not dip from one side to the other. If they do, it could indicate movement in the adjacent construction;
- Where the side strings are attached to the internal faces of external walls or internal walls at low level, test the timber for signs of moisture. Repairing a stair

string properly will often involve rebuilding the entire staircase, an expensive operation;

- If the underside of the staircase is open to view, inspect the following:

  - Check all wedges and blocks are well glued and fixed into position, so treads and risers are secured;
  - Look for wood-boring beetles (common in this location);
  - Check the rear face of the bottom riser and the base of the adjacent strings for wood rot and moisture content. Moisture in adjacent walls and floors often affects stairs here;

- If the underside of the stair has been plastered or boarded, check the soffit for soundness. Where the staircase is in a common hallway serving two or more flats, the underside of the staircase should be fire-protected. Where there is an under-stairs' cupboard, it should be fitted with a fire-resisting door (see Section 9.4.5).

10.5.2 Other joinery items

This will include a wide range of components, including internal doors and linings, skirtings, architraves, window surrounds, dado panels, picture rails, fitted cupboards, etc. Many of these have been mentioned in previous sections when specific defects were discussed. Because most are made of timber, special care should be taken to look for signs of wood rot and beetle, especially on the internal faces of external walls. The level of detail reported will depend on the type of survey, but the following tips might be useful:

- Open and close all internal doors to check for warps, twists, binding and general operation. Check ironmongery to see if it is working properly;
- Check all doors, windows and other openings for squareness. Distortion might suggest structural movement;
- Are there any joinery items that have special aesthetic or historic features? For example, panelled doors or original dado features might not only be attractive but also protected through any 'listing'.

10.5.3 Internal features in listed properties

In listed properties, you should assume joinery and other items have the same special protection as the externally visible roofs, guttering, windows and walls, etc. Internal features might include:

- Staircases including handrails on landings;
- Fitted cupboards, kitchen dressers, wardrobes and bookcases;
- Doors, frames and skirtings;
- Fire surrounds, plaster, cornices; and
- Internal parts of windows such as shutters, box frames, sash cords, weights and sills.

Such features may be considered a fundamental part of the architectural interest and history of the property. Indeed, they may be included in the listing description. Removal or alteration of these elements may significantly affect the character of the listed building. The surveyor's client will assume the liabilities of the previous owner and could be fined or, in extreme circumstances, find themselves in prison for the criminal offence committed by another. The surveyor should consider the following:

Key inspection indicators:

- As part of the desktop study, carefully read the listing and note any specific references to internal joinery (and other items);
- During the inspection, confirm all those items are present and those that have been either been removed nor materially altered;
- Confirm any issues as appropriate for the legal adviser.

10.6 Internal decorations

10.6.1 Introduction

Condition of internal decorations rarely affects the value of a property and will usually have little impact on condition. There are some general aspects to consider when advising on internal decorations:

- Most new owners want to redecorate to their own tastes even if the vendor spruced up the house before it went on the market. This is usually done on a DIY basis and many clients will see this as almost a 'fixed' home move cost;
- Internal decorations are features many people feel able to assess for themselves;
- 'Taste' is a very subjective concept and should be left to the client to judge. The surveyor's role should therefore be focused on the following:

  - What general condition are the decorations in?
  - Are there signs in the decorations that suggest defects in other parts of the fabric?
  - Does the type of decorations affect the safety of the occupants? (For example, are the decorations likely to be flammable?);
  - Will the costs of redecorating be high due to the particular nature of the existing decorations?

These are discussed in more detail in the next section.

10.6.2 General condition of decorations

The quality of the decorative finish mostly depends on:

- The quality of the original materials used; and
- The standard of workmanship including preparation.

Although many clients will be able to make this assessment, an objective view will help them. They will expect to do some redecoration; the key question is how much. The answer may affect their decision to buy. A room-by-room schedule would not be appropriate for most standard surveys. Instead, an overall judgement should be given.

10.6.3 Following the trail

Comments should be made where the decorations have been affected by other building defects, such as moisture staining, mould growth, cracking and so on. These signs are possible 'trails of suspicion' in themselves and the surveyor should make sure they have been properly followed. Cross-reference to other parts of the report might be appropriate.

The surveyor should make special note of circumstances in which the property has been decorated very recently. This can make analysis of cracks and other imperfections in the internal surfaces, such as walls and ceilings, difficult. It might also indicate the vendor has deliberately tried to conceal defects – textured coatings are a favourite method of concealing cracks and moisture.

Eradication of wood-boring insects from painted joinery is very difficult and sometimes impossible due to the presence of paint oils in the wood. Any timbers affected by such infestation may need to be renewed. This may not be possible, for example, in a listed building, so the client should be warned.

10.6.4 Safety implications of decorative finish

Over the years, several different finishes may have been applied that might be potential hazards to occupants. These include:

- Surface finishes that allow excessive fire spread and release toxic gasses when ignited. The most common example would be polystyrene tiles (especially when painted with a gloss finish) but could also include fabric wall hangings, and so on;
- Textured finishes that may contain harmful substances such as asbestos. Older 'Artex' finishes may be particularly problematic;
- Any property built before 1970 is likely to have paint that contains lead, which is hazardous. Sanding down can create dust particles, so clients should be warned of the hazard;
- In blocks of flats, communal areas should have the benefit of an asbestos register, confirming whether asbestos is present.

See Chapter 13 (hazard chapter) for further discussion.

10.6.5 Relative costs of redecoration

Where decorations require renewal, the relative cost will depend largely on the nature of the existing finish. For example, imagine a magnolia and white painted room that is a little grubby. If the new owner wishes to repaint it in similar colours, a quick wash down and one coat of paint may suffice. Alternatively, a poorly decorated room with a heavy lining paper painted in a deep colour will require much more work. The client should be given an impression of what work might be involved. The following may help:

- Dark colours will be far more difficult to cover up with lighter shades, so may take many coats of paint;

- Where wallpaper is removed, the condition of the plaster beneath becomes very important. If the surface of the existing wall shows signs of unevenness, cracks and loose and bulging sections of plaster, these areas are likely to be damaged even more when the wallpaper is removed. A large amount of filling or even plaster repairs might be required;
- Heavily textured surfaces will be very difficult to remove. Artex and sculptured plaster may have been fashionable several years ago, but now most people want to remove them. This can be very costly and time consuming.

These are just a few examples that are commonly encountered. People do weird and wonderful things in the privacy of their own homes. Although it might suit their unique tastes, it could be very off-putting to those with more conventional opinions. One of the most dramatic examples encountered by the authors was when the teenage daughter of the previous owner had created a work of art in the corner of her bedroom. Although the sculpture was very artistic, not everyone appreciates over 300 cola tins stacked and glued from floor to ceiling in a black-painted bedroom.

10.7 Basements, cellars and vaults

10.7.1 Introduction

These are terms for the lowest space in a dwelling. The nature of this space will vary from a simple sub-floor void, a full-height room with natural daylight with potential for conversion to a habitable room, to an entire home, including several bedrooms and all modern facilities. The main issues surveyors must consider arise from the fact that these parts are either totally or partially below ground, hence:

- The walls and floors can face particular problems from lateral forces imposed by retained ground and sometimes very high levels of moisture; and
- There are some specific or enhanced hazards in basements.

These and other issues are now considered.

10.7.2 Strength and stability

Key inspection indicators include:

- Where the basement walls are retaining, is there evidence of horizontal displacement such as bulging or cracking?
- Is there evidence of relatively recent extensions or other buildings close to the basement walls? Loads from new foundations may put pressure on old basement walls. Information from Party Wall etc. Act 1996 notices and supporting drawings may be helpful;
- Do internal load-bearing partitions appear stable and well founded? Do they line up with the internal partitions above? Have they been altered in a way that might affect their stability? Internal partitions at basement level may be taking

a considerable load from above and play an important role in helping external walls resist the thrust of retained ground;

- Is there evidence the lowest floors have been renewed and possibly lowered? Look at the lower part of the existing walls – are there still signs of a former floor level? Renewing and lowering a floor may increase the head height and cut down on dampness levels but may affect the stability of walls by reducing the effective depth of foundations.

10.7.3 Moisture problems, beetle infestation and wood rot

Basements can be poorly ventilated, damp places where conditions are often perfect for wood rot and insect attack. The following checklist might be helpful:

- Is the basement adequately ventilated? Use the rule of thumb described in Section 9.4.2 as an initial guide. Pay special attention to stagnant areas, as many basements are subdivided into small spaces;
- Are there any timbers built into damp walls such as major timber supporting beams, trimmers or joists. Wet rot can cause a slow deterioration and compression of timber, leading to the settlement of any supported structure above. It is important to test these features with the moisture meter and prodding with a sharp implement for signs of rot. Other joinery items can also be vulnerable, such as door frames and linings, cupboards, shelf brackets and fixing plates. Stored timber can also start a dry rot outbreak;
- Is the basement ceiling underdrawn? If yes, what is its condition? Any finish to the underside of the floor above will restrict the inspection, which must be reported to the client. It will also restrict ventilation to the floor structure, increasing the risk of wood rot. It is common practice to recommend underdrawing be removed. This has two main implications:

  - It may affect the fire resistance of the entire building. This is important for flatted accommodation and houses with three storeys or more. If a fire starts in a basement, it can quickly spread upwards and prevent escape for occupants at higher levels. Therefore, in these cases, removal of any underdrawing needs to be carefully considered;
  - The amount of uncomfortable draughts and heat loss may increase with the removal of the under-drawing. It is sensible to advise draughtproofing the floor with hardboard and insulating from beneath.

10.7.4 Moisture in refurbished, extended and waterproofed basements

Introduction

The last 20 years have seen an increase in the conversion of former cellars and inadequate basements to provide additional living space. Creating additional floor space without increasing the footprint of the building can bring great benefits, but if the work is not done properly, then serious issues can result. Tracing and repairing faults with waterproofing systems can be disruptive and very costly.

BSI and PCA guidance

The Property Care Association (PCA) suggests the most important basement guidance is British Standard 8102: 2009, 'Code of practice for protection of below ground structures against water from the ground' (BSI 2009; PCA 2013). This created three different types of waterproofing for rooms below the ground:

- Type A (barrier protection) – Gives protection against water ingress into the basement dependent on a separate barrier system applied to the structure (either on the inside or outside). A cement:sand tanking barrier is most commonly associated with this type. Figure 10.17 shows a different approach where a bituminous membrane is sandwiched between the existing wall and a new concrete inner skin;
- Type B (structurally integrated protection) – The structure itself gives protection against water ingress. Structural concrete walls and floors with an added waterproofing additive are common examples;
- Type C (drained protection) – Protection against water ingress into the useable spaces is provided by an internal water management system. This usually consists of a studded membrane directing water to a series of drainage channels and a sump where an electric pump removes the water.

Older basement conversions were usually waterproofed with cement and sand tanking or a DPM held in position with a masonry inner skin (type A protection). The walls and floors of purpose-built basement dwellings usually exclude moisture themselves (type B). More recent conversions, however (within, say, the last 10–15 years), usually incorporate a type C barrier.

Maintaining type C protection systems

Where type C barriers have been used, the structure itself provides the primary resistance to water penetration. This type of protection should also incorporate a drained cavity and there is a permanent reliance on this cavity to collect groundwater seepage that penetrates through the structure. Drainage channels along the base of the cavity direct the water to a sump for removal by an electric pump to the nearest suitable drain. In this way, type C systems do not provide a 'hydrostatic' barrier against the flow of water but achieve the same effect by means of water management. The internal face of the studded plastic membrane is usually dry lined with foil-backed plasterboard fixed to a metal track system. Although some contractors use timber battens, the timber will always remain susceptible to wood rot and so should not be used.

Effective drainage is an important part of this system, and it can be particularly vulnerable to blockages caused by free lime and other waterborne debris that may get through the structure. These impurities can be deposited behind the membranes, within the drainage channels and sump chamber and around the pump itself. This can cause the pump to break down and the rest of the system to fail, allowing water to get into the basement. To minimise the risk of this happening, the drainage channels and sumps must be cleared out, tested and properly commissioned after completion of the building work. The system should then be inspected and serviced at regular intervals (normally annually), although this could be more frequent in areas where levels of free lime and/or impurities in water are high. Some owners also install backup pumps where the basements are deep and/or water levels high.

Information from the vendor

Where a basement has been converted or extended, the surveyor must ask the vendor (or their agent) several questions:

• What was the date of the work?
• Does the development have planning permission? Where an existing basement is converted, it is unlikely planning permission would have been required. However, permission may be needed in the following situations:

  • Where the new space was created by excavating down to a lower level;
  • The building was listed at the time of conversion;
  • Conversion created a separate unit and/or materially changed the use of the basement;
  • A light well was added that altered the external appearance of the property;
  • Ensure you know the local planning authority's requirements and policies;

• Does the development have building regulations approval covering aspects such as fire escape routes, ventilation, ceiling height, damp-proofing, electrical wiring and water supplies and is there an appropriate final completion certificate?
• Were any necessary Party Wall Act 1996 notices served and an Award prepared?
• Was the work carried out by an appropriately qualified and experienced contractor? A typical example would include organisations that are members of the Property Care Association (PCA) and employ staff with a Certificated Surveyor in Structural Waterproofing (CSSW) qualification;
• Is the work covered by an insurance backed guarantee? A typical insurance policy normally lasts for up to 10 years – establish how many years are left;
• If a type C system, has it been maintained and serviced? If yes, who did this and are there proper records?

Even if you receive satisfactory responses from the vendor (although our experience suggests this is rare), you should still advise your client to refer these matters to the legal adviser, who should confirm their validity and offer formal confirmation and advice.

The surveyor's inspection

Inspection of a waterproofed basement must allow for the fact that the waterproofing is usually under hydrostatic pressure from water. At ground level, although moisture can rise by capillary action, it is always fighting against gravity; however, below ground level moisture can move laterally and hydrostatic pressure increases with depth. Positive water pressure tests the effectiveness of any waterproofing system and the standard of workmanship will be critical in this respect. The following features are most vulnerable:

• The junction between adjacent sheets of the studded membrane material;
• Where pipes, wires and other services pass through the membrane; and
• At the critical junctions in the basement construction such as internal corners, around staircases, within services cupboards and so on.

In high water table areas, failure of the waterproof system can result in water flowing into the basement, so look out for standing water and water staining down walls and across floors. More usually, the tell-tale signs include damp patches, water staining, salt deposits and high moisture meter readings. Water pressure can sometimes push a type A barrier on the inside wall face away from its background, creating uneven and bulged areas. Tapping with the handle of a screwdriver will produce a hollow sound. Ideally, locate the rodding, cleaning points and sump cover. These can give insight into the system's effectiveness – a poorly designed or maintained system will show excessive amounts of lime scale and other debris in the cleaning points, channels, sump and around the pump.

Assessing converted and refurbished basements

In most basements, the waterproofing layer will usually be concealed by the lining material and this will limit assessment. Consequently, surveyors must decide based on what can be seen and deduced. Assessing a concealed basement waterproofing is a complex process. Most surveyors are not structural waterproofing specialists, so they do not have enough technical knowledge to carry out a full appraisal. Therefore, they should decide whether the system is either satisfactory (CR1) or unsatisfactory (CR3) (usually requiring further investigation) using the following protocol described in figure 10.16.

Making decisions about complex parts of a building using simple criteria like these must be done with great care. You will have to make the final professional judgement using the information you have managed to collect. However, using a simple but objective decision-making framework like this will help you produce a balanced report and explain the complex issues to your client.

10.7.5 Hazards specific to basements

Many basements in older dwellings have living accommodation that fails to comply with modern safety standards – rooms have often been converted or adapted many years ago without approvals. Examples include:

- Moisture-related problems are a safety hazard, together with associated issues such as mould;

Basement moisture assessment protocol	
Assessment	Documentary and condition criteria
Satisfactory – CR1	There is clear documentary evidence the system has been properly designed with all necessary approvals, installed, maintained and serviced within the last 12 months. There are no visual indications of a problem and the basement spaces are free from any signs of moisture problems.
Unsatisfactory – CR3	There is no documentary evidence the system was properly designed, installed and/or maintained. And/or there are actual or potential problems and defects. This would typically include evidence of direct moisture penetration, high moisture meter readings, water and crusty salt staining and so on.

Figure 10.16 Basement assessment protocol.

Figure 10.17 A typical type A barrier protection to an existing retaining wall. The original wall surface was levelled up with a cement render and a self-adhesive, bitumen-based membrane applied. A blockwork wall was built up in front of the membrane and the 30 mm gap filled with cement slurry in stages. The vertical membrane was linked to the DPM in the new floor.

- Fire – the surveyor should use the benchmarks in Approved document 'B' to assess basements and hazards should be reported on as appropriate. See Section 9.4.5;
- Radon – there are particular problems relating to radon in basements – see Chapter 11 and Chapter 13 for more information;
- Services – many basements contain gas and electricity meters and installations, often inadequately maintained. However, the presence of these services, if visible, provides an excellent opportunity to inspect the condition of these elements. See Chapter 10 for more details.

References

BRE (1983). 'Suspended timber ground floors: Chipboard flooring – specification'. *Defect Action Sheet 31*. Building Research Establishment, Garston, Watford.

BRE (1984). 'Suspended timber ground floors: Joist hangers in masonry walls – specification'. *Defect Action Sheet 57*. Building Research Establishment, Garston, Watford.

BRE (1986a). 'Plasterboard ceilings for direct decoration: Nogging and fixing – specification'. *Defect Action Sheet 73*. Building Research Establishment, Garston, Watford.

BRE (1986b). 'Suspended timber ground floors: Remedying dampness due to inadequate ventilation'. *Defect Action Sheet 73*. Building Research Establishment, Garston, Watford.

BRE (1987a). 'Suspended timber floors: Notching and drilling of joists'. *Defect Action Sheet 99*. Building Research Establishment, Garston, Watford.

BRE (1987b). 'Chimney stacks: Taking out of service'. *Defect Action Sheet 93*. Building Research Establishment, Garston, Watford.

BRE (1988). 'Solid floors: Water and heating pipes in screeds'. *Defect Action Sheet 120*. Building Research Establishment, Garston, Watford.

BRE (1990). 'Assessment of damage to low rise buildings'. *Digest 251*. Building Research Establishment, Garston, Watford.

BRE (1996). 'Acid resistance of concrete in the ground'. *Digest 363*. Building Research Establishment, Garston, Watford.

BRE (1997a). 'Domestic floors – assessing them for replacement or repair'. *Good Building Guide 28, Part 3*. Building Research Establishment, Garston, Watford.

BRE (1997b). 'Domestic floors – assessing them for replacement or repair: Concrete floors, screeds and finishes'. *Good Building Guide 28, Part 2*. Building Research Establishment, Garston, Watford.

British Standards Institution (BSI) (2009). *BS 8102: 2009, Code of Practice for Protection of Below Ground Structures Against Water from the Ground*. BSI, Davy Avenue, Knowlhill, Milton Keynes MK5 8PP.

Burkinshaw, Ralph and Parrett, Mike (2003). *Diagnosing Damp*. RICS Business Services Ltd, Surveyor Court, Westwood Business Park, Coventry CV4 8JE.

Department for Communities and Local Government (DCLG) (2005). *Regulatory Reform (Fire Safety) Order 2005*. DCLG, London.

Department for Communities and Local Government (DCLG) (2006). *Guidance for Landlords and Property Related Professionals, Housing Health and Safety Rating System (HHSRS)*. DCLG, London.

Douglas, J. (1997). 'The development of ground floor constructions: Part II'. *Structural Survey*, Vol. 15, No. 4, pp. 151–156.

HAPM (1991). *Defect Avoidance Manual – New Build*. Housing Association Property Mutual, London.

Harris, D.J. (1995). 'Moisture beneath suspended timber floors'. *Structural Survey*, Vol. 13, No. 3.

Historic England (2016a). *Energy Efficiency and Historic Buildings – Insulating Suspended Timber Floors, v1.1*. Historic England, London.

Historic England (2016b). *Energy Efficiency and Historic Buildings – Insulating Solid Ground Floors, v1.1*. Historic England, London.

LACORS (2008). *Housing – Fire Safety, Guidance on Fire Safety Provisions for Certain Types of Existing Housing*. Local Government Association, London.

Marshall, D., Worthing, D. and Heath, R. (1998). *Understanding Housing Defects*. Estates Gazette, London.

Melville, I.A. and Gordon, I.A. (1992). *Structural Surveys of Dwelling Houses*. Estates Gazette, London.

MiTek Industries Ltd (2020). *The World of Posi Technology, the Posi-Joist Technical Handbook, Issue 7*. Dudley. Available at https://irp-cdn.multiscreensite.com/b9e44cf6/files/uploaded/Posi%20Joist%20-%20Technical%20Handbook%20%28Issue%207%29.pdf.

Office of Public Sector Information (OPSI) (2013b). Building Regulations approved document 'C'. *Site Preparation and Resistance to Contaminants and Moisture.* OPSI, Richmond, Surrey.

Office of Public Sector Information (OPSI) (2013c). Building Regulations approved document 'K'. *Protection from Falling, Collision and Impact.* OPSI, Richmond, Surrey.

Office of Public Sector Information (OPSI) (2015a). Building Regulations approved document 'E'. *Resistance to the Passage of Sound.* OPSI, Richmond, Surrey.

Office of Public Sector Information (OPSI) (2016). Building Regulations approved document 'L1A'. *Conservation of Fuel and Power in New Dwellings.* OPSI, Richmond, Surrey.

Office of Public Sector Information (OPSI) (2018). Building Regulations approved document 'L1B'. *Conservation of Fuel and Power in Existing Dwellings.* OPSI, Richmond, Surrey.

Office of Public Sector Information (OPSI) (2019). Building Regulations approved document 'B'. *Fire Safety, Volume 1 – Dwellings.* OPSI, Richmond, Surrey.

Property Care Association (PCA) (2013). *Summary of BS 8102:2009 Code of Practice for Protection of Below Ground Structures Against Water from the Ground,* 11 Ramsay Court. Kingfisher Way, Hinchingbrooke Business Park, Huntingdon, Cambridgeshire PE29 6FY.

Ridout, Brian (2000). *Timber Decay in Buildings – The Conservation Approach to Treatment,* By E and F. N. Spon, 11. New Fetter Lane, London EC4P 4EE.

Rock, Ian, (2006). *The Victorian House Manual.* Haynes Publishing, Sparkford, Yeovil, Somerset BA22 7JJ.

RoSPA (Royal Society for the Prevention of Accidents) (2020). Available at www.rospa.com/Home-Safety/Advice/General/Facts-and-Figures.

Smith v. *Bush* (1990). 1 AC 831.

Sneesby and another v. *Goldings* (1995). 36 EGCS 137.

Watt, David S. (2007). *Building Pathology – Principles and Practice,* 2nd edition. Blackwell Publishing Ltd, 9600 Garsington Road, Oxford OX4 2DQ.

11 External and environmental issues

Contents

DOI: 10.1201/9781003253105-11

11.1 Introduction

This chapter looks at those features on or around the property. Some of these may not directly affect value, but most will be important to the client. A word of warning – it is common for many of these elements to be inspected last when surveyors are often tired, lacking concentration and considering their next appointment rather than carefully evaluating external and environmental issues.

11.2 The site

11.2.1 Garages

The most important outbuilding on most properties is usually the garage. This has a direct impact on the value of the property, so its condition is important. The type of construction will vary and could include:

- A substantially built structure, detached or attached, that uses the same constructional methods as the main house. This may include 225 mm brick walls, pitched and tiled roof with well-designed and manufactured doors and windows;
- An integral garage that is incorporated into the footprint of the main property and built using the same constructional methods;
- An old sectional building made up of prefabricated metal and asbestos-cement panels. Timber components are also common. Neglect over the years may have resulted in a poorly maintained structure that is close to collapse. If asbestos has been used, many of the components may pose a real danger to the occupants and so might need expensive removal;
- A garage in a terrace of similar structures, with brick and concrete block walls, concrete floor, timber roof structure and flat felt covering. These can be close to the property or some distance away. Legal issues might include party walls, shared access, parking and repairing liabilities.

Garages may have developed other uses over time, containing animals, bird sanctuaries, DIY workshops, storage areas and even small home offices. They can present a variety of problems, including unhealthy conditions, places where potentially explosive or hazardous materials are kept, and serious breaches of planning regulations. The importance of correct reporting on garages was illustrated in Allen and another v. Ellis and Co (1990) where a surveyor failed to warn his client the garage had an asbestos roof. After purchase, the owner climbed on the roof to carry out repairs, fell through and was injured. The judge felt the surveyor should have warned the client that walking on asbestos could be dangerous.

Key inspection indicators include:

- Can the structure described by the estate agent as a 'garage' be used as such?

    - Older drives are sometimes not wide enough to give access for modern vehicles, for example, between the house and the boundary. In this case, the 'garage' may only be capable of being a store; and
    - Many garages are too small for a modern car and that does include some recently constructed garages;

- Ease of operation of vehicle door – some up-and-over types are prone to buckling. Electrically powered doors are often short-lived and have to be lifted with a variety of safety features. In these cases, it is important to check whether the mechanism has been regularly checked and maintained by an appropriately qualified person;
- Many pitched roof garage structures suffer from:

  - Lack of adequate triangulation (see Chapter 8), with consequent bulging of walls;
  - Wood-boring insect attacks, as the timbers are often exposed to the open air (often a good indicator the main property has a similar infestation);

- Flat roofs – inspect carefully above and below and use moisture meter for signs of water ingress or wood rot;
- Inspection pits – are they safely covered or full of water? You should warn the client of the potential safety hazard;
- Boilers of all types and other heat producing appliances (gas, oil and solid fuel) may be subject to special regulations and so you should checked/serviced by an appropriately qualified person (see volume 3 for further detail);
- The same applies to electrical installations. Does this part of the system match with current standards? The key is how the system is linked to the main house electrical system and should be protected by a residual current device and appropriate protection to any externally run cable;
- Does the garage have necessary planning and building regulation approvals and, or has there been a change of use?
- For an attached or integral garage, is there satisfactory fire separation between the garage and the home? This is especially important for the ceiling, dividing walls and doors;
- Many garages are attached to other garages and there will be one or more party walls, which should be identified and reported to the legal adviser;
- If there is no garage, is there space for one? The lack of vehicle parking for a property can significantly affect its value and saleability.

11.2.2 Outbuildings

Other outbuildings and structures in the grounds of residential dwellings are much more variable than most garages and they are seldom maintained with the same degree of care as the main building (Noy 1995, p. 299). Consider the following range:

- Outside toilets;
- Greenhouses and sheds;
- Summerhouses and 'follies';
- DIY-built tool sheds and workshops;
- Old air-raid shelters;
- Animal cages and runs;
- Large timber-framed barns and many more.

Although outbuildings are unlikely to affect a purchasing decision, their condition can impose significant liabilities on an owner. Outbuildings are prone to rapid deterioration, as few regulations control their design and construction. Most owners spend little time and money on maintenance.

If the structures are unstable or contain hazardous materials, they could give rise to claims from neighbouring owners or third parties. If a client suffers injury after moving into the property, the surveyor's PII policy may suffer. Therefore, the outbuildings should be clearly described, carefully inspected and their condition assessed. Clear warnings should be given if the structures pose a danger to users. Outbuildings typically present safety hazards, including:

- Non-safety glass in doors, windows and especially greenhouses;
- Asbestos materials in roofs and cladding; and
- Inadequately maintained electricity installations under leaky roofs.

Clients should be told that repairing or removing dilapidated or hazardous outbuildings can be very expensive, especially if they contain asbestos.

The extent of inspection will depend on the type of survey, but take particular care, even in simple sheds. This is because experience shows that this is where the vendor may store what could be described as incriminating evidence. Over the years, the following items have prompted successful 'following the trail' processes in other parts of the property:

- Tubs of rodent poison;
- Bottles of toilet and drain blockage remover;
- Tins of sealant for preventing water penetration through walls and roofs;
- Containers with rot and 'woodworm' treatments;
- An entire shelf of weed-killer (for amateur knotweed treatment remover);
- Rolls of roofing felt and 'flashband';
- Drain rods (well used).

11.2.3 Retaining walls

Many properties have retaining walls within or forming their boundaries. Whether a small, informal rubble wall holding back a flower bed or a tall, properly engineered retaining structure, it needs to be very carefully considered. This is particularly important, where the retaining wall also marks the boundary line. Neighbouring properties may have acquired a right of support and should the wall collapse, the owner of the subject property may be liable for damages to their neighbour as well as for rebuilding a very expensive feature. A retaining structure must resist very high lateral loading from the soil and water behind it. A rule of thumb for brick retaining walls is that the thickness at the base should be one third of its height (see Figure 11.1). As retaining walls often reduce in thickness, the width of the wall at the top need not be the same as the width at its base. A well-designed wall should be drained through the wall or by a land drain at its base to prevent excessive loading during wet periods.

Key inspection indicators include:

- Are there any cracks? Is the wall vertical? Is there any bulge? If so, these can be assessed using the middle third rule outlined in Chapter 5;

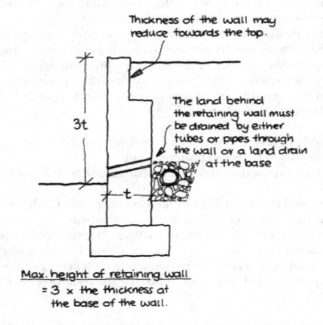

Thickness of the wall may reduce towards the top.

The land behind the retaining wall must be drained by either tubes or pipes through the wall or a land drain at the base

3t

t

<u>Max. height of retaining wall</u>
= 3 × the thickness at
the base of the wall.

Figure 11.1 Sketch section through a typical retaining wall.

- Condition of the wall such as the masonry, copings, pointing and bedding mortar. Are the bricks or stone breaking down through frost or chemical action and is there a good coping with DPC to provide protection? Such defects can progressively weaken the wall;
- Does the wall appear well built using the same materials, or is there evidence the wall has been constructed using different methods at various times? This could suggest poor building practices or suggest the land has been further added to since the original retaining wall was constructed;
- Is there any provision for drainage through the wall? Reasonably sized drainage pipes at 1.5–2 m spacings would be appropriate. If they are present, is there evidence that they are functioning properly? For example, is there any staining or damp patch below them suggesting a flow of water? If these drainage points do not relieve the build-up of groundwater behind the wall, it could cause collapse. Some retaining walls may be drained by a hidden land drain, so absence of drainage holes should not be a definitive indicator;
- Is there a building on the retained side close to the wall? Providing this is at least the height of the building's wall away from the retaining wall, this should be satisfactory (see Figure 11.2). If closer, it may be imposing an excessive load on the retaining wall and further advice may be needed. The same principle applies to imposed loads such as piles of bricks, soil or other heavy materials on the retained side. If the neighbouring land can be inspected, confirm any large ponds, streams or other water features. They may be the cause of potential problems – see Rylands v. Fletcher (1868) in Chapter Volume 1, Chapter 5.3.

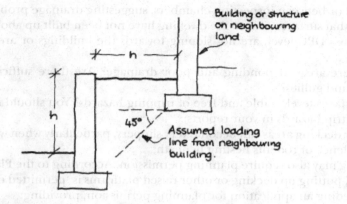

Figure 11.2 Relationship between imposed loads and retaining walls.

If there are any such concerns, further investigation by a structural engineer or other professional with geotechnical experience might be appropriate.

11.2.4 Drives, paths, patios, decking and steps

The vehicle drives on a property can have a considerable impact on its usability. If there is a long distance from the building to the entrance, it is important that the drive is safe to use by occupants and delivery and service vehicles that are usually much bigger than normal cars. Paths can allow good access to the garden, and patios can be delightful features adding to the quality of the accommodation if well designed and built, but both can be dangerous if uneven. Steps must be stable and safe to use. The range of materials used for paths and drives include (Noy 1995, p. 290):

- Flexible paving such as tarmacadam and other bitumen-based materials;
- Concrete paving and paviors;
- Tiles and setts;
- Gravel and hogging;
- Timber decking.

In all cases, these features should be described, their condition assessed, hazards pointed out and legal matters flagged up.

Key inspection indicators include:

- Are the drives, paths and other hard surface areas in good condition with a sound, even and durable finish? Are there excessive numbers of weeds pushing up through the path material? This could indicate poor workmanship and inadequate thickness of the path itself;
- Are there any signs to suggest the drives or paths are shared by neighbours? For example, an unadopted road with significant maintenance liabilities;
- Is there root damage from nearby trees or shrubs?

- Is there evidence of cracks or areas of settlement to surfaces? For example, around or between inspection chambers, suggesting drainage problems;
- Check that surfaces next to the dwelling have not been built up above the subject property's DPC level, are not sloping towards the buildings or are blocking air bricks.
- Are there areas of ponding and poor drainage? Are there sufficient drainage points and gullies?
- Are surfaces level, stable and free of tripping hazards? You should include warnings of trip hazards in your reports;
- Timber decking areas can become very slippery, particularly when wet; also check for evidence of rodents nesting beneath;
- Decking may also require planning permission. According to the Planning Portal (2022), putting up decking or other raised platforms is 'permitted development', not needing an application for planning permission, providing:

  - The decking is no more than 30 cm above the ground;
  - Together with other extensions, outbuildings and so on, the decking or platforms cover no more than 50% of the garden area;
  - None of the decking or platform is on land forward of a wall forming the principal elevation.

  Anything different to this is likely to require planning permission. These regulations will be more restrictive in national parks, areas of outstanding natural beauty, conservation areas and within the grounds of a listed building.

- In localities where there are significant slopes, some drives and paths may be very difficult to use in icy conditions;
- Other hazards include:

  - Poor vision splay where the vehicle exits onto the highway;
  - Having to reverse onto the highway because there is no turning circle on the property; and
  - Significant drops (more than 600 mm) on steps and at unguarded edges.

11.2.5 Trees and other vegetation

The effect of trees on buildings has been covered in Section 5.6 and will not be repeated here. With respect to large trees that are in the grounds and are not within influencing distance of the building, they should be reported and the variety identified if possible. Although they might not undermine the foundations, trees can still be a liability – the presence of any trees or shrubs should prompt a warning to the client that future maintenance (pruning and so on) will be required.

Key inspection indicators:

- Are the trees mature and/or in poor condition?
- Are there overhanging branches in danger of collapsing?
- Are there trees close to boundary walls and likely to undermine any walls?
- Are trees:

  - Within influencing distance of buildings on neighbouring properties?
  - Blocking light from the property or neighbouring buildings?

- Close to drain runs where the roots might cause damage?
- Possibly protected, such as with Tree Preservation Orders?

- Are there trees on adjoining properties that might affect the subject property?

Any recently planted trees should also be noted. A row of 30 leylandii may seem to solve a privacy problem when they are 1.75 m tall, but because they have a mature height of some 20 m, they can soon become a maintenance issue and/or the source of a 'right of light' neighbour dispute.

11.2.6 Japanese knotweed

The first edition of this book did not include the term 'Japanese knotweed'. Although it was as widespread as it is now, in 2001, it simply had not affected the residential market in the same way. At the time of writing this second edition, it is a different story.

Scope of this section

We do not want to recreate material that has been more than adequately covered in other excellent publications (for example, RICS 2012 and the Property Care Association 2021 in particular). Instead, we want to focus on the issues that are most important to residential practitioners who provide condition reports for their clients. These topics include:

- How to identify Japanese knotweed;
- How desktop research can help identify the risk of Japanese knotweed in the vicinity;
- The scope and extent of the inspection of the grounds;
- How to assess the scale of any Japanese knotweed infestation; and
- What to tell the client.

Please note: at the time of writing (November 2021), RICS were due to publish their revised guidance note for inspecting and reporting on Japanese knotweed.
   We will go through each of these headings.

How to identify Japanese knotweed

While residential practitioners are not expected to have a specialist knowledge of Japanese knotweed, we should be able to identify the plant where it is clearly visible and is displaying the typical visual characteristics. In general, the appearance of Japanese knotweed will vary through the seasons:
   Early in the growing season (March to April): Small red/purple shoots or 'spears' begin to emerge from the ground. These can be seen growing amongst the tall stems of the previous year's growth. The leaves begin to unfurl and can be red or green with red veins (see Figure 11.3).
   Main growing season (April to September): As the plant develops, the leaves develop into large shield or heart shapes with a flattened base and have a distinctive lush green colour. To begin with, the stems are initially green but develop distinctive

Figure 11.3 Japanese knotweed early in the growing season.

purple speckles as the growing season progresses. The stems are hollow and never woody during the growing season and this helps distinguish knotweed from many other types of plant.

Where the plant is growing uninhibited, it will form dense and tightly packed clumps or 'stands'. The overall form of these can be very distinctive (see Figure 11.4).

In our opinion, the most distinctive feature of Japanese knotweed is the way the leaves grow on alternate sides of the leaf stem, forming a clear 'zig zag' pattern (see Figure 11.5).

Figure 11.4 Stands of Japanese knotweed in the mid-growing season.

Figure 11.5 Japanese knotweed leaves in the mid-growing season.

End of the growing season (September/October). Towards the end of the growing season, spikes of small creamy white flowers appear. The length of the spikes is usually 100 mm (Figure 11.6).

Winter (November through to spring): In the winter, the leaves fall off as the plant turns orange and brown in the early part of winter. The stems eventually peel, leaving them a pale 'straw' colour. Although the stems are hollow, they can also be resilient. It is normal for the 'stands' to be standing the following spring. During the winter, remnants of the flowers can often be seen on the stems (see Figures 11.7).

Figure 11.6 Japanese knotweed late in the growing season, showing spikes of creamy white flowers.

Figure 11.7 Japanese knotweed in the winter.

If you have not encountered Japanese knotweed before, then it is important you aim to improve your identification skills. This section will act as a good primer, but you should supplement this with additional study and research. The information and guidance provided by the invasive team at the Property Care Association will help deepen your knowledge.

How desktop research can help identifying the risk of Japanese knotweed in the vicinity

The Home Survey Standards (RICS 2019) states that all RICS members must be familiar with the type of property to be inspected '. . . and the area in which it is situated'.

This is taken to include general environmental issues where the '. . . information is freely available to the public (usually online)'. Although we consider this knowledge of locality in Chapter 3 it will be useful to highlight the type of features that may put you on notice that the risk of Japanese knotweed is high.

The Property Care Association focus on this process and points out that JK is spread almost exclusively by soil movement (PCA 2021). This could be associated with features such as:

- Water courses;
- Public rights of way,
- Railways, trams and the underground;
- Roads;
- Public car parks; and
- Commercial/industrial land especially derelict sites that are prone to fly tipping.

The PCA also suggests that burrowing animals may also be significant vectors, as rhizomes could be redistributed around fox and badger sets, for example.

Consequently, when carrying out your desktop research, you should be alert to these features that can often be identified on:

- Satellite and mapping websites and apps;
- Google Street View can provide historic photographs that in some cases, go back to 2008–2009. We have discovered a number of knotweed stands on older images that have been concealed by current owners;
- Historic maps – the presence of former quarries, railway sidings and the like are important clues;
- In some areas where Knotweed infestations are common, local authorities may have useful information.

Over the last few years, online map resources have been developed, and although some may give a rough indication of the local frequency of Japanese knotweed in an area, their accuracy is often untested and often not updated, so use this information with particular care.

You should clearly note any features of concern on your sites notes/property file so that the information can inform your inspection.

Where relevant and practical, the owner and/or seller or their agent should be asked whether the property or any neighbouring properties have been affected by Japanese knotweed and, if applicable, for details of any Japanese knotweed management plan or guarantees/warranties.

The scope and extent of the inspection of the grounds

The scope and extent of a typical inspection at the different levels of service are described in Appendix B of the Home Survey Standard (RICS 2019). As described in the introduction to that Appendix, the document does not provide a comprehensive listing of what is or what is not inspected. Instead, it provides critical benchmarks '. . . around which an RICS member's service can be built'.

In relation to the inspection of the grounds, although benchmarks are always open to interpretation, the pertinent parts of the description of the different levels are as follows:

- **Level one:** involves '. . . a cursory inspection . . .'
- **Level two:** at this level '. . . a thorough inspection . . .' should be carried out.
- **Level three:** '. . . a comprehensive inspection . . .' would be appropriate.

Clearly, such broad descriptions are open to interpretation, but in our view, inspections for Japanese knotweed at the different levels should include:

For all levels of inspection

The following is applicable:

- Visual inspection of the grounds from within the boundaries of the subject property and where necessary, from the adjoining public property. In our view, 'public property' is taken to mean public footpaths, passageways, parks, recreation areas and so on to which members of the public have access. These spaces must also share a boundary with the subject property;
- Visual inspection of the grounds from an upper storey window, balcony or other part of the subject property. Such a view may also allow an inspection of the neighbouring property – important when Japanese knotweed is straddling a boundary.

Additionally, as described in the new RICS Guidance Note (due to be published in late 2021 or early 2022), the inspection is '. . . not a plant-by-plant check for Japanese knotweed'.

For level one services

The inspection is described as a '. . . cursory inspection of the grounds during a general walk around' and in keeping with the spirit of HSS at this level, the inspection is less than for levels two and three. It would typically include walking around the grounds using formal access features such as patios, paths, driveways and so on. Although you should endeavour to view all boundaries, this will often be from a distance and maybe restricted by heavy planting.

Where the property is relatively small and the gardens compact, it will be easy to inspect the grounds, even within the level one definition. It is with larger properties where the grounds are more extensive where you will have to take some professional judgements.

For level two services

The HSS benchmark changes to a 'thorough inspection of the grounds'. In our view, this involves a more extensive inspection using both formal access features as well as easy-to-access areas such as lawns, rough meadows, open borders and the like. At this level, you should anticipate viewing the majority of each boundary, but because level two services provide '. . . a professional opinion at an economic price . . .', there is a limit. For example, we don't think you should be pushing into heavily planted borders where foliage has to be parted to provide access.

For level three services

For level three services, RICS use terms like 'detailed assessment' of the property and the inspection is '. . . more extensive than a level two service'. Consequently, at this level, the inspection should indeed be 'comprehensive' and include all features described for a level two service. In addition, you should consider more intrusive inspections too. For example, where a large planted area could be more effectively inspected by gently parting some of the foliage, then we think you should do that even if it means stepping off a path or lawn and getting your feet muddy. In these situations, you should ask the vendor for permission and make sure it is safe to do so.

Following the trail during the inspections

The concept of 'following the trail' has been discussed in Volume 1 and Chapter 3 and in relation to JK, we think it is important to provide a little more context. The start of any trail is the visual trigger and would typically include:

- JK leaves/stems growing amongst or protruding above established shrubbery/ planting on the subject property;
- JK showing above the other side of a boundary fence/hedge suggesting it is growing on a neighbouring property; and
- Recently replanted/landscaped areas or other commonly used concealment methods such as decking, paving, tarmacadam and so on.

In these circumstances, the extent to which you follow the trail will depend on the level of service provided:

- At level one you should follow the trail sufficiently to confirm that JK is present on or near the property so a recommendation for further investigation can be included in the report;
- At level two not only should you seek to confirm the presence of the plant but you should also make a broad assessment of its extent. Although the primary advice would be a recommendation for a further investigation by an appropriate specialist, you will also be able to give your client an outline of the scale of the problem;
- At level three you will still be recommending a further investigation but also be making a more detailed record of the knotweed's location and extent so a more detailed explanation of the likely extent of remedial work can be included in the report.

Site notes and JK

Although many practitioners make an adequate record of the subject property during the inspection (including written notes, plans and photographs), they often fail to adopt a similar approach when recording the inspection the grounds. Consequently, site notes for a typical level two inspection might include:

- A sketch of the property in its grounds showing the essential features such as boundaries, block plans of the property and outbuildings and other notable features such as formal paths, drives, retaining walls, major trees along with the location and extent of any Japanese knotweed;
- Where JK is identified, you should note the following:

  - Proximity to built structures, hard landscaping, underground drain runs together with a description of any damage and disruption;
  - Location, approximate height and area of all stands of JK; and
  - Where the infestation is likely/has already crossed boundaries.

- A satisfactory range of photos/videos that record what was visible at the time of inspection. This is particularly important with JK because of its rapid growth rate. If JK is not picked up on an inspection either because of concealment or a practitioner's omission, a revisit a few months later can reveal verdant green stands of the plant where none were visible just a few months before. Without an adequate photographic record, there is very little defence.

Concealment

Due to its notoriety, property owners will often attempt to conceal the presence of JK. Although practitioners do not yet have x-ray vision, you should still look out for those signs that could indicate infestations may have been concealed, especially in areas you know are at risk of JK. Typical signs include:

- Recently turfed areas;
- Parts of the garden covered with landscape fabric and covered with gravel;
- New areas of decking and patio paving and so on.

If you notice any signs of potential concealment, you should specifically ask the vendor if the property has been affected by JK, recommend the legal adviser also ask similar questions during the conveyancing process and report the likely risks to the client in a balanced way.

Inspecting large properties

Where larger properties have extensive grounds, it can often take a considerable amount of time to inspect the grounds to the extent described above. To properly manage this issue, you may want to consider the following:

- During the initial stages of the instruction process, you should be establishing the nature of the property so you can draw up suitable terms of engagement that match the nature of the property and the client's requirements. This is usually achieved by studying the agent's details, the potential client's description of the property and/or a quick look on Google satellite/Street View. This will reveal the relative size of the property and its plot;

- If the grounds are considered extensive then you should discuss with the client the possible implications. There are usually two options:

  - You negotiate an additional fee for the time involved in a more thorough inspection; or
  - You limit the extent of the inspection. In this case, it is important you explain to the client the risks associated with this approach.

Discussing these matters early in the process can help avoid later misunderstandings.

11.2.7 Boundaries

Introduction

Establishing the ownership of the boundaries of the property is not the duty of the surveyor. The client's legal adviser should be able to establish this. However, the surveyor should inspect and report on apparent legal discrepancies and on the nature and condition of the boundaries in order to:

- Identify any encroachments on the land;
- Help the client assess financial implications;
- Highlight any current or future neighbour disputes.

More information on the legal aspects of boundaries can be found in Volume 1, Section 5.1.6.

As Anstey (1990, p. 22) states, 'there is an awful lot of trouble to be found in the suburban back garden'. Disputes with neighbours over the position of fences, condition of walls, trespass of dogs, cats, chickens, sheep and even children because of inadequate fences have all been well documented. The surveyor must be the eyes and ears of the client's legal advisor, highlighting specific concerns that need to be checked further.

Ownership

Any surveyor experienced in boundary disputes will testify that there are no clear rules in deciding the ownership of a boundary from visual evidence. Over the years, pragmatic repairs by a frustrated owner may contradict the information shown in the deeds. Despite this, Noy (1995, p. 294) has attempted to identify some general rules (see Figure 11.3):

- Garden walls

  - Buttress or pier on one side only – the wall belongs to the owner with the buttresses;
  - Symmetrical piers – boundary runs down the middle – shared ownership;
  - Plain brick boundary walls – boundary runs down the middle – shared ownership.

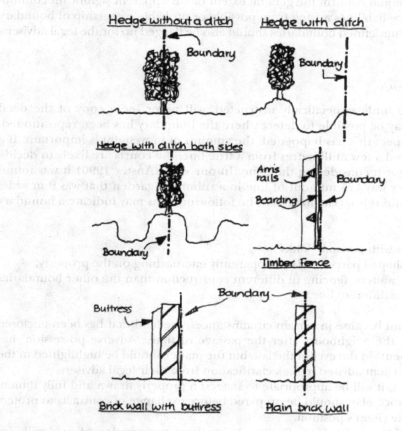

Figure 11.8 Typical boundary positions.

- Fences
  - Posts, arris rails and boarding – the latter usually faces the person that owns the fence;
  - Post and wire fence – usually constructed on the boundary itself so no presumptions can be made about ownership, also applies to modern ranch fencing, concrete panels dropped into concrete posts and so on;
- Hedges and ditches
  - Ditch with bank and possibly hedge on the top – boundary on the far edge of the ditch;
  - Ditch without hedge – no presumption can be made; likely that boundary is in the middle, although look for remnants of the bank which could indicate ownership;
  - Hedge – usual for ordnance survey to assume ownership runs down the middle.

The surveyor should confirm the general extent of any apparent significant continuing maintenance liabilities arising from possible or probable ownership of boundary structures. Any undefined boundaries should also be flagged up for the legal adviser's attention.

Position of the fences

Most surveyors, unless specifically instructed, will never see a copy of the deed plans, but it may be possible to detect where the boundary has been repositioned. Where you suspect this has happened, the amount of movement is important. If a fence has strayed a few millimetres from a true line, the courts are likely to decide that it is not worth considering the issue. In one case (Anstey 1990) it was found that a boundary was 175 mm out of line in a suburban garden that was 9 m wide. The courts found this was actionable. The following signs may indicate a boundary has 'moved':

- Boundaries with unusual 'kinks';
- Strangely shaped parcels of land apparently encroaching on the property;
- Lengths of walls or fencing of different construction than the other boundaries and follow a different line.

This is important because in certain circumstances, the land that has been enclosed might become the neighbours after the passage of time. 'Adverse possession' has strict requirements in the eyes of the law, but the matter should be highlighted in the report and the client advised to seek clarification from their legal advisers.

In some cases, it will be appropriate to suggest a properly drawn and fully dimensioned conveyance plan should be prepared before exchange of contracts to protect and confirm the client's position.

The Party Wall etc. Act 1996, Section 1, regulates construction of new walls (of buildings and garden walls) on a 'line of junction' or, in other words, on or astride a boundary. If there is evidence of such work on the property or adjoining properties, the legal adviser should be notified in the report.

Condition of boundaries

If the boundaries are at the end of their useful life, replacement can be expensive, since they can be very long. Holes and gaps allow animals to get between properties. If parts of the boundary wall are unstable, they might fall on people or property, exposing the owner to legal claims, especially if the boundary faces a public right of way. The surveyor must inspect and report on their condition as carefully as for the main property. BRE has published an excellently illustrated guide to condition of boundary walls (BRE 1992).

Key inspection indicators:

- Brick and stone walls

    - Are the walls vertical or leaning? Is the wall sufficiently thick for its height and reinforced with buttresses?

- Is there any evidence of cracking? Boundary walls often have shallow foundations susceptible to subsidence. Many do not have expansion joints, so thermal and moisture movement can be a problem;
- Are the copings loose and liable to become dislodged?
- What condition is the masonry and the pointing and bedding mortar? Over
  the years, these can be eroded by the elements leading to further accelerated
  deterioration. Sulphate attack can be common;
- In old rubble stone walls, has the inner core begun to come apart and slump
  so the wall is unstable? Look out for bulging to either side of the wall;
- Is there a DPC below the coping and towards the bottom? (However, this is
  rare in our experience.)

- Timber fences

  - What type of posts support the fence? Concrete posts generally last much
    longer than timber equivalents that can rot. Are the posts stable? Is there
    any evidence of remedial supports being fixed at the bottom of the posts
    that would suggest past failures? (For example, iron angles, spur posts and so
    on.);
  - Is the fencing of good quality and properly fixed? Some fences can be an
    inadequate 'DIY' hotchpotch of panels and timber.

- Other fences

  - Concrete fence posts and panels – still a common fencing method. Panels or
    planks slot into rebates in the side of the posts. These can be very robust but
    can become very unstable if the posts have not been set deep enough into
    the ground. Older installations can suffer from spalling of the concrete cover
    caused by corrosion of reinforcement;
  - Post and wire fence – this can include steel angle or concrete posts with wires
    stretched between, sometimes clad with plastic. If the straining wires corrode
    or are broken, fencing can easily distort and be vandalised.

11.2.8 Rights of way

As described in Section 5.1.2 of Volume 1, as with adverse possession, a right of way
can be acquired over someone else's land through regular and continued use. This
tends to be a problem associated with larger properties, but rights of way issues have
been known to affect the sale of 'normal' properties.

Key inspection indicators:

- A well-trodden path or drive across the property not leading to the property itself;
- Stiles or even gates in the boundary walls or fences;
- Evidence of a footpath on an ordnance survey map of the area.

If another party appears to be using or passing through the property, the client must
be notified so the deeds can be checked by the legal adviser. A right of way may affect
the enjoyment of the property, in some cases affect its value and be very difficult to
'extinguish'. Similarly, if there appears to be a right of way or similar in favour of

the subject property over an adjoining property, this should be included in the legal adviser's section of the report.

11.2.9 Highways

How both vehicles and pedestrians get access to the property is very important, as it can have significant financial implications for the client. Before we go further, it will be useful to briefly outline the legal status of the different types of roads that can serve a property:

- **Adopted road or highway** – this is usually a road that is managed and maintained by the highway authority. In most areas, this will be the local authority;
- **Unadopted road** – this usually refers to roads which do not have to be maintained by the Highway Authority under the Highways Act 1980. A legal duty to maintain these roads still exists, but it falls to the owners of the road. This usually consists of the owners of properties fronting the road but may also include those where the side or length of their property fronts the unadopted road. In most cases, the public have a right of access over unadopted roads;
- **Unadopted estate roads** – although these are not strictly another legal category of a road/highway, we think it is important to mention them. Over the last few years, an increasing number of housing developments/estates have not had their roads adopted by the local authority. Instead, the developers have included a range of charges on the freeholders and/or leaseholders (see explanation ground rents section below);
- **Private road** – this is usually a road or driveway that is closed to the general public. It is on a privately owned property and can only be used by the owner (or owners) or those with specific permission. The owner (or owners) is responsible for the maintenance of the private road.

Although it will be the legal adviser's responsibility to establish the legal status of the road, in your role as the 'eyes and ears' of the legal adviser, the following indicators may give some important clues:

- Adopted roads tend to conform to the local authority 'design standard' for the local area including road markings, curbs/dropped curbs, road signs and road lighting;
- Unadopted roads may not conform to this 'design standard' as the various owners may have carried out piecemeal repairs and improvements over the years. Although standards of maintenance can vary between both publicly and privately owned roads, it is our experience that unadopted roads tend to be in a poorer condition;
- Unadopted estate roads in more modern developments are difficult to spot. This is because a number of connected roads in the development may share many of the design features of adopted roads and especially in their early years, the condition of the surfaces may be satisfactory.
- Private roads will often be gated or have signs like 'private road' or 'no access to the public' fixed at their entrance.

Ground rents and other types of charges

Over the last few years, the issue of ground rents payable on leasehold properties has hit the media headlines. Properties have seen their ground rents increase dramatically and in some cases, this has doubled every 10 years – increases that far outstrip inflation. Not only does this have serious financial implications for leaseholders, it can also affect the property's saleability. It is not the residential practitioner's role to identify and advise on ground rents, but an understanding of some of the different types of charges associated with property can help us provide more holistic advice.

Service charge: Commonly found in shared multi-occupied buildings, such as blocks of flats. The aim of this charge is to cover the running and maintenance costs of the building. This may incorporate a sinking fund to cover larger items of expenditure in the future (such as new roofs) and some may have an escalation mechanism. Whatever the particular circumstances, the landlord/freeholder has certain legal duties to consult how these service charges are calculated.

Ground rent: Rent payable to the freeholder for the use of the land. Although some ground rents may be modest (often known as a 'peppercorn rent'), some may include an escalation mechanism which may be set as a percentage of the retail price index (RPI), based on the property value or some other assortment of complicated calculations. In some cases, the cost of ground rent can quickly reach several thousands of pounds a year.

Estate rent charge: A mechanism by which a third party (usually a developer or a management company) can get contributions from freeholders towards the cost of upkeep of shared areas on an estate (such as unadopted roadways and green areas). In some ways, the estate rent charge is similar to the service charge for leaseholders but without any of the legal protection. The freeholders have no right to see how the estate rent charges are formulated and the penalties for non-payment are severe. For example, if the estate rent charge is not paid within 40 days, it could result in the third party taking possession of the property. If this happens, the lender's security would be compromised and so many lenders are reluctant to take that risk.

We know of a number of blocks of modern flats with all three of these charges that can result in a high level of expenditure.

Other important features associated with vehicular access to properties Dropped kerbs

At the time of writing, Section 184 of the Highways Act 1980 makes it an offence to drive a vehicle across a footway or verge and into a property where there is no proper vehicle crossover or 'dropped kerbs'. New drop kerb vehicle crossings were historically called 'carriage crossings' and it includes both the physical lowering of the kerb (the dropped kerb) and a permission to allow a vehicle to cross the public footway. On adopted roads, a property owner must apply to the local/highways authority for permission to install a dropped kerb and its suitability will depend on a range of factors, including adequate sightlines along the road and path; proximity to junctions, traffic lights or crossings; extent and depth of services within the footpath; the speed of the road, bus and cycle lanes and the existing parking or bus stops. In other words, installing a dropped kerb is a complex process. If your client inherits an unapproved dropped kerb, it could leave them liable to any damages issues that may arise.

Grass verges

In many areas, the pavement that runs along the side of the road may also include a grass verge or, in some cases, a grass verge alone. In most cases, there is a legal presumption that the grass verge is part of the highway and should be maintained by the same entity that is responsible for the road. However, in some rural areas, these verges may still be owned by the parish council or even the 'Lord of the manor'. Whatever the legal situation, you should clearly identify the arrangement and bring the matter to the legal adviser's attention. You should be particularly vigilant for any access drives built across the verge and any attempts to enclose the verge.

Parking restrictions or permits

The usability, value and saleability of a property may be affected if it does not have an immediately adjacent parking provision. Even if the parking is unrestricted, due to the rise in car ownership and subdivision of larger properties into separate flats, it can be very difficult to find a convenient parking spot. This is often down to your local knowledge and where this is the case, you may want to let your client know, especially if they are new to the area.

11.2.10 Gates, including automatic gates

All vehicle and pedestrian entrance gates should be inspected and included in the report, especially if there are any condition, safety or legal issues associated with the features.

Electrically operated entrance gates are increasingly common in flat developments and private homes. In 2010, following the death of two children and injuries to several other people, new legislation was introduced applicable to gates with electrically operated, automatic opening systems. The law specifies safety requirements to help prevent accidental crushing of people, especially elderly people and children. As a result, the current recommendation is that automatic gates should be inspected and serviced every 6 months. As the Health and Safety Executive confirms, 'a powered gate must respond in a safe way when any person interacts with it . . .' and a typical example of this would be children playing in close proximity (HSE 2020).

You should ask the vendor for information about the gates and when they were last tested/serviced. If no records are available, this should be reported to the client as a hazard.

11.2.11 Easements, wayleaves and other matters

Section 5 of Volume 1 describes the legal issues of which a surveyor must be aware, should note during inspection of the site and include in the report where appropriate. These include:

- Shared areas – this includes a wide range and large number of features such as common access staircases, landings, corridors in flats, entrance drives, parking areas, garages;
- Overhead power, phone masts or other service cables that cross over the property to serve other properties. Alternatively, cables that serve the subject property but

cross over other properties, including all associated poles and other equipment. Be particularly vigilant for electrical substations;

- Possible rights of access enjoyed by service companies such as a drainage company maintaining a dyke or stream;
- A possible practical requirement for access by an adjoining owner onto the property. For example, to maintain a wall on a boundary and a similar requirement for the subject property over an adjoining property;

See also 'rights of way' in the preceding section.

11.2.12 Other matters

There is a broad range of other issues you should be looking out for during your inspection of the grounds. Although some of these have been covered in other parts of this volume, they have been summarised here for convenience:

- Evidence of flooding or waterlogging on site or in the vicinity. This can include large areas of flatten vegetation and piles of water-borne debris such as dead trees, large branches and other debris associated with flooding. You should also look out for any obvious flood defences, such as dams, raised levees on rivers and moveable gates and barriers on either the subject property or neighbouring land (where these can be easily seen from the subject property itself);
- The presence of protected species. Although this might include a range of species that will be difficult to spot (for example dormice, newts and red squirrels to name but a few), some animals like badgers and otters may leave physical evidence of their presence and this may be very damaging. Local knowledge will be very important in this respect;
- Hedgerows and/or trees protected by preservation orders;
- Presence, or suspected presence, of archaeological remains, scheduled monuments and so on.

The desktop study will provide very important clues for the last two points.

11.3 Contaminated land

11.3.1 Introduction

Properties built on contaminated land can present two problems:

- Hazards to human health;
- A possibly significant, detrimental effect on property value and, in some cases, condition.

In the UK, there have been few pollution incidents on contaminated land that have caused loss of life. However, incidents of properties exploding owing to the presence of landfill-derived methane gas are causes of major concern (Viney & Rees 1990). Therefore, it is very important for the surveyor to pick up this possibility.

11.3.2 Definition

A 'contaminant' is defined in the Building Regulations as 'any substance which is or may become harmful to persons or buildings including substances which are corrosive, explosive, flammable, radioactive or toxic' (OPSI 2013a, p. 5).

New buildings and some material alterations are subject to strict requirements to prevent contaminants from affecting the health of occupants, including the necessary remedial measures to help reduce the risks associated with those contaminants. You must be aware of these current safety benchmarks and report on them accordingly, including the possibility that old and recently constructed properties may be affected.

11.3.3 Recognising contaminated sites

Before the site inspection, many contaminated sites will have been identified through planning records, other publicly available information or local knowledge during the desktop exercise (see Chapter 3). This process is fundamental to identifying possible contamination, since spotting contamination on a site is difficult. Many liquid and gaseous contaminants are mobile and can affect neighbouring land. For example, the building regulations Approved Document C suggests gaseous substances can migrate by as much as 250 m (OPSI 2013a, p. 19). You therefore need to take a broader view of the neighbourhood and its history. Sites that are likely to contain contaminants (for a full list, see OPSI 2013a, p. 14) include:

- Asbestos and chemical works;
- Gas, coal and coal by-product plants;
- Landfill and other waste deposit sites;
- Metal mines, steel-making plants and metal-finishing works;
- Oil storage and distribution sites;
- Railway sidings and depots;
- Power plants and nuclear installations;
- Garages, petroleum fuel stations and scrap yards;
- Former sewage works.

The local authority environmental health officer might help identify such sites. Another good source of information is old ordnance survey maps that show the sites of old industries, including brickworks, railway sidings and chemical works. Caution must be exercised when interpreting these clues. You should ideally have good knowledge of materials used in the former process and where they were stored. This could mean the difference between isolated pockets of contamination and the need for whole-site treatments. In these cases, employing a specialist may be vital. You can only be expected to identify potential uses and then it should be referred to a specialist who can determine where further investigation may be required to prove a contaminant exists and what action is required.

Key inspection indicators:

If desktop studies fail to turn up any information about an area, visual clues should be checked. These include:

- Large open sites with uneven or unnatural topography;

- Surface features of open ground with the following characteristics:

  - Absence of vegetation, or growth that is poor or unnatural;
  - Unusual surface materials that may be strangely coloured and contours that suggest wastes and residues;

  - Fumes and odours suggesting organic chemicals at very low concentrations;
- Road and street names can give important clues. For example, Pit Street, Cemetery Road, Pond Street, Tannery Row, Gas Street Run and Railway Terrace are just a few taken from the A-Z map of a typical town. In these (actual) cases, the use suggested by the name was no longer visually evident.

Where evidence is noted, further desktop studies or enquiries by the surveyor may be required. If contamination is confirmed or suspected, a referral to a specialist company would be appropriate.

11.3.4 Implications of contamination

Making the site safe if a site has been contaminated can be expensive. There are three broad options:

- Treatment – Application of biological or chemical compounds to remove toxicity;
- Containment – Encapsulation of the contaminant using plastic or similar layers and soil;
- Removal – Disruptive excavation and carting soil away to a licensed tip.

These options are usually expensive, and in many cases it is impossible to identify the polluter, let alone get them to pay. Therefore, contaminated land can seriously affect the value of dwellings.

11.4 Radon gas

11.4.1 Introduction

Radon is a naturally occurring, radioactive, colourless and odourless gas. It is formed in small quantities by the radioactive decay of naturally occurring uranium and radium in the ground. Radon can move through the subsoil and collect within the voids and living accommodation of buildings. It has been described as 'the second leading cause of lung cancer in North America and the leading cause of lung cancer for individuals who have never smoked', with similar assertions for the UK (Field 2011, p. 1).

11.4.2 Identification of risk

All properties

In 2007, following collaboration between the Health Protection Agency (now 'Public Health England' covering England and Wales, with equivalent organisations covering Northern Ireland and Scotland) and the British Geological Survey, a revised map of areas affected by radon was published. Maps covering the entire UK are now available online (the 'radon atlas') and an example is reproduced in Figure 11.9 (Miles 2007).

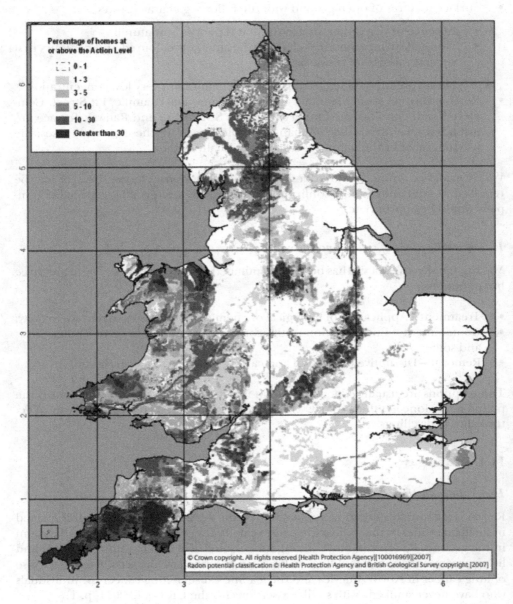

Figure 11.9 Map of radon-affected areas in England and Wales.

PHE recommends existing homes in what it calls 'affected areas' should have radon measurements carried out to establish whether there is significant risk for occupiers and therefore whether remedial work should be implemented to reduce the hazard. An 'affected area' is defined as having a '1% probability or more of present or future homes being above the Action Level' (Miles et al. 2011, p. 6). Such action includes reducing concentrations of radon in the home to levels as low as is reasonably practicable below the 'action level', which is 200 Bq m^{-3}. Such action is confirmed in Approved document 'C' of the Building Regulations, which requires some new homes, conversions and extensions in affected areas to be built with radon protection – usually a thicker plastic membrane at ground floor level to help prevent radon entering the property.

Radon levels in existing properties can be measured by detectors supplied by PHE or other certified laboratories for a small charge. Detectors are usually placed in a bedroom and living room, monitoring takes place for three months and detectors are returned to the laboratory for confirmation of results. This is not something that can usually take place before a transaction is completed and the surveyor needs to explain the potential risks.

Radon in basements and cellars

BRE has described basements and cellars as 'major contributors to the radon problem in a building' (Scivyer & Jaggs 2010, p. 1). This is because the parts of the basement in contact with the soil (from where the radon emanates), the walls and floor, have a surface area in contact with the soil three times that of a 'normal' property without a basement. In addition, the gas collects in the basement, which acts like a sump – retro-fitting a sump into a property is one of the methods of collecting radon. BRE has therefore concluded that every basement and cellar in the UK is 'at increased risk of elevated levels of radon, regardless of geographical location' (Scivyer & Jaggs 2010, p. 29).

Given the cancer risk, you should therefore recommend radon testing for all living accommodation and occupied spaces below ground. Current advice from PHE and BRE is that if the property has a basement with no ventilation, or the basement is used for habitable purposes, it should be provided with either a sump with a fan or a positive ventilation system (BRE 2015).

Key inspection indicators:

- Identify basements and cellars, *especially* those with living accommodations, as radon hazards can affect *all* parts of the UK;
- Consider how remedial work might be carried out, if required – where might the sump be located beneath the ground floor slab, and can the vent terminate in a satisfactory location, away from openings into the property.

11.4.3 Remedial measures

Examples of works and costs to reduce radon levels in a property include (based on 2021 prices):

Works	Typical cost £	Normal range £
Installing a 'radon sump', described as the most effective method of reducing levels of the gas, especially when fan powered. A sump is a small, bucket-sized space just under the floor slab, vented to the outside (if fan powered, usually terminating at roof level, away from openings).	800.00	Up to 2000.00

Figure 11.10 Radon remedial measures and costs, based on PHE online advice.

Works	Typical cost £	Normal range £
Positive ventilation, where a fan is installed to blow air into the property. This dilutes the radon concentration inside and increases internal air pressure, thereby helping to prevent radon from getting into the property.	550.00	Up to 1000.00
Natural under-floor ventilation – through the vents usually provided for lower floors with a gap beneath.	200.00	Up to 600.00

Figure 11.10 (Continued)

11.4.4 Typical action and advice

The RICS Home Survey Standard (HSS) indicates that the surveyor should have knowledge of 'high' radon levels and similar statements are made in other RICS documentation (RICS 2019, p. 31). When preparing for a survey, in all cases the surveyor should check the radon atlas to identify whether the property is in one of the areas where radon may be an issue, particularly where there is an occupied area below ground or in a basement. In the report, the surveyor should confirm a potential hazard and flag the issue for the legal adviser when the property:

- Is situated in any area that has radon levels above the 'action' level; that is, more than 1%; and/or
- Has a basement, cellar or underground living space below ground.

The need for a test need not interrupt the sale process. The legal adviser can agree arrangements for a 'radon bond' or 'retained fund'. This sum of money, typically around £2,500, is set aside to cover the possible need to carry out remedial works in the future. After testing, if no radon is found, the money is paid over to the seller.

11.5 Flooding

11.5.1 Introduction

At the time of writing, climate change is on the international agenda as extreme weather events now affect all parts of the globe and the UK is no exception. According to the Meteorological Office, climate change is already being felt across the UK, with 2020 being the third warmest, fifth wettest and eighth sunniest on record.

In terms of rainfall, the UK has been on average 6% wetter over the last 30 years (1991–2020) than the preceding 30 years (1961–1990). Six of the ten wettest years for the UK in a series from 1862 have occurred since 1998. Add to this the storms Ciara and Dennis that hit the UK only one week apart helped make February 2020 the wettest on record and brought devastating flooding, affecting many homes and businesses (Kendon et al. 2021).

The impact of flooding on local areas is easy to find. A simple internet search for 'flooding' in most towns or cities will normally produce various accounts of dramatic local incidents qualifying this information as 'local knowledge' of which residential practitioners should be aware. The impact on property owners can be significant:

- Flooding can be life threatening;
- The financial cost of protecting a property from flooding and repairing it after it has flooded can be prohibitive;
- In some flood prone areas, buildings and contents insurance can be very high and, in some cases, not available at all. This will affect property values and saleability;
- Research has found people who have been flooded in the UK are nine times more likely to experience long-term mental health problems than the general population. Post-traumatic stress disorder (PTSD) was the most commonly reported condition, with a prevalence rate of between 7.06% and 43.7%. Anxiety and depression were also common. As one homeowner stated, 'We never sleep when it's raining'. Cruz et al. 2020.

This is not a problem that affects a small minority of properties that happen to be on the banks of major rivers. Current estimates indicate that 1 in 6 houses is vulnerable to flooding. This means that a typical residential practitioner is likely to inspect at least one or two flood risk properties a week.

In our view, this makes the risk of flooding an important factor about which buyers should be aware. Although residential practitioners are not experts on flooding matters and our standard levels of service do not include a formal flood risk assessment, we have an important duty to bring these issues to our client's attention and, as a minimum, where a flood risk exists, at least point them in the right direction so they can be properly advised during their purchase decision.

11.5.2 Different types of flooding

According to the Flood Guidance, there are a number of different types of flooding in the UK:

River flooding: Sometimes called fluvial flooding, this is very common in the UK. This is where a river's flow exceeds the bank sides and causes damage or obstruction to a nearby area. In some established urban areas, water courses may have been 'culverted' where the water passes through a tunnel. Although these may not conform to the typical image of a stream, fluvial flooding can result in the water bursting out of its culvert with very damaging results.

Coastal flooding: This type of flooding affects communities situated close to the sea. With high tides, stormy weather and climate change all contributing to an increased risk of coastal flooding, this is a major hazard for many areas of the UK. Seawater over-topping coastal defences can cause significant damage and disruption to communities, often requiring people to retreat further inland. With the added problem of salt in seawater-damaged buildings, coastal flooding is a serious issue and should be addressed with appropriate flood protection.

Surface water flooding: This is also known as pluvial flooding. This can affect properties all around the country, even if they are not situated near a river or the sea. Surface water flooding occurs after periods of heavy rainfall where excess water cannot drain away, urban drainage systems are overwhelmed and water flows out into streets and nearby structures. This may be due to a range of reasons, including blocked drains or even rainwater running off roads. Just because a property has not flooded before does not mean it won't flood in the future.

Flash floods are normally associated with pluvial flooding and are characterised by an intense, high-velocity torrent of water triggered by torrential rain falling within a short amount of time within the vicinity or on nearby elevated terrain. They can also occur through the sudden release of water from an upstream levee or a dam. Flash floods are very dangerous and destructive, not only because of the force of the water, but also the hurtling debris that is often swept up in the flow.

Reservoir flooding: Reservoir flooding is extremely rare in the UK due to very strict regulations and mandatory assessments and most residential practitioners did not consider the risk of this type of flooding. That was until the Whaley Bridge dam incident. In August 2019, a dam spillway partially collapsed and this resulted in about 1,500 people being evacuated from the town of Whaley Bridge over fears the Derbyshire town would be flooded. Reservoir flooding will cause very fast-flowing water to flow down the natural water path in large quantities and can threaten lives and properties many miles away from the reservoir itself (Flood Guidance 2021).

11.5.3 Assessing the flood risk of a property

As mentioned above, a residential practitioner is not expected to carry out a detailed flood risk assessment of a property, but we must be familiar with the nature and complexity of the region/area in which the property is situated. This is reinforced in RICS's Home Survey Standard, which states:

> RICS members need to be familiar with the nature and complexity of the locality in which the property is situated. This includes general environmental issues where the information is freely available to the public (usually online).
> (Appendix C, RICS 2019)

The HSS goes on to specifically list flooding (including surface, river and sea) as one of the issues to be considered.

As a consequence, we think an appropriate assessment suited to all levels of service can be split into a number of specific stages:

- Pre-inspection 'desktop' research;
- Information from the vendor/occupier;
- The inspection;
- Reporting to the client.

Pre-inspection desktop research Using the government's website

Most practitioners have been using the government's own web-based information sources for a number of years. Although the nature and format have varied, the website allows users to check:

- The long-term flood risk for an area;
- The possible causes of flooding;
- How to manage flood risk.

(UK Government 2021)

The service uses computer models to assess an area's long-term flooding risk from rivers and the sea, surface water, reservoirs and some groundwater. The website states that the results are an indicator of an area's flood risk, particularly the likelihood of surface water flooding. It is not suitable for identifying whether an individual property will flood and it does not include the flood risk from sources such as blocked drains and burst pipes.

Despite these limitations, we think it is important that you use this website as part of the desktop research **before** you leave the office for **every** instruction. This is important because:

- Where a property shows flooding in close proximity, it is important to adjust your inspection routine to see if there are any physical indications of the problem. For example, this could include waterlogged ground, flood prevention devices fitted to the subject and neighbouring properties, and even piles of old sandbags near the front gate. These will all provide evidence that the property could have flooded before;
- The information on the flooding website is regularly updated so it is not safe to assume the nature and extent of flooding in any one area remains static even where you know the area well. Therefore, it is important to check every time.

Currently, the most useable information is on a section of the website called '*Check the long-term flood risk for an area in England*' (https://flood-warning-information.service.gov.uk/long-term-flood-risk/risk). There are equivalents for Scotland, Wales and Northern Ireland.

After entering the property's postcode and address, the software generates a flood risk summary for surface water, rivers and sea. These are in categories ranging from 'very low risk', 'low risk', 'medium risk' and 'high risk'. This information can be viewed on a map that shows the areas at risk from flooding more graphically. The flood risk from reservoirs can also be selected from a drop-down list.

Additional facilities on the government website

In addition to identifying the flood risk, at the time of writing, the website also contains useful information on how to manage a flood risk, including a section on insurance and moving house. These additional facilities include:

- Email link to the Environment Agency to find out if a property is in an area which has flooded in the past and request a flooding history report;
- Email link to the Environment agency to obtain evidence of the flood risk for insurance purposes by requesting an 'insurance-related report' that should be received within 20 days; and
- If flood-related work has been done on a property, a proforma 'flood risk report' can be downloaded so a surveyor can complete the appropriate sections and confirm the extent and nature of the work.

These additional features show how this particular website has developed over the last few years and we are sure this increased functionality will be useful to a number of property owners. However, in our view, residential practitioners must be cautious about using these expanded services because:

- The accuracy of the information is unknown;
- The level of support to the web-based service will not be assured; and
- It is inevitable that development of internet-based facilities like this will continue to evolve. This can outdate current usage.

Consequently, residential practitioners have to offer information and advice that is helpful, accurate and sustainable. In relation to desktop research, we recommend you restrict your use of the government's website to identifying the risk of flooding from rivers and sea, surface water and reservoirs.

Using the internet for general searches

As described above, a general internet search for flooding in specific roads, neighbourhoods, villages, towns and cities can often identify previous flooding problems not otherwise captured by more formal sources of information. Like all information on the internet, the output has to be treated with caution, but online articles/features from local media, councils, community groups and even social media can help build a picture of the flooding problem around a subject property.

Information from the vendor/occupiers

This is an important part of the process. Although the reliability of information from the selling agents, owners, occupiers and tenants is treated with caution by many, it is important to ask whether the property and its grounds have suffered from flooding. This is because it can help 'triangulate' information from desktop research and provide information early in the sale and purchase process.

Although the legal adviser will formally ask the seller similar questions (normally on a form called TA6 Property Information form), this usually comes at the end of the sale process when all parties are normally focused on getting the sale completed. It is far better to try to get this information earlier so any information can be objectively evaluated.

The 'Flood Re' scheme

Before we consider how residential practitioners can advise clients on flooding matters, it is important to explain what 'Flood Re' is, how it works and the implications it has for property owners in flood risk areas.

Flood Re is a joint initiative between the government and insurers. Its aim is to make the flood cover part of household insurance policies more affordable. Every insurer that offers home insurance in the UK must pay into the Flood Re Scheme and this raises £180m every year that is used to cover the flood risks in home insurance policies (Flood re 2021).

The Flood Re scheme works with insurers behind the scenes. When a home insurance cover is bought, the insurer can choose to pass the flood risk element of the policy to Flood Re for a fixed price. Then, if a valid claim for flooding is made, the insurer will initially pay the claim but later claim it back from the 'Flood Re' fund. In this way, the scheme helps to keep the premiums down.

Eligibility for Flood Re

Flood Re does not apply to all properties at risk from flooding. To come under the scheme, properties must meet a set of criteria and the main ones are as follows:

- It must have a domestic Council Tax band A to H (or equivalent);
- It must be used for private, residential purposes;

- It must be a single residential unit or a building comprising of two or three residential units;
- It must be insured on an individual basis or have an individual premium;
- It must have been built before 1 January 2009; and
- It must be located within the UK comprising England, Wales, Scotland and Northern Ireland (excluding the Isle of Man and the Channel Islands).

The types of properties that can be covered will usually include:

- Bed and breakfast premises paying Council Tax and insured under a home insurance contract;
- Some farmhouse dwellings and cottages;
- Holiday/Second Homes;
- Properties occupied by home workers;
- Individual leaseholders protecting their own property/flat;
- Leasehold blocks if they contain 3 units or fewer and the freeholder(s) lives in one of the units to be insured;
- Single unit leasehold properties where the leaseholder insures the structure of the property;
- Residential 'buy to let' properties;
- Static Caravans/homes if in personal ownership.

In some cases, Flood Re will also cover a tenant's and/or individual's contents in rented or leasehold properties, even where the buildings risk would not be eligible (such as in large blocks of flats).
   Properties which would not be covered include:

- Bed and breakfast premises paying business rates;
- Blocks of more than three residential flats;
- Company houses/flats;
- Properties covered by contingent buildings policies (for example held by banks);
- Farm outbuildings;
- Properties used by freeholders/leaseholders in deriving commercial income insuring blocks/large numbers of properties in a portfolio;
- Housing association's residential properties;
- Multi-use properties under commercial or private ownership;
- Some residential 'buy to let' that do not meet core criteria;
- Social housing properties; (eligible for Contents cover but not eligible for Buildings cover); and
- Static caravan site owners (for commercial gain).

Like all complex schemes, it is important to check the detailed eligibility criteria on Flood Re's website.

The inspection

As briefly mentioned in Chapter 3, pre-inspection research and enquiries can help inform the property inspection itself. Where potential problems associated with flooding have been identified, you should be alert for the following features:

- Proximity to water courses, drainage ditches, large ponds and lakes, culverts, waterlogged grounds, areas that have been scoured by water flow and so on;
- Evidence of flood prevention measures including demountable flood barriers at gates, doors and sometimes windows; sandbags; airbrick covers; non return valves in inspection chambers and so on;
- Properties where ground floor rooms and been fully refurbished (possibly following a flood);
- Outside walls that show signs of extensive past moisture problems such as bands of salts on masonry;
- Dwellings that sit lower than neighbouring properties and/or public highways. A typical example would be a sloping drive from the main street down to the property.

This information can be added to the previous data to provide the client with more useful advice.

Reporting to the client

Although this will depend on the level of service provided, you should take care not to exceed the terms of engagement in relation to any advice you give, even though you may want to be as helpful as you can. However, as the HSS reminds us (RICS 2019, p. 15), the report should be 'property specific' and not simply regurgitate publicly available information. Therefore, the report should endeavour to:

- Objectively present the outcome of your investigations including the desktop research, vendor enquiries and physical inspection;
- Where you have concerns about past, present or possible future flooding matters, recommend that a full flood risk assessment should be carried out before commitment to purchase;
- Recommend the client should make appropriate enquiries with their chosen insurer to see if building and content insurance will be available for that property;
- Where the property is unlikely to be covered by the Flood Re scheme, clearly explain the implications;
- Recommend the client discusses any highlighted issues with their legal adviser and make sure these issues are clearly and individually highlighted in the legal section.

Where the indicators suggest flooding could affect a property, it is important to recommend the client commission a full flood risk report from a commercial organisation. Although many legal advisers may include this as a standard part of an environmental assessment report, it is important to be specific and not leave it to chance.

11.6 Other environmental issues

11.6.1 Introduction

At the time of writing, the HSS includes an appendix (RICS 2019, p. 28) dealing with local environmental issues, emphasizing that the list is not exhaustive. Most of such information will be publicly available, usually through the internet. You must not assume the client will carry out such research because the information is freely accessible – this will be no legal defence to a complaint.

11.6.2 Typical issues

In addition to the environmental issues discussed in this chapter, a surveyor's list for consideration might include:

- Possible nuisances, such as noise or smells from:
  - Nearby military or civilian airports;
  - Local transport networks such as roads and railways;
  - Schools, universities, hospitals and other local institutions that may create a nuisance;
  - Agricultural land or premises;
- Locally known incidences of local pollution (for example, water, air, and so on);
- Local land uses such as industrial, commercial and similar properties, e.g., public houses;
- Risks from local soil or ground conditions, e.g., mundic, sulphates, etc.;
- Situation in a smoke control area;
- Proposed local developments likely to affect the property, such as a new block of flats on adjoining vacant land.

The surveyor's role is to inform the client regarding all local matters that might reasonably affect his or her decision to purchase.

11.7 Deleterious materials

11.7.1 Introduction

The dictionary definition of 'deleterious' is 'harmful or injurious to the health'. The presence of these materials in a property could affect the health of occupants and, in extreme circumstances, visitors. The full extent of this might not be appreciated because the materials are often performing an adequate function. For example, asbestos pipe lagging may still be as originally designed and present no immediate danger. The problem is the hidden danger of the future deterioration of the material. The other complicating factor is the attitude of the occupiers. Some will refuse to share their home with a potentially dangerous material and demand complete removal. Others will be prepared to maintain the material in a safe condition. The latter approach leaves an uncertainty as ever-increasing standards and research and public perception of the risks may result in the need for remedial work in the future. However, if this doubt is accepted, such a minimalist approach can provide the surveyor and client with initial guidance.

The typical substances included under this heading include:

- Asbestos;
- Lead; and
- Urea formaldehyde foam.

Some aspects of these have been covered under other sections and references are made to the relevant text.

11.7.2 Asbestos

Introduction

Breathing in asbestos dust can cause the development of asbestos-related diseases such as cancer of the chest and lungs – mesothelioma (Barreiro & Katzman 2007). The material is responsible for 5,000 deaths annually in the UK. Most people currently dying from these diseases were exposed to the substance during the 1950s and 1960s, when the material was more widely used. Surveyors should be able to:

- Identify the presence of possible asbestos-containing materials (ACMs);
- Assess the condition of the component;
- Give broad advice to the client on the implications of having it in their future home; and
- Give guidance on what should be done about it.

Identifying asbestos

Asbestos risks should be considered proportionately. Materials containing a high percentage of asbestos are more easily damaged. Sprayed coatings, laggings and insulating boards are more likely to contain the more harmful blue and brown asbestos and can be made up of 85% asbestos fibres. These materials pose the greatest risk. However, asbestos cement products can contain only 10–15% asbestos, often of the white variety. These fibres are tightly bound into the material and will only give off dust if damaged or broken – these represent a lower risk.

   Key inspection indicators:

- ACMs are usually present in properties built between 1950–1980, especially in non-traditional steel-framed dwellings; and remained in use in buildings until 1999, albeit on a decreasing basis;
- High-risk component include:

    - Sprayed coatings or laggings as thermal insulation to boilers and flues, perforated ceiling tiles;
    - Insulating boards used for thermal insulation, fire protection to floors and service ducts, etc;

- Low-risk component:

    - Asbestos cement products including corrugated roof sheets; wall claddings; 'vinyl-asbestos' floor tiles, linings to the soffits of external features such as roof eaves, balconies, porches, covered ways, etc.;
    - Ceiling lining in garages;
    - Gutters, rainwater pipes and water tanks.

The HSE points out that ACMs were used in many building components and, if in doubt, the default is to assume it is present until proven otherwise (HSE 2012b, p. 3).

In properties of multi-occupation such as flats or houses in multiple occupations, the Control of Asbestos Regulations 2012 imposes a duty on the 'duty-holder' (usually the owner, who can devolve the responsibility to the managing agent) to identify the presence and condition of ACMs in those parts defined as 'non-domestic', which in practice are communal or shared areas. Such parts typically include entrance halls, stairways, lifts, landings, storage areas, plant rooms, roof spaces and common external parts such as communal areas and bicycle stores.

Most duty-holders therefore arrange for an asbestos survey to be carried out on the property to:

- Identify the presence of ACMs or possible ACMs;
- Assess the risks from the material;
- Prepare a management plan for the material and the risks; and
- Ensure that persons who may be at risk are informed about the risk; such as flat owners and visitors and persons who use the premises as a workplace, that is building contractors, but also surveyors.

This document, or documents, (sometimes called an 'asbestos register') is therefore a potentially good source of information for surveyors inspecting such properties to help identify the location of ACMs and assess any residual risks to both the surveyor during their inspection and their client. However, if an asbestos register exists, it does not remove your obligation to look out for obvious materials and features that are commonly known to contain suspected asbestos materials.

Given the legal obligation on the duty holder to ensure availability of this document, you should flag up whether this document exists for the legal adviser; otherwise, the client will be faced with the cost of paying for an ACM survey themselves.

Assessing the condition of the asbestos

The next stage is to identify the risk of asbestos fibres being released into the air. The risk is high if:

- The material is being disturbed, e.g., is in a position likely to be knocked and scraped by normal usage;
- The surface of the material is damaged, frayed or scratched;
- Surface sealants (paint, etc.) are peeling or breaking off;
- The material is becoming detached from its base, e.g., sectional insulation is falling away from a flue, etc;
- There is dust or debris in the immediate area.

Whereas, the risk is low if the asbestos component:

- Is in good condition; and/or
- Has been 'encapsulated' in some way, for example with proprietary sealant; and
- Is not likely to be:
    - Damaged; and
    - Worked on.

Advising the client on possible action

If the asbestos is in satisfactory condition, it is likely to be safest to leave it in place and introduce a management system to keep it safe. This could include labelling the asbestos product and keeping a record of where the asbestos is. This should be left with the deeds so it can be passed on with the sale of the property to protect future owners and users. Slightly damaged asbestos can be made safe by repairing it and sealing it with a special coating.

If the asbestos is in poor condition and/or is likely to be disturbed during routine maintenance work or normal use of the property, it must be removed completely.

Because of the special health risks, the Health and Safety Executive (HSE) confirms removal of high-risk asbestos materials such as pipe insulation, asbestos insulating board or sprayed coatings should be carried out by a licenced contractor. Working on lower-risk ACMs does not currently require a licence but is nevertheless likely to require special training and procedures and therefore increased costs when compared with working on similar components not containing ACMs.

Depending on the extent, this could affect the value of the property and certainly endanger the client's health if the work is not done properly. Therefore, the hazards involved should be included in the report if possible ACMs are identified.

11.7.3 Lead

Lead water pipes are discussed in Volume 3 of this publication. Lead can also pose a health risk in older paint coatings. There is a risk of inhaling lead dust during DIY paint stripping and sanding, a particular danger to young children. Studies have shown household dust is contaminated with deteriorated lead paint and is one of the main sources of lead exposure. Research confirms that 'there is no safe level of lead exposure for young children and, although small, these effects are enduring and possibly permanent . . . effects appear to persist at least into early adolescence' (Canfield et al. 2015, pp. 1 & 11). Pregnant women are also at risk (HSE 2012a, p. 4).

Lead in paint has been gradually phased out since the 1950s, but the HSE suggests it remains a problem in properties built until the early 1980s and enough remains to present a potential hazard in possibly 50% of UK housing stock (HSE 2014). Like asbestos, the potential risk should be kept in perspective. The main triggers for concern are:

- Pre-1960 properties; the older the property, the greater the risk;
- Large areas of loose and flaking paint creating residues of dust and smaller particles.

Lead-based paint can be treated by overcoating sound material with modern lead-free paint. Where the original coating is unstable, removal may be the only option. It is suggested that people working on paint that includes lead are at most risk during several procedures, including burning lead paint from components and stripping paint from joinery (HSE 2012a, p. 1). Because of the health risks, stripping of lead-based paint should involve extra precautions, including:

- Planning the work – protecting the home with plastic sheets, wearing overalls and special mask (known as a FFP3 mask);

- Stop dust getting into the air – wetting sandpaper, or using special dust extraction devices;
- Stop lead fumes being produced – setting hot air guns at lower temperatures, not using blow torches; and
- Washing arms and forearms before eating and drinking (HSE 2014, p. 2).

Although these safety measures are not as rigorous as with asbestos, they will add to the cost and inconvenience of the work. Lead is also used in other parts of a property, including roofs, some glazing and flashings; these also represent a risk, particularly to a DIY homeowner. Clients should be warned of these factors, the additional hazards and the need for any further advice before undertaking work. Additional information is available from the HSE.

References

Allen v. *Ellis & Co* (1990). 1 EGLR 170.

Anstey, J. (1990). *Boundary Disputes and How to Resolve Them.* RICS Books, London.

Barreiro, T.J. and Katzman, P.J. (2007). 'Malignant Mesothelioma: A Case Presentation and Review'. *The Journal of the American Osteopathic Association,* Vol. 106 No. 12, pp. 699–704.

BRE (1992). 'Surveying brick or blockwork freestanding walls'. *Good Building Guide 13.* Building Research Establishment, Garston, Watford.

BRE (2015). Report BR 211 *Radon: Guidance on Protective Measures for New Buildings (Including Supplementary Advice for Extensions, Conversions and Refurbishment).* Building Research Establishment, Garston, Watford.

Canfield, R.L., Jusko, T.A. and Kordas, K. (2005). 'Environmental lead exposure and children's cognitive function'. *Rivista italiana di pediatria,* December, pp. 293–300.

Cruz, J., et al. (2020). 'Effect of Extreme Weather Events on Mental Health: A Narrative Synthesis and Meta-Analysis for the UK'. *International Journal of Environmental Research and Public Health,* Vol. 17, No. 22. MDPI journals.

Field, R. William (2011). *Radon: An Overview of Health Effects.* University of Iowa, Iowa. Available at www.researchgate.net/publication/288034741_Radon_An_Overview_of_Health_Effects.

Flood Guidance (2021). 'Types of flooding'. Available at www.floodguidance.co.uk/what-is-resilience/types-flooding/. Accessed 29 September 2021.

Flood Re (2021). Available at www.floodre.co.uk/. Accessed 30 October 2021.

Health and Safety Executive (HSE) (2012a). *Lead and You – Working Safely with Lead.* The Stationery Office, PO Box 29, Norwich NR3 1GN.

Health and Safety Executive (HSE) (2012b). *Managing Asbestos in Buildings: A Brief Guide.* The Stationery Office, PO Box 29, Norwich NR3 1GN.

Health and Safety Executive (HSE) (2014). *Old Lead Paint – What You Need to Know as a Busy Builder.* The Stationery Office, PO Box 29, Norwich NR3 1GN.

Health and Safety Executive (HSE) (2020). Available at www.hse.gov.uk/work-equipment-machinery/powered-gates/safety.htm. Accessed February 2020, HSE.

Kendon, M., et al. (2021). 'State of the UK Climate 2020'. *International Journal of Climatology,* Vol. 41. Royal Meteorological Society. London.

Miles, J. C. H., et al. (2007). *Indicative Atlas of Radon in England and Wales, Health Protection Agency, Dicot.* Oxfordshire and British Geological Survey, Keyworth, Nottinghamshire.

Miles, J.C.H., et al. (2011). *Indicative Atlas of Radon in Scotland, Health Protection Agency, Dicot.* Oxfordshire and British Geological Survey, Keyworth, Nottinghamshire.

Noy, E.A. (1995). *Building Surveys and Reports.* Blackwell Science, Oxford.

Office of Public Sector Information (OPSI) (2013a). Building Regulations approved document 'C'. *Site Preparation and Resistance to Contaminants and Moisture.* OPSI, Richmond, Surrey.

PCA (2021). *Guidance Note Japanese Knotweed – Guidance for Professional Valuers and Surveyors, June 2021.* Property Care Association, Huntingdon.

Planning Portal (2022). Available at https://www.planningportal.co.uk/permission/common-projects/decking/planning-permission. Accessed 6 September 2022.

Public Health England and Building Research Establishment (PHE & BRE) (2015). *Quick Guide 7, Reducing Radon, Underground Rooms – Cellars and Basements, Public Health England,* pp. 133–155. Waterloo Road, Wellington House, London SE1 8UG and Building Research Establishment, Garston, Watford.

RICS (2013). *Japanese Knotweed and Residential Property, RICS Information Paper,* 1st edition (IP 27/2012). RICS, London.

RICS (2019). *RICS Professional Statement, Home Survey Standard,* 1st edition. (HSS) Royal Institution of Chartered Surveyors, Great George Street, London.

Rylands v. Fletcher (1868). L.R. 3HL.

Scivyer, C.R. and Jaggs, M.P.R. (2010). *BRE 343, A BRE Guide to Radon Remedial Measures in Existing Dwellings – Dwellings with Cellars and Basements.* Building Research Establishment, Garston, Watford.

Viney, I.F. and Rees, J.F. (1990). 'Contaminated land: Risks to health and building integrity'. *Buildings and Health. The Rosehaugh Guide to the Design, Construction, Use and Management of Buildings.* RIBA Publications, London.

UK Government (2021). Available at https://flood-warning-information.service.gov.uk/long-term-flood-risk/. Accessed 29 September 2021.

12 Non-standard forms of construction

Contents

DOI: 10.1201/9781003253105-12

12.1 Introduction

In the first edition of this book, we focused on what was then called 'non-traditional housing' that covered prefabricated housing built soon after the Second World War and extending into the 1960s and 70s. In the early 1980s, the increasing popularity of the 'right to buy' policy resulted in thousands of non-traditional dwellings entering the private housing sector for the first time. Because many of these house types came with their own inherent problems, the housing market took some time to adjust.

Over the last 20 years, innovation and technological developments have resulted in many other forms of construction entering the marketplace. Although brick and block houses are still the most common type of construction in England, new materials and techniques are now used, presenting residential surveyors with a new set of challenges.

Consequently, we will widen the scope of this chapter to include the following categories of non-standard forms of construction:

- Precast reinforced concrete;
- In situ concrete;
- High rise blocks (50s and 60s especially);
- All types of timber-framed properties; and
- Modern methods of construction.

12.2 Historical development of non-traditional housing

The First World War had slowed down house building and a shortage of skilled labour had resulted in an urgent need for new properties. The Tudor Walters Report in 1918 tried to co-ordinate a centralised view of the industry but largely failed, resulting in only a few industrialised buildings developed by contractors and manufacturers. These included:

- Waller System – precast storey height slabs;
- Dorlonco System – promoted by Dorman Long, used a steel frame with rendered ribbed metal lathing externally and 50 mm clinker blocks internally;
- Atholl House – steel plate externally supported by steel 'T' frame;
- Boswell House – precast reinforced concrete columns at corners with cast in situ concrete clinker cavity walls.

The Second World War led to a similar dislocation of the construction industry. Special legislation was enacted that aimed to produce large numbers of houses quickly. The 'prefab' was the first to be built, followed by the Aluminium Bungalow, the Arcon, Uni-seco and the Tarran types of prefabricated dwellings. The politicians encouraged the use of non-traditional systems to speed up the post-war reconstruction and initially approved the BISF and the Airey House types. Many different varieties followed some with national distributions, while others were limited to specific regions, often in small numbers.

As the 1950s progressed, these non-traditional houses still could not satisfy the demand for house building. Local authorities were encouraged to build upwards through generous government subsidies. Prefabricated high-rise blocks soon overtook low-rise development. As a result, hundreds of different types of non-traditional dwellings were developed and fall into the following categories:

* Precast reinforced concrete (PRC);
* Steel framed;
* Steel frame with brick/concrete/block cladding;
* Timber-framed;
* Cast in situ concrete types.

12.3 Non-standard housing and the market

12.3.1 Housing Defects Act 1984

In the early 1980s, the Building Research Establishment investigated a fire in a precast reinforced concrete (PRC) house type. During this study, they discovered that the concrete had deteriorated through a process called carbonation that had nothing to do with the fire. The extent was so surprising that further investigations were carried out in other similar dwellings. This revealed that a large number of PRC properties either suffered or had the potential to suffer from the same defect. By this time, thousands of people had bought their homes under the right to buy legislation, unaware of the extent of the physical problems with their homes.

After early attempts to award grants to homeowners, the government brought in the special provisions under the Housing Defects Act 1984 (consolidated into the Housing Act 1985). This allowed the Secretary of State or the local authority to designate a particular type of dwelling as 'defective by reason of their design or construction'. Once designated, the owner of the dwelling could apply for a grant to fund either remedial work or repurchase by the original landlord, depending on the financial viability of each option. Each type has a cut-off date before which the owner must have acquired the interest.

A total of 24 systems were designated, which included:

> Airey, Boot, Boswell, Butterly, Cornish, Dorran, Dyke, Gregory, Hawksley SGS, Myton, Newland, Orlit, Parkinson, Reema, Schindler, Smith, Stent, Stonecrete, Tarran, Underdown, Unity, Waller, Wates, Wessex, Winget, Woolaway.

This list contains 26 types, but often 'Schindler and Hawksley SGS' and 'Unity and Butterly' are combined to give a total of 24. Some lenders have a longer list, as they include those dwelling types that were designated in Scotland and may include:

> Lindsay, Blackburn-Orlit, Boot, Dorran, Myton-Clyde, Orlit, Tarran, Tarran Clyde, Tee Beam, Unitroy, Whitson-Fairhurst, Winget.

Only Orlit was designated in Northern Ireland.

The Smith and Boswell Houses were the last to be designated in 1986 and 1987, respectively. This left a large number of other house systems that were not designated, including all of the steel framed types. Figures 12.1 to 12.5 show just a few of the most common types. The BRE have produced a large number of Information Papers and reports on prefabricated house types that are widely available. A few

Figure 12.1 An Airey house. The shiplap concrete panels are the distinctive feature of this house type.

Figure 12.2 Unity house. This type is faced externally with concrete panels.

Figure 12.3 Smith house. This house is clad with clay tiled panels that resemble traditional brick, so it is easy to mistake for a traditional dwelling.

Figure 12.4 Boot house. The externally-rendered exterior can make them difficult to identify. Vertical cracking at the corners is a very common feature.

publications have been identified in the reference section of this chapter under 'Further reading'.

Homeowners had ten years to claim assistance under the scheme. Although local authorities had the discretion to extend the period by a further 12 months, no further grants have been available since March 1998.

Figure 12.5 Boswell house. The Boot and Boswell houses can appear very similar. When they are well maintained and painted, they can be mistaken for a 'Wimpey no fines' house which is not designated.

12.3.2 The PRC scheme

The National House Building Council (NHBC) set up PRC Homes Ltd to administer the licensing and insurance of repair systems associated with the passing of the Housing Defects Act 1984. Various structural engineering firms and other organisations applied and obtained licences for the repair of the different standard housing types. There were three parties involved with the design and execution of the repair work carried out under the PRC Homes Ltd scheme (Kettlewell 1988):

- The **Designer** – a professionally qualified architect, engineer or surveyor who has designed the repair system;
- The **Repairer** – the builder who carried out the repair;
- The **Inspector** – a professionally qualified architect, engineer or surveyor approved to check that the work is in accordance with the original repair system.

If the repair was completed in accordance with the scheme, then PRC Homes provided Repair Insurance, which covered:

- Loss due to repairer bankruptcy;
- All defects that occur during the first two years;
- Structural defects that occur between the 3rd and 10th years after completion.

It is important to distinguish between two terms:

- **Licensed systems** – This is the technical repair scheme of the particular housing system;
- **Licensed scheme** – This is the framework that the repair is carried out within, that is, the design, supervision and insuring of the repair project.

In September 1996, PRC Homes officially ceased approving new schemes and licences and only administered 'run-on' insurance until September 2006. The ownership of the original licences still remains with the agencies that obtained them, so they can still be offered to dwelling owners requiring the repair, but it will not:

- Be carried out by a registered builder (the repairer);
- Be checked by a registered inspector; and
- Will not have a 10-year insurance cover.

Consequently, repair projects carried out using licensed systems since September 1996 must be judged on their own merits. They will not necessarily have the overall assurance of projects carried out under the original licensed scheme but can offer some comfort to prospective purchasers.

12.3.3 Other repair schemes

In addition to the official licensed schemes, non-traditional houses may have been repaired by other means:

- PRC types repaired before PRC Homes Ltd came into operation (1984);
- Local authority designed and supervised schemes;
- Privately organised contracts;
- DIY repairs.

All of these approaches must be judged on their own merits, but many will prove to be inadequate if judged by the PRC Homes criteria. Even when local authorities have used licensed systems, they may not have followed the procedures fully. For example, financial restrictions may have led to a reduction in the scope of the specification, resulting in a different standard of repair. Consequently, when assessing non-licensed schemes, the following criteria should be taken into account:

- Who designed and supervised the scheme? Were they recognised professional people with experience of repairing non-traditional buildings?
- Who carried out the building work? Did they have experience of this sort of work? Were they appropriately supervised?
- Does any certificate of final completion/structural worthiness exist? Is there any form of insurance-backed guarantee?

If these inquiries fail to confirm the quality of the repair work, then the suitability of the property must be seriously questioned.

12.3.4 Attitudes of the lending institutions

If you are providing a condition report that does not include a valuation, the attitude of lenders may not be of a central concern. However, as most properties will be placed on the open market at some time in the future, it is important to make the client aware of the implications of owning a non-standard property.

Lender attitudes appear to have changed over the last 20 years. When the first edition of this book was published, as long as a non-traditional property had been designated and repaired using a licensed scheme, then most lenders would find the property acceptable. This approach seems to have changed. For example, at the time of writing, one major lender lists the following types of property as unacceptable:

- Properties listed under the Housing Defects Act;
- Steel clad houses;
- System built concrete construction;
- Prefabricated reinforced/poured or shuttered concrete construction;
- Easi-form construction (except by Laing from 1945 onwards);
- Mundic block property;
- Timber-framed property with cavity wall insulation unless installed during construction.

This lender clearly had little to do with 'non-traditional housing'. Other lenders may be more selective about particular types they want to exclude, but as long as the dwelling has been properly repaired, they will see it as acceptable risk. Whatever the current view of the lenders, it is important to warn the client about the possible restrictions they may have to face in the future.

12.4 Inspecting and advising on non-traditional properties

Particular care is needed when inspecting non-traditional houses, as many may have been altered or improved. New roofs, rendered finishes and windows can easily mask the main characteristics of non-traditional dwellings. Residential surveyors will be expected to be familiar with the published material of the main systems and have some knowledge of local types and their locality. Many local councils publish useful guides that help with identification.

12.4.1 Inspection strategy

The inspection strategy can be broken into a number of stages:

Stage one

- Identify it as a non-traditional dwelling. Look for the following clues:
- Chimney stack above and below the roof covering. Often constructed of mass concrete or precast concrete blocks;
- Cracks to joints of large wall panels;
- Loft space inspection of party walls. Even if the external elevations have been covered over, concrete sections may still be seen in the loft;
- Size of windows – often quite small in comparison to modern standards because they need to fit between structural members;

- Contained within an area of well-established housing (60 to 80 years old);
- Neighbouring properties. Non-traditional dwellings were always built in groups. Unless all neighbouring properties have been demolished or improved, then a simple drive around the local area might reveal a few dwellings in their original state.

Stage two

- Identify the particular system. If necessary, take photographs and identify later;
- Establish whether the system is designated or not;
- Make a broad assessment of condition by following the trail of suspicion. For example, are there any defects that may cause concern? For example:

  - spalling, cracking, splitting of concrete cover revealing the reinforcement below;
  - bulging or leaning of walls;
  - corrosion of the cladding;
  - poor seals at panel junctions and so on;

- Make a note of any repairs, renovations or extensions that have recently been completed. Ask the vendor for full details of the schemes, including planning and building regulations;
- Obtain any guarantees, reports, specifications that have been produced on the dwelling.

12.4.2 Disguised non-traditional housing

Local authority repair schemes, unlicensed contractor packages and DIY alterations may result in the masking of the essential visual characteristics of many non-traditional dwellings. Figures 12.6 and 12.7 show typical examples. Surveyors have been known to mistake this type completely. Whatever the extent of the improvement work, the best indicators are still the chimney, the loft space inspection and the presence of non-disguised dwellings in the neighbourhood.

12.5 Giving advice on non-traditional housing

When the first edition of this book was published, a significant number of dwellings were still under the insurance-backed guarantee of the original PRC scheme, so establishing whether the repair was covered by an appropriate scheme had practical value. When the second edition is published, the last guarantees would have come to an end some years ago. As a consequence, although any warranty/guarantee will no longer be effective, asking for documentation associated with the repair scheme is a way of establishing whether the work met the appropriate benchmark at the time the contract was carried out.

The following guidance will therefore still have value.

If the dwelling is a PRC-designated type and it has been repaired:

- Ask the vendor for all the documentation associated with the PRC Homes scheme and advise the client that their legal adviser should check the validity of the evidence;

Figure 12.6 A reclad Unity house. The new cladding is easy to spot when the other semi-detached house is unimproved, but what if the whole street has been altered?

Figure 12.7 This Orlit house has been completely refurbished. Not only has it been reclad but extended as well. This changes its proportions completely. The give-away is the pre-cast concrete chimney stack to the left-hand side of the photograph.

- If the repair was not under the PRC Homes scheme, then ask for whatever documentation is available. This should describe a scheme equal to the PRC approach. Even if this is obtained, you should point out that this may not be acceptable to some lenders;
- If no certification is available then it is unlikely to be acceptable in the marketplace until an appropriate structural assessment has been carried out and the appropriate repairs completed.

If the dwelling is a PRC-designated type but is unrepaired:

- For most pre-cast reinforced properties, a comprehensive structural assessment will have to be carried out and an appropriate repair scheme equal to the former PRC Homes scheme should be organised. An appropriate certificate should be kept with the title deeds. Even when this has been done, you should advise the client that this repair may affect the future saleability of the property;
- If the dwelling is an in-situ concrete type the situation is slightly different. Many lenders will often consider this more robust form of construction acceptable for lending if it does not suffer from obvious defects. If significant cracking is present or it was built before 1950, then some lenders require a more detailed assessment of the property.

Other PRC systems

For properties built from a PRC system that was not designated, each one should be assessed on its own merits taking account of:

- Resaleability;
- Location and exposure – some systems are particularly vulnerable to rain penetration for example; and
- Condition and state of repair.

In most cases, a full structural assessment and repair will be required and even then it will be appropriate to warn your client about possible future restrictions on saleability.

12.6 The special case of steel-framed dwellings

Despite representations to the government by groups of owners at the time, no steel-framed dwellings were designated under the Housing Defects Act. One of the main reasons given was that even if a steel-framed dwelling began to corrode, the repair is fairly straightforward. The same could not be said of concrete buildings, where the only way of properly resolving carbonation defects was to remove the concrete components completely.

Therefore, most steel-framed houses have to be assessed on their own merits. Some lenders automatically call for a full structural assessment, while others ask for this only if the original inspection identifies defects that give cause for concern. Further investigation would have to include an assessment of concealed parts of the steel frame

Figure 12.8 The front elevation of a typical BISF house.

and this will require the removal of small areas of internal lining. Not only will this be disruptive, but it will also be expensive.

The main type of steel-frame dwelling is the British Iron and Steel Federation (BISF) type. Assuming a typical example, worrying visual signs might include (see Figure 12.8):

- Cracking to any rendered finishes in the position of underlying steel stanchions or columns. This could suggest corrosion and expansion of the structural frame;
- Excessive corrosion to the steel sheeting to the walls.

12.7 Timber-framed dwellings

12.7.1 Introduction

In this country, houses have always been constructed using timber framing in one form or another. Before we describe how to assess this type of structure, we think it will be helpful to define what we mean when we say 'timber-framed buildings'. This is because it has become a generic term that includes a range of different types of timber structures. For example:

Historic timber-framed properties. These include those properties built using centuries-old techniques and technologies. Cruck construction is a typical example that is made up of pairs of inclined timbers which meet at the apex and are tied together by some form of collar. This is a specialist form of construction and their assessment is beyond the scope of this book.

The 'classic' timber-framed property. Although not a universally recognised classification, we hope you know what we mean. This refers to timber-framed properties developed after World War II by the Timber Research and Development Association (TRADA). This system uses 'open panels' of timber studs, rails and lintels that are either built on site or prefabricated in a factory. Insulation is usually placed between the studs and then sheathing and breather membranes are fixed to each side. The outside can be clad with a number of different materials.

Timber-framed dwellings using modern methods of construction. These versions use innovative techniques and technologies, such as:

- Prefabricated wall panels are insulated with polyurethane with oriented strand board sheathing (or OSB for short) fitted on the outside and plasterboard liner fixed to the inside in the factory; and
- Similar to above but using timber 'I' section vertical sections rather than softwood studs. This allows for thicker timber wall panels that can accommodate greater levels of insulation.

The one thing all these different types have in common is the vertical timber components support the imposed loads. This distinguishes it from other innovative methods such as structurally insulated panels described in Section 12.10.

12.7.2 The change in timber-frame construction

Figure 12.9 shows how this type of construction has changed over the years. The top sketch shows an early timber frame common in the 1960s and 1970s with minimal thermal insulation – this could be as little as a 25-mm layer of fibre-board in some cases. A layer of breather paper was fixed to the external face of the studs with the cladding fixed over. Foil-backed plasterboard was fixed to the internal side of the studs to provide a vapour check.

Changes in the building regulations and general development of the various systems resulted in the 'classic' modern timber-framed construction consisting of the following (see Figure 12.9 lower sketch):

- **Cladding** – this is usually brick but can be any masonry material including stone;
- **Cavity** – usually 50 mm;
- **Breather membrane** – a micro-porous layer that allows water vapour to move outwards but will not let water in;
- **Sheathing plywood or other boarding** – usually nailed or stapled to the timber studs. Very important for the stability of the structure as a whole;
- **Timber-frame** – usually ex 50 x 100mm sections set at 400–600 mm centres;
- **Insulation** – this usually fully fills the cavity within the timber frame. Commonly mineral fibre quilt, although polyurethane foam is sometimes used in factory-built components;
- **Vapour control layer** (VCL) – usually a robust polythene sheet. Because it is punctured by nails and other fixings, the name 'vapour barrier' has been dropped. 'Vapour control layer' is a more accurate description;
- **Interior lining** – usually plasterboard but more recently built properties may include additional layers of insulation to meet the thermal requirements of the building regulations.

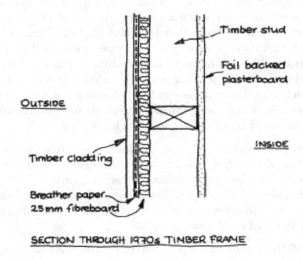

SECTION THROUGH 1970s TIMBER FRAME

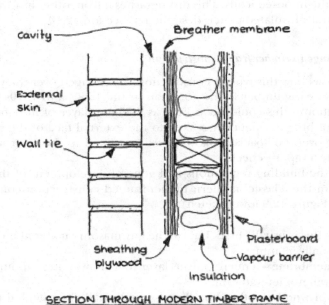

SECTION THROUGH MODERN TIMBER FRAME

Figure 12.9 Sketch section through two different types of timber-framed walls.

The BRE's Good Building Guide titled 'Timber-framed construction: an introduction' has a more detailed review of this method of construction if you want to deepen your knowledge (BRE 2003).

Cavity fire barriers (this can go in an information box)

Another feature of timber-frame construction is the importance of non-combustible fire barriers. These are installed in the cavities to reduce the risk of flames and hot gases spreading around the property and to the neighbouring dwellings in a terrace or block of flats. Typical locations include corners of the building, party wall positions, intermediate floors, roof/wall head junction and around openings (BRE 2003).

Despite the importance of these fire barriers, recent investigations have revealed that a major house builder failed to install the barriers in a large number of new properties, creating 'an intolerable risk' (Independent 2019). At the time of writing, the house builder is undertaking a programme of investigations and replacing the barriers where they are found to be missing.

During your visual inspection, it is impossible to check whether cavity fire barriers have been properly installed. The only outward sign on some timber-framed properties is the presence of weep holes that drain the cavity tray above a long horizontal cavity fire barrier between floors. Even if the weep holes are present, it gives no indication whether the fire barriers are actually there.

So what do we do? This is yet another challenge to our role of giving balanced advice. On the one hand, the issue is in the public domain, so we need to take account of it during our assessment. But on the other hand, we do not want to cause unnecessary concern. The final judgement will always be yours, but here are a few considerations you may find helpful:

- Recently built timber-framed properties seem to present the biggest problem;
- Local knowledge is important. All residential surveyors should be building up information on their locality and keeping a watching brief on new developments is part of that process. Ideally, you should be making a note of the developer and constructional methods used on new sites in your area;
- You should also use the internet to search for media coverage of complaints/campaigns about defects in housing on particular estates/areas;
- Where possible ask the vendor directly if the property has been subject to any defects investigations and in particular the effectiveness of any cavity fire barriers. You may want to recommend to the client that they ask their legal adviser to seek answers to the same information, too.

12.8 Assessing timber-frame properties

Although timber-frame construction is increasing in popularity and modern forms are fully approved by local building control departments and the majority of warranty providers, they have to be designed and constructed correctly. They are dependent on satisfactory construction practice because of the vulnerability of the timber components to deterioration through the effects of two main sources of moisture: directly penetrating moisture from the outside or interstitial condensation caused by water vapour on the inside

There are two important stages in the assessment of timber-framed buildings:

- **Identifying that the dwelling is timber framed in the first place**. This is important for a number of reasons:
    - Timber-frame structures perform very differently from masonry buildings and confusing the two types could result in you missing some important problems during the inspection;
    - Some lenders and building insurance companies can have very particular attitudes towards some forms of timber-frame properties. Therefore, it is important for the client to know the true nature of the property's construction;

- **Assessment**. The inspection and assessment of this type of property is different to masonry buildings and shows the importance of identifying the difference between the two.

These will be considered in turn.

12.8.1 Identifying timber-framed dwellings

It is very easy to miss that a property is timber-framed, especially when carrying out an inspection for a valuation or a level one service. Changes in building regulations and technological developments have made this even more difficult. Consider the following:

- 20 years ago, one of the simplest ways to identify a timber-frame construction was by lightly tapping the inside face of the outside walls. Timber frame would give a hollow sound, while a masonry wall was usually plastered directly and would give a solid sound. However, many modern brick and block masonry properties are now dry lined internally, giving a similar sound to timber-frame walls when tapped;
- In 2008, when we wrote the 'Domestic Energy Assessor's Handbook' (Parnham and Russen 2008), measuring the thickness of external walls was considered the best way of helping determine the nature of an outside wall. There is merit to this approach if construction methods remain largely unchanged. This is not the case today. Brick slips instead of outer leaf of cavity walls; wider cavities and thick aerated concrete inner skins can result in wall thickness ranging from as little as 150 mm through to the 350–400 mm range.

In other words, it is not as easy as it used to be. Faced with this variation in constructional methods, it is important to consider a range of indicators (Marshall et al. 2014). A number of these are described in the following pages, but you should also bear in mind you could see a number of these on masonry-built dwellings too:

Externally

- If the outside of the wall is covered with a lightweight cladding (such as timber boarding, timber shingles or cement render on lathing) the wall will be much thinner when compared to a masonry wall;
- Windows tend to be fixed to the timber stud inner skin resulting in deep reveals to the openings;
- Movement joints with mastic infills are often provided around all the windows, beneath the eaves and sometimes at the party wall positions. On three-storey properties, these gaps can be up to 15–18 mm wide;
- Weep-holes at all the normal locations (for example, above the openings, just above damp- proof course and so on) but also at first and other intermediate floor levels. This may indicate a cavity tray DPC over a horizontal fire stop.

Internally

- The party walls in the roof space may have a plasterboard or mineral board finish. The internal face of the gable wall may also be similarly lined. In older timber-framed properties this boarding could be asbestos-based;

- Where a property has a gable wall, this is usually formed from a spandrel panel;
- Some older dwellings may also have 'cross-walls' of brick or block with only the front and rear walls consisting of timber frame;
- If it can be seen, the wall plate of a timber-framed dwelling that supports trussed rafters will be in 'planed' or finished timber. This is because it is usually made in a factory. Wall plates of traditionally built properties tend to be in 'sawn' or rough timber;
- When knocked, the internal linings of the external walls will give a hollow sound. Care must be taken, as dry-lined solid walls can also have a similar response.

12.8.2 Assessment

There are a number of publications that advise on inspection routines that should be followed when assessing the condition of timber-framed dwellings (BRE 1995, BRE GBG 11 and 12 and TRADA 2021 Survey of timber-framed houses, wood information sheet, number 10). Many of these involve a variety of destructive tests, including drilling into sole plates, removing plasterboard and skirtings and inspecting fire barriers in cavities. All of these are beyond the scope of most standard inspections. Therefore, once a dwelling has been identified as timber-framed, you should look for a number of visual key indicators that may trigger the need for further investigations.

Key inspection indicators

EXTERNALLY

- Establish whether the cladding is fixed directly to the frame. These can be more vulnerable than those that have a cavity between the cladding and the frame;
- Determine the age of the dwelling and if possible the system used. Older timber-framed dwellings will be more likely to have problems (say before 1985);
- Make a note of the orientation of the dwelling and the exposure of the surrounding terrain. Fully exposed sites warrant greater levels of attention during the inspection, especially the elevations that face the worst of the weather;
- Look out for evidence that the cavities of the external walls have been filled with insulation. Large 10–12 mm circular drill holes at the junction of the vertical and horizontal mortar joints are the typical signs. Insulated cavities will dramatically affect the build-up of moisture within the wall, creating ideal conditions for rot to develop. Some lenders will exclude timber-frame properties that have had their cavities insulated.

Weather tightness of the building envelope. Critical areas include:

- Chimney to roof details – especially important as frame shrinkage may cause maximum movement here. Check inside the loft as well;
- Junction of different claddings;
- Deteriorated claddings – any damaged, decayed or displaced cladding panels or boards. Prod for decay, especially on the most exposed elevation at corners or near foot of wall;
- Moisture affected patches, green staining or staining on cladding – this suggests excess moisture and should be investigated further;
- Seals around all openings – check that the mastic seals are in good condition and still effective. This also applies to any expansion joints.

Structural integrity – Most problems are associated with brick-clad housing and are attributable to inadequate allowance for movement between the frame and cladding. This can result in cracking, bowing and leaning in the cladding, especially around openings. Consider the following:

- **General cracks on the face of the cladding that are not necessarily related to openings or junctions.** These can be caused by inadequate foundations, particularly in the vicinity of new additions such as extensions and porches. If this is the case, further investigations as described in Chapter 5 will be required;
- **External cracks at the eaves, verge, generally around windows and doors and at junctions between different claddings.** These can be due to inadequate allowance for differential movement between the frame and cladding. For a conventional timber-frame building, the timber-frame itself will shrink as it dries out, while a brick out skin will expand as it takes on moisture. For a three-storey dwelling, this could result in movement of between 20 and 30 mm and result in sealant displacement and/or sill and lintel rotation.

  Depending on the scale of the movement and the age of the building, retrofit expansion joints may be required and sills and lintels may need to be rebedded.
- **Bowing/bulging/leaning in the wall.** When sighting across the outside face of the walls, distinct areas of bowing/bulging/leaning can be seen. This can have a number of causes:

  - Inadequate number or incorrect positioning and fixing of wall ties between the frame and cladding;
  - Sulphate attack on the masonry. If this was the case, then you would expect to see some of the signs described in Chapter 5;
  - Differential movement between the frame and cladding as described above. In this case, the outer skin could be restrained by the eaves/verge above forcing the brickwork to bow outwards;
  - Leaning outwards of the wall especially on the gable wall. This could be caused by the overturning effect of the trussed rafters, which are inadequately braced.

According to the BRE (BRE 1993, p. 3). If the bowing from any cause exceeds 20 mm, then remedial work may be required and could typically include:

- The number and distribution of the wall ties will have to be assessed and new ones retrofitted if required;
- For sulphate attack, some amount of rebuilding may be inevitable;
- As previously described, differential movement may require retrofitted expansion joints; and
- If an unbraced roof is the cause, then a structural assessment may be appropriate.

INTERNALLY

Linked to the external inspection, visible internal signs can help to confirm a potential problem and justify the need for further investigation. There are three main defects to look for:

   Moisture-related – Look for evidence of water staining, mould growth, breakdown of plaster surfaces and decorations and so on. You should pay particular attention to the lower parts of the outside walls and the adjacent skirtings. The sole plate at the base of the wall panel will be just behind the skirting and can be vulnerable to moisture problems. Although the BRE recommend drilling holes through the skirtings into the sole plate so the 'deep probe' attachments to moisture meters can be used, this goes well beyond the scope of most inspections. Consequently, you will be restricted to taking moisture meter readings to the skirtings and the lower part of the wall lining.

Structural problems – cracking or bowing of the floors and ceilings, splitting along plasterboard joints and popping of nail heads. According to the BRE, nail popping is usually due to movement in the frame in the early life of the building, but can also be caused by poor workmanship. In the worst cases, the plasterboard may need refixing but is unlikely to reoccur, as future movement should be small and unlikely to disrupt the refixed plasterboard.

Evidence of alterations or repair – timber-framed dwellings are sensitive to inappropriate repair and alteration work. For example, if a vapour control layer is punctured or a plywood sheathing board is cut through, the integrity of the whole unit may be affected. Look for the following indicators:

- Plumbing leaks and inferior plumbing installations;
- Changes to services that may have damaged the vapour control layer or affected the fire resistance of the party wall. Examples would include new electrical sockets, TV aerials and so on;
- Installation of new windows that may not be properly sealed with DPCs or provided with suitable movement joints to replace the ones that would have been disturbed;
- Structural alterations that may have weakened internal load-bearing partitions. For example, 'through' lounges, new door openings, serving hatches, enlarged bedrooms and so on;
- Sagging or springy floors;
- Any gaps or spaces through or over the separating walls in the loft that may compromise the fire resistance.

12.9 Giving advice about timber-framed properties

Like all assessments, giving advice about the condition of a timber-framed house is a matter of judgement. No one sign or symptom should automatically result in a recommendation for further investigation. It will often be a matter of balancing the inadequate features against the damage that is likely to arise.

12.10 Modern methods of construction

According to the National House Building Council Foundation (NHBC 2018) 'Modern Methods of Construction' or MMC for short is a wide term, embracing a range of offsite manufacturing and onsite construction techniques that provide alternatives to traditional house building. MMC ranges from whole homes being constructed from factory-built volumetric modules to the use of innovative techniques for laying concrete blockwork onsite.

Seen as a possible solution to the housing shortage, the government has sought to encourage the house building industry to adopt modern methods through a number of trials and initiatives. Although developers appear to be engaging with some aspects of these innovative techniques, there is yet to be a breakthrough in terms of housing completions (NHBC 2018).

12.11 Drivers for change

Some of the drivers for a change in the way we construct our dwellings include (NHBC Foundation 2006):

Skill shortage: Underinvestment in training in the building industry in recent years has led to a decrease in skill levels in site-based operatives, and this has affected the quality and quantity of the housing produced. As most of an MMC dwelling is 'built' in the factory where operatives are usually directly employed, there is a greater incentive to invest in training and this can result in higher quality products.

Housing shortage: Estimates have put the number of new homes needed in England at 345,000 a year (House of Commons 2021). Library briefing paper *'Tackling the under-supply of housing in England'* (14 January 2021). Yet in 2019/20 the housing stock increased by only 244,000 – well short of the government's own targets. Many commentators see that an invigourated MMC sector can help fill this gap.

Poor quality of new housing: The National Consumer Satisfaction Survey in recent years has shown low levels of satisfaction with new housing, especially around snagging issues. Construction in factory-controlled conditions is seen as one way to improve these figures.

Tougher building regulations: In recent years, the building regulations have broadened to cover the performance of buildings, particularly in relation to thermal and acoustic standards. These indicators can only be tested after construction and if they fail to come up to standards, costly remedial work can be the result. Many builders look to MMC as a potential way of providing a more predictable performance in the completed dwelling.

Types of MMCs

Although there are no formal definitions of what constitutes a modern method of construction, most commentators seem to agree on the following classification:

- **Offsite volumetric:** Building systems composed of three-dimensional units (volumetric modules), produced in a factory and fully fitted out before being transported to site. These modules can be stacked onto prepared foundations to form the dwellings.
- **Offsite panelised:** Building systems composed of two-dimensional units that are typically manufactured offsite and assembled onsite. Panelised systems include timber frame, light-gauge steel frame, cross-laminated timber and structural insulated panels.
- **Offsite Hybrid types:** Construction that combines a volumetric pod system with a panelised system.
- **On site sub-assemblies and components:** An assembly of building components to form a building element which is then incorporated with other elements to form the building. Examples include prefabricated chimneys, porches and dormers.

12.12 Assessing MMCs – the challenge of change

This may sound like a pretentious heading, but if the last 20 years is anything to go by, then the next few years will see a remarkable change in how we build houses in this country. Already many builders have been using new on-site sub-assemblies and components for a number of years. Examples include metal web floor joists, prefabricated chimneys and porch roofs, to name but a few. This has already blurred the distinction between 'traditional' and 'modern methods' of construction. If the construction industry does move towards a greater use of MMCs, then this will be a natural evolution from past techniques.

However, commercial, political and environmental pressures are likely to accelerate this rate of change. For example, the current concern about global climate change may speed up the development of MMCs in an effort to reduce the carbon footprint of the construction process even further. The challenge for residential surveyors is to be able to provide a professional service in this rapidly changing environment.

In the eyes of the many commentators in the construction sector, residential surveyors are often seen as a conservative profession (with a very small 'c') that is resistant to change and new developments and acts as a brake to innovation. Although there is some justification for adopting a cautious approach (PRC housing, Ronan Point and Grenfell Tower, to mention but a few), we think we have no option but to engage more fully with these new and developing technologies while continuing to provide an independent and objective service to our clients. Although we do not have a crystal ball, we think the following pointers will help you build an appropriate strategy.

12.13 The main types of MMCs

A quick search on the internet will reveal a growing number of modern methods of construction. Here is a description of just two that are commonly used at the time of writing:

Insulated concrete formwork (ICF): This system is based on large hollow polystyrene components that interlock together without intermediate bedding materials to provide a formwork system into which concrete is poured. Once set, the concrete becomes a high-strength structure and the formwork remains in place as thermal insulation, with very low U-values.

Structurally insulated panels (SIPs): This is a sandwich construction comprising two layers of sheet material bonded on to a foam core. This panelised system does not rely on internal studs for structural performance. The SIP panels are used for both wall and floor elements.

We do not have the scope to cover the increasingly numerous types of other MMCs. Normally, we would have recommended Keith Ross's excellent book 'Modern methods of housing construction' (BRE 2005, FB 11) as a starting point for improving your knowledge in this area, but this has been 'archived' by the BRE and is no longer available. Useful sources of information include the Building Research Establishment Bookshop, as well as the increasing number of publications from the National Housing Building Council's 'Foundation' (see www.nhbcfoundation.org/).

12.14 MMC and the marketplace

12.14.1 Lender attitudes to MMCs

Although this book seeks to address condition-related matters, it is important to consider the context in which our services are delivered, and the attitudes of lending organisations is an important influence. New and developing technologies have always presented a challenge for lenders: How can they assess the lending risk on a property built using a construction system that does not have a track record? The Council for Mortgage Lenders (now UK Finance) proposed a set of criteria that should be used to judge MMC schemes:

- Durability: it should achieve a life span of at least 60 years;
- Whole life costs: the system should be comparable to traditional construction;
- Reparability: no undue repair costs, and ability to use a range of local repair services;
- Adaptability: the property should, without difficulty, support the usual range of adaptations/extensions such as a porch and conservatory;
- Insurability: buildings insurance should be available on normal terms.

Assessing a property to this level of detail is outside the scope of most point-in-time' residential services. However, picking up just three aspects will show how problematic the assessment of MMCs can be.

12.14.2 Insurance

From a building insurance point of view, modern methods of construction are often classified as 'non-standard' properties and put into the same category as Cob, wattle and daub and straw bales and will attract an additional premium. Additionally, if the nature of the dwelling's construction isn't properly declared to the insurance company in the first place, there is a risk that the insurance policy could be nullified. Consequently, it is important not only to identify the true nature of a property's construction but also to make sure your client understands the possible insurance implications of owning such a home.

12.14.3 Repairability

When 'traditionally' built properties are damaged or deteriorating, the construction technologies are generally well understood (by traditional we also include the 'classic' modern timber frame). Repairs can usually be procured in a straightforward manner in terms of operative skills, components and materials. By their nature, MMCs are new and unfamiliar and use innovative materials and techniques in novel ways. In some cases, unsuitable repairs will not only be ineffective, but may also undermine the performance of the system as a whole. Take structurally insulated panel systems as an example.

In conventional timber-framed construction, the primary load paths are the vertical timber studs and the exterior sheathing board and internal linings serve as lateral bracing for the panels. If a portion of a timber-framed wall is damaged (say, a localised fire or excess moisture), normally the affected area can be removed and repaired as the adjacent studs can take the weight.

SIP systems work differently. The panels intentionally eliminate the timber studs for reasons of thermal efficiency and the loads are instead carried by the two OSB skins. The adhesion between the skins and the foam core braces each panel, preventing buckling and allows the panel to function as a composite member. From a structural point of view, SIPs are designed to be uniformly loaded. Where concentrated loads occur (for example, purlins in the roof), supplemental vertical timber supports are usually provided.

If a portion of an SIP panel is similarly damaged, it cannot simply be cut out because it only works as a composite. Anything that affects the bond between the OSB and the foam core will directly reduce the wall/roof/floor strength and stiffness. Consequently, where a panel has been damaged, it should be initially assessed by a structural engineer or other suitably qualified person. Unlike with classic modern timber frames, it is unlikely that a general building contractor will be able to make the assessment. If the damage is structural, temporary support may be required while repairs are carried out. If the repair includes the use of conventional timber studwork, then the SIPs thermal efficiency would be compromised.

Manufacturers of some modern steel-framed properties also recommend that any repairs and alterations should be designed and supervised by a structural engineer.

The outcome of this added complexity can be summarised as follows:

- MMCs often use unfamiliar technologies and it is unlikely the local building repair sector will not have the knowledge and experience of this type of system;
- Compared to traditionally built properties, the involvement of suitably qualified professionals in the assessment and design of repairs schemes is more likely with MMCs; and
- This added complexity may result in more costly repairs that are more difficult to arrange.

Additionally, where you see evidence of past repairs, you should ask how these were procured in an attempt to make sure they were properly carried out.

We think it is important that you highlight these matters to the client in a balanced way so they understand the implications of owning a dwelling that includes an MMC component.

12.14.4 Adaptability

The owners of all properties will want to make changes over time. Adding an extension or conservatory, removing an internal partition to form a through lounge or refitting a kitchen or bathroom are all common projects even in relatively new homes. Although these will all be possible in dwellings built using MMCs, they may have to be done in accordance with manufacturers' instructions. For example, how would an opening through an external wall built with insulated concrete formwork be formed? It may not be rocket science, but it will have to be done properly. Similarly to the repair process described above, not only should you check all the necessary approvals are in place, you will have to try and find out if the constructional techniques used for the alteration are suited to that particular system.

12.15 Identifying MMCs

Although the development of modern methods of construction is fast changing, the most important first step in properly advising clients is recognising that the property been built using an MMC in the first place. This process is similar to that described for distinguishing between the 'classic' modern timber-framed property and those built with masonry but with added complications. For example:

- Many MMC dwellings are clad with brick slips and depending on the levels of workmanship, it can be challenging to tell them apart from conventional brickwork;
- In SIP structures, the panels are also used to form the roof creating a 'room in the roof' habitable space. Consequently, there will be no roof space or void that will give an easy insight into the building's structure.

As mentioned previously, if you do not identify that a property is formed by an MMC, it could have serious implications in terms of breaching lender guidance and creating insurance underwriting issues.

Using some of the excellent advice put forward by Ross (BRE 2005) and our own observations, we have identified some of the visual clues that may help you identify whether an MMC has been used.

12.15.1 General

The following pointers may help you identify those properties that are most likely to have been built using a modern method of construction:

- The likelihood of MMC use increases after the 2000–2002. In this initial period, local authorities and housing associations were the principal users of the systems;
- Smaller, one-off developments and single dwellings commonly use MMCs. We call it the 'Grand Design' effect, as self-builders are often more prepared to use innovative techniques;
- On larger sites and estates, developers often use a variety of construction methods including traditional technologies. So where you have strong evidence one dwelling was built using MMCs, then don't assume the rest of the development follows suit.

12.15.2 Desktop studies

For all instructions, residential surveyors should carry out appropriate research before carrying out the inspection (see Chapter 3 for more information). The process for MMCs is no different, but there should be an emphasis on particular aspects.

- **Seller's questionnaire.** Although many surveyors use the outcomes of this process with caution, it can be very important for newer properties. You may want to ask the vendor 'whether a modern method and innovative form of construction was used to build the property' and make a note of the response in your site notes. In an effort to be fair to property owners, they might not know their house is an MMC, as many developers do not mention in their marketing and sales particulars;

- **Local knowledge.** As previously discussed, you should be keeping a note of what is being built in your area for future reference. Occasional 'drive-by' visits during the construction phase can often help you identify the structural system. It is important to keep this information in an accessible database so that you can quickly check the information if you are instructed to inspect one of the properties in the future;
- **Google Street View.** We would like to add the phrase 'other mapping systems are available', but to our knowledge none are as useful as Google Street View. In many areas of the country, Street View has a number of historical images that go back to 2007/8. In some cases, we have found eight or nine images spread over 12 or 13 years. In one particular instruction, these images revealed the whole development process for a small site that used MMCs. The images showed the site being cleared, after the ground floor SIP panels had been installed, with most of the roof SIP panels in place and one showing the cladding being fixed – invaluable information that helped us precisely identify the system (we even got the phone number of the contractor). Not all properties will have this number of images, but it should be part of your standard research.

12.15.3 *External inspection*

Like conventional construction, MMCs can be clad with a whole variety of materials. Rather than go through a list of possible alternatives, we will focus on the various types of brick claddings. This is because brick slips fixed directly to the structural panels can render a property unmortgageable, as many lenders require a drained and ventilated gap between the cladding and the structural panel.

The following visual clues may help you distinguish between the two:

Weep holes: In a wall with a drained cavity, weep holes are usually positioned over window and door openings, at ground level and over other penetrations such as meter boxes and flues. Weep holes are not needed on brick slip claddings fixed directly to the structural panels because there wouldn't be a cavity.

Reveals to opening: In general terms, where a half brick thick outer skin and cavity has been used, the reveals will usually be at least 100 mm deep. With a brick slip cladding, the depth is often less.

Lintels and sub sills: Where stone and concrete lintels sub sills are used, the cladding is likely to be a separate brick skin with a cavity behind. This is because it would be impractical to incorporate these relatively heavy features in brick slip cladding.

Damp-proof courses: If conventional brickwork skin is used to clad the building, then you should be able to see a DPC at the usual level. Where brick slips are fixed directly to the frame or panel, DPCs are not usually required.

12.16 MMC and certification

Since the judgement in the court case Hart v Large (see Volume 1, Chapter 5), residential surveyors have sought ways of helping their clients get the necessary reassurances that a newly built or refurbished property has been constructed to a satisfactory standard. This is particularly relevant to dwellings incorporating modern methods of construction because of the wide and constantly changing products in the market. For example, few lenders will give blanket approval to a construction system and will want each property independently assessed. This is because although the main structural system might have the necessary accreditations, the roof, walls and other secondary elements may not. One lender requires the following information:

- The principal constructional material (for example, insulated concrete formwork (ICF), structurally insulated panels (SIPs), steel frame and so on);
- Details of the roof and wall finishes; and
- Details of a mainstream warranty that covers the development.

Although condition reports do not have to follow lenders' criteria, this framework does provide a useful benchmark for assessment and also sets the 'norm' for the sector. The information covering the first two points should be gathered during the inspection, so the rest of this section will focus on the available warranties/guarantees that can cover a property.

12.16.1 Warranties for MMC dwellings

Warranty schemes for newly built properties have been around for some time and are generally well understood. However, these apply to traditionally constructed dwellings and many are not suitable for the range of novel and unusual construction methods that have been developed over the last 20 years. This has often resulted in some types of MMCs not being accepted by lenders.

In an attempt to encourage innovation and avoid this type of impediment, a range of key stakeholders launched the 'Buildoffsite Property Assurance Scheme' (or BOPAS for short). In essence, BOPAS is a risk-based evaluation which demonstrates to funders, lenders, valuers and purchasers that homes built from non-traditional methods and materials will stand the test of time for at least 60 years. Put simply, it works like this:

- Manufacturers apply to BOPAS and their systems are checked to ensure the construction method meets agreed lender standards for mortgage purposes. If they are successful, the system is placed on the BOPAS database;
- When residential surveyors are inspecting a property and discover the form of construction is unconventional, they can access the BOPAS database and see if the construction method/system has been approved under the scheme;
- Valuers can then provide a valuation taking into account the normal criteria, but confident that the construction technology has been approved for lending purposes and, additionally that the Latent Defects cover on the completed unit meets lender requirements (www.bopas.org/about/how-the-scheme-works).

Over the last few years, a number of other warranty schemes for MMCs have become available, including:

- **NHBC Accepts** – according to the NHBC, this is a comprehensive review service for innovative products and systems. It offers a fast-track route for acceptance of these products and systems for use in homes covered by all NHBC warranty and insurance policies; and

- **LABC Systems Acceptance** – provided by the Local Authority Building Control (LABC), this is aimed at developers intending to use modern methods of construction. They have an Innovations Team that can provide guidance on the suitability of systems before work starts on site. In this way, as long as the construction system meets with the LABC's technical requirements, then the development will be covered by an LABC warranty. However, as the LABC state on their website, the 'LABC Warranty system acceptance does not constitute a third party product approval from a UKAS accredited testing body or any form of building regulation approval. The LABC Warranty acceptance is purely to recognise that the system can meet our warranty requirements and must not be considered to be anything else'.

This last issue is important. Although a property that incorporates an MMC system may have a warranty from a major provider, you should not assume it covers all parts of the construction. Although this may be difficult to assess during a typical instruction, you should make it clear to your client of the importance of having the appropriate warranty/guarantees in place so that any potential risks are minimised.

To help you advise your client, we have outlined an informal 'hierarchy of certification':

- **Third party certification:** This will typically include a British Board of Agrement certificate or certification from the Building Research Establishment. Although this will vary between providers, the certification must cover the structural system AND the roofing and walling elements used in combination;
- **Warranty schemes**: The property should be covered by 10-year, insurance-backed warranty from one of the major providers that specifically covers the MMC system as well as all the other elements used in the dwelling's construction;
- **Building regulation approval:** This should include the final completion certificate;
- **Manufacturers quality schemes:** In some cases, manufacturers can offer their own quality schemes such as 'Q-Mark' which is the UK Timber Frame Association Quality Scheme. Although these quality schemes can provide some reassurance, on their own, they are not sufficient;
- **Competent person scheme:** Like more conventionally built properties, a relatively large number of elements of the building may have been signed off by contractors who are part of government approved competent persons schemes. Common examples will typically include heating, electrics, hot water, and microgeneration installations, to name but a few.

If an MMC-type property has the majority of the approvals/warranties/certificates, it may lower the likely risk of future problems, but this assessment must be done on a case-by-case basis. You should:

- Clearly advise your client to instruct their legal adviser to investigate the status of the property, its associated guarantees and warranties and explain the limitations if the full range of approvals/warranties are not in place; and
- Check with their chosen building insurers to make sure insurance will be available for this type of dwelling at market rates.

PROFESSIONAL CONSULTANTS CERTIFICATES (PCCS)

In place of a warranty, lenders will often state that a Professional Consultants Certificate, in certain circumstances, will be an acceptable alternative. The status and reliability of a PCC has been hotly debated. In this section, we have attempted to outline our opinion of what can be a less-than-effective substitute for a conventional warranty.

Lenders' attitude to PCCs   As UK Finance include the following description of a PCC in the Lenders Handbook:

> Lenders will generally only lend on a newly built, converted or renovated property where the property is covered by a warranty scheme or a Professional Consultant's Certificate (PCC). The PCC is for use by professional consultants when designing and/or monitoring the construction or conversion of residential buildings.
>
> (UK Finance 2021)

They go on to say that the purpose of the PCC is to confirm to the lender or its conveyancer that a professional consultant:

- Has visited the property during construction to check its progress, its conformity with drawings approved under building regulations and its conformity with drawings/instructions issued under the building contract;
- Will remain liable to the first purchasers and their lender and subsequent purchasers and lenders for the period of 6 years from the date of the certificate;
- Has appropriate experience in the design and/or monitoring of the construction and conversion of residential buildings; and
- Will keep a certain level of professional indemnity insurance in force to cover their liabilities under the certificate'.

UK Finance also provide a template of a typical form that can be used as a Professional Consultants Certificate. This is a very simple form that essentially requires the professional to confirm a number of essential requirements and is usually completed by hand. The Building Societies Association (BSA) follow a similar approach.

Retrospective PCCs   In our opinion, this extract provides us with a clear rule – UK Finance and the BSA will only accept PCCs that have been issued by the professional who '. . . visited the property during construction'. It follows that PCCs issued after the building is complete by professionals who were not involved in the design and/or visiting during the construction phase are not normally acceptable to lenders. Anecdotal feedback from lenders' representatives confirms this.

Despite this unambiguous position, a cursory internet search for 'PCC' quickly reveals a large number of organisations that offer to provide retrospectively issued professional consultants certificates that are approved by '. . . the Council for Mortgage Lenders'. In our view, these are not acceptable because a professional issuing a 'retrospective PCC' would not have inspected the construction at the different stages of the build process. Consequently, it would **not** conform to UK Finance and BSA requirements and would **not** give any reassurance or comfort to clients about the quality of the construction. Although retrospective PCCs are usually based on a visual inspection of the property, important parts of the building will remain unseen, such as the depth of the foundation trench, the damp-proof course and the nature of concealed parts of the structure, such as beams and columns. In these situations, retrospective PCCs merely act as an insurance policy designed to provide the lender with enough reassurance so that the loan and sale can go ahead. This is far from providing the buyer with reassurance that the building work is satisfactory.

Consequently, it is our view that you should advise your client of the disadvantages of relying on retrospective PCCs and accept only those issued by the professional who inspected the building work at the appropriate stages.

12.17 Defects in properties built using modern methods of construction

It is difficult to concisely describe common defects with modern methods of construction for two reasons:

- There are so many different types in the marketplace (and often used in combination) ranging from those systems based on the use of mass concrete through to relatively lightweight steel frames. Each type will perform differently to others;
- At the time of writing many of these systems are less than ten years old, so there is little information on performance in the public domain that can help the assessment of performance.

The other challenge is that these building systems can react very differently to the normal agents of deterioration when compared to traditionally built properties. For example, consider a brick and block cavity wall. When affected by subsidence, this type of wall will usually show characteristic diagonal cracking and other signs that can help in the diagnostic process (see Chapter 5). When excessive levels of water get into the cavity and cross over to the inner skin, the signs on the internal face are usually predictable and closely associated with the source of the problem (see Chapter 6).

Now consider a property that has been built using insulated concrete formwork walls and clad with brick slips glued to a backing board that has a 15 mm ventilated and drained cavity behind. Additionally, it has a structurally insulated panel roof structure covered with concrete tiles. The way this structure reacts to a loss of support at foundation level or rainwater getting past the cladding will be very different. Because the walls are of mass concrete, subsidence cracking patterns will be related to points of weakness (for example, between openings) and possibly remote from where the foundations have actually failed. Rain water penetration into the cavity could go unnoticed for long periods of time, whereas the junctions between ICF walls and the doors/windows will be vulnerable to water penetration because they are not usually protected by a deep reveal and rely on a bead of mastic as the sole protection against rain water penetration.

Consequently, we have focused on problems normally associated with modern methods of construction. Mostly these are generic, but some are specific to particular types of systems. What is common to them all is that we have focused on problems that can usually be seen during a normal inspection of a residential property without any 'opening up' work or the use of specialist equipment such as borescopes, infrared cameras and the like.

Important note: where you are assessing an MMC that incorporates timber panels, then much of your assessment will follow a similar process described for modern timber-framed buildings described in Sections 12.7 to 12.9 of this chapter.

12.17.1 Workmanship and extemporisation ('making things fit')

Most MMC systems rely on accurate setting out of the building, as the tolerances between components can be relatively small. Although anecdotal, our research has revealed that the constructors often have to adjust and alter the various parts to make them fit together on site. In most cases, this may involve only minor adjustments, but if excessive alterations could affect the integrity of the system as a whole.

When inspecting the property, try to find those spaces where you can get an insight into how the system has been put together, such as accessible roof spaces, services ducts and spaces, storage cupboards, undercrofts, cellars, basements and so on. In these areas, look out for:

- Components that have little or no bearing;
- Parts of the system that have been cut or shaped so they can fit together; and
- Sides of adjacent panels that do not fit together properly leaving gaps that can be penetrated by moisture and fire.

Depending on their extent, these makeshift alterations can act as a 'trail of suspicion' and cast doubt on the quality of the build. In this case, you should inform your client and to ensure appropriate warranties and guarantees are clearly in place.

If you consider that the problems are so serious, then a further investigation may be required. To be done properly, parts of the construction may have to be 'opened up' and will clearly have an impact on the client's purchase decision.

12.7.2 Moisture-related problems

The BRE 1993 consider the condition of the 'waterproof envelope' that includes the chimney, roof, walls, windows and doors and much of this assessment is similar to that for conventionally built properties and very similar to those matters we have described for conventional timber-framed properties in Section 12.9 above. In this section, we will concentrate on MMC buildings and in this respect, you should pay particular attention to the following:

- **Chimney to roof details.** Not only should you check the waterproof junction between the chimney and roof covering, you should also remember that many chimney stacks on MMC properties are prefabricated. Some are used for aesthetic reasons, while others may also incorporate a flexible flue liner from a heating appliance below.

Even though the prefabricated chimney stack might be made out of glass rein-forced plastic (GRP), when clad with brick slips and in some cases flaunched with cement mortar, the component can be reasonably heavy. Although most types of prefabricated chimneys are supported by the trussed rafters of the roof structure, they also require additional strengthening. This might typically include additional noggins or battens through to a thick plywood board fixed to the top of the trusses. If this is not provided, the roof structure beneath the chimney may distort, the roof covering can sag and the chimney itself may move out of vertical.

- **Cladding systems.** MMC systems will use a variety of claddings ranging from the conventional (outer skin of brickwork with a cavity) to the new (brick slips through to 'brick effect' thin renders). You should inspect for problems in the following locations:

  - At the junction of different cladding systems. Look out for incorrect detailing at the junction between the two. This could result in a concentrated flow of water getting behind the cladding;
  - This concentrated flow could result in saturated areas of cladding, green sur-face growths or water staining on the surface;
  - Also note any deteriorated cladding materials such a poor pointing, spalled or cracked render, broken or missing brick slips or tiles and damaged timber cladding.

If you noticed any of these signs, then 'follow the trail' to the inside of the prop-erty to see if the problems have affected the internal surfaces.

- **Windows and doors.** Check the sealants around all the doors and windows to see if any are missing, poorly applied, deteriorated or damaged as a result of differ-ential movement. This is particularly important with some types of MMCs where the windows and doors are not protected by deeper reveals and, as a result, are very exposed. In these cases, the sealant may be the only barrier to water penetra-tion. Where you suspect this is the case, check around the internal reveals for signs of leaks.

You should also check that the doors and windows have appropriate sills and weatherbars that have an adequate drip and are correctly positioned so the water is thrown clear of the structure.

- **The base of the outside walls.** In Chapter 6, we described moisture problems that affect the base of walls and the importance of appropriate detailing around the DPC position. The same principles also apply to MMC buildings, but are par-ticularly important for those that use timber panels in their construction. This is because the system is usually built off several important timber components that are close to ground level. These could include the sole plate, the baseplate of the closed panel itself, as well as any timber sheathing skins.

Although this is similar to the problems faced by 'classic' modern timber-frame systems (see Section 12.9), in our view, structurally insulated panel construction is more vulnerable to damage. This is because the strength and integrity of the panels as a whole rely on the adhesion between the stressed skins and the rigid insulation. This adhesion could be undermined if the sole plate, the base plate of the panel and the base of the panel skins are affected by excessive levels of moisture.

Consequently, you should look out for the following:

- External evidence that indicates the base of the wall may be exposed to high levels of moisture. This may typically include high ground levels that reduce the difference between internal floor level and external ground level;
- Areas of hardstanding that slopes towards the base of the outside walls;
- Areas of staining/green algae that suggests the base of the cladding remains wet for most of the time; and
- Evidence of previous flooding including old sandbags close by or flooding barriers fixed across door openings, air brick openings and so on;
- Internally, you should check for moisture problems to the base of the walls affected by the problems described above.

If you note any of these potential problems you may want to recommend to your client, the problems should be further investigated. However, because of the special nature of modern methods of construction systems, it is important that the person investigating the problem is familiar with the method of construction used. For example, if the base of the outside wall to a SIP built property is affected by a moisture-related defect, then it is unlikely a damp-proofing specialist more used to repairing traditionally built dwellings will come up with the correct solution.

References

BRE (1993). 'Supplementary guidance for assessment of timber-framed houses: Part 1 examination'. *Good Building Guide 11*. BRE, Garston, Watford.

BRE (1995). 'Supplementary guidance for assessment of timber framed houses. Part 1: Examination'. *Good Building Guide 11*. Building Research Establishment, Garston, Watford.

BRE (2003). 'Timber framed construction: An introduction'. *Good Building Guide 60*. BRE, Garston, Watford.

BRE (2005). *Modern Methods of House Construction: A Surveyor's Guide (FB 11)*. BRE Trust, Garston, Watford.

House of Commons (2021). Library briefing paper, *Tackling the Under-Supply of Housing in England*, 14 January 2021. House of Commons Library, London.

Independent (2019). Available at www.independent.co.uk/news/business/news/persimmon-homes-fire-barriers-safety-risk-report-a9250411.html. Accessed 6 September 2022.

Kettlewell, D. (1988). 'NHBC's licensing scheme for PRC repairs'. *Structural Survey*, Vol. 6 No. 2, pp. 102–107.

Marshall, D., Worthing, D. and Heath, R. (2014). *Understanding Housing Defects*. Estates Gazette, London.

NHBC (2006). *A Guide to Modern Methods of Construction*. NHBC Foundation, Milton Keynes.

NHBC (2018). *Who Is Doing What – Modern Methods of Construction (NF 82 MK)*. NHBC, Milton Keynes.

Parnham, P. and Russen, L. 2008. *Domestic Assessor's Handbook*. RICS, London.

TRADA (2021). *Survey of Timber Framed Houses*, Wood information sheet, number 10. TRADA, High Wycombe.

UK Finance (2021). Available at https://lendershandbook.ukfinance.org.uk/lenders-handbook/pcc/. Accessed 6 September 2021.

Further Reading

BRE Reports on individual house types:

no. 34 (1983). 'Structural condition of Boot pier and panel cavity houses'.
no. 35 (1983). 'Structural condition of Cornish unit houses'.
no. 36 (1983). 'Structural condition of Orlit houses'.
no. 38 (1983). 'Structural condition of Unity houses'.
no. 39 (1983). 'Structural condition of Wates houses'.
no. 40 (1983). 'Structural condition of Woolaway houses'.
no. 52 (1984). 'Structural condition of Parkinson framed houses'.

BRE INFORMATION PAPERS

10/84 (1984). 'The structural condition of some prefabricated reinforced concrete houses'.
14/87 (1983). 'Inspecting steel houses'.
15/87 (1983). 'Maintaining and improving steel houses'.

13 Assessing safety hazards for occupants

Contents

DOI: 10.1201/9781003253105-13

13.1 Introduction

When the first edition of this book was published, hazards were mentioned throughout each chapter, mainly by reference to the Building Regulations Approved Documents (BRADs). To a certain extent, this reflected guidance from the professional bodies. RICS suggested in 1997 that any 'defect judged to be an actual or developing threat . . . to personal safety' constituted an urgent repair and should be included in the report (RICS & ISVA 1997, p. 14). However, hazards were only included within each building element, whereas most current survey products have an additional hazard summary section, so the client has a complete hazard list.

This chapter reflects the way personal safety has increasingly become more important in survey and valuation reports, driven by public awareness of personal safety and liability owed by professionals to their clients. Perhaps the most recent example has been the Grenfell fire, of which more later. We therefore discuss possible liability, current guidance and address hazards in the context of current best survey practice, suggesting possible methodologies or protocols to help practitioners report appropriately on hazards.

13.2 Guidance on hazards

13.2.1 Mortgage valuations

Surveyors acting on behalf of lenders must reflect significant hazards where they materially affect value and report on them as appropriate, given their contractual and tortious liabilities to lenders and potential homeowners. As a result of the Housing Acts, where a property is let, some issues take on more significance and could make the property unlettable until the defect has been repaired. Of particular concern are the services, such as electricity or gas, which, if incorrectly serviced, could cause serious health issues or even death. Similarly, condensation leading to mould growth can also seriously affect health. Such issues are of great concern to lenders who may lose revenue during the period in which the property cannot be occupied, as the landlord does not have revenue to pay the mortgage. This is a differentiator between a let property and one that is owner occupied, where the owner can take on the responsibility for such repairs and may continue to occupy a property and pay the mortgage.

Failure by the landlord to have current servicing certificates is a significant matter a Valuer would need to address, since even though the cost of repairs may be small, the consequences could have more significance, such as prompting the tenant to take legal action against the landlord.

RICS has issued Guidance to Valuers in their publication The Valuation of buy-to-let and HMO properties. At the time of writing, the 2016 edition was in force. In Volume 1, Chapter 5, we have indicated that it is likely the lender's valuer will have no liability to the purchaser of a buy-to-let residential property – *Scullion v Bank of Scotland plc* [2011].

The Housing Health and Safety Rating System (HHSRS) is used by local authority surveyors and assessors to judge whether hazards in a property that is rented should be removed or reduced. These judgements are based on a complicated risk assessment process. Residential surveyors carrying out inspections and reports on rented (and owner-occupied) property should be aware of this guidance, associated legislation and understand basic concepts of risk assessment.

Mortgage Valuers are not expected to undertake a full risk assessment as set out in the Housing Health and Safety Rating System under the Housing Acts, but they should be aware of the types of risk that could affect whether a landlord can let the property, especially those that could take time to make good and could result in significant costs. Some issues, such as steep stairs, cannot usually be corrected at reasonable cost, so this may limit the type of occupant who can take the tenancy.

Some lenders specialise in buy-to-let lending, and their guidance and reporting formats prompt the Valuer to consider specific hazards the lender may have concerns about. However, some lenders use a standard format and, in such circumstances, Valuers need to be careful to detail the consequences of the repairs not being undertaken. The consequences are outlined in the following sections.

13.2.2 Level two surveys

The recently superceded HBR Professional Standards (PS) provides current guidance for the RICS level two survey product. The PS's definition of a CR3 is a defect that is 'serious' and/or requires repair 'urgently'. One of the tests for 'urgent' is a defect 'which presents a safety threat', including 'a visible broken power point, missing/broken stair handrail' (RICS 2016, pp. 11 & 12). Surveyors should not assume *any* safety threat should be a CR3, since defects that have 'existed for some time and which cannot reasonably be changed' should not be condition rated (RICS 2016, p. 26). The PS furthermore confirms that 'risks to people' should not be 'too remote', giving examples of 16 hazards (RICS 2016, p. 62).

To summarise, this guidance suggests a safety threat:

- Must arise from a defect;
- Must not be too remote;
- May be condition rated for action; unless
- It has existed for some time and it is unreasonable to change it.

The writers agree that a test of 'reasonableness' should be applied to all hazards noted (i.e., hazards must not be too remote). However, the public's attitude to risk in their homes has changed over the years, especially due to the Grenfell fire and it is suggested hazards that are reasonably foreseeable and could cause harm to an occupant and/or visitor should be considered and 'condition-rated'; whether they have existed for some time or if they exist without constituting a 'defect'.

13.2.3 RICS Home Survey Standard (HSS)

This 2019 guidance lists some 'common safety hazards' (most are in the HBR PS list referred to in the previous section) a surveyor might encounter in a residential property, emphasising that the hazard list 'is not intended to be exhaustive' (RICS 2019, p. 30). Given that advice, Tables 13.1, 13.2 and 13.3, based on the HHSRS, are included to inform surveyors of other hazards they should consider. Even this list is not conclusive – any hazard the surveyor sees must be noted, considered and

Table 13.1 Comparison of HHSRS and HSS hazards, 'physiological requirements – hygrothermal conditions' and 'pollutants (non-microbial)'. Hazard number in HHSRS Regulations, Schedule 1 in column 1.

No.*	HHSRS hazard	HSS closest equivalent	Notes (page numbers, DCLG 2006)
	Physiological requirements – hygrothermal conditions		
1.	Damp and mould growth – exposure to house dust mites, damp, mould or fungal growths.	–	HSS example of 'damp' basements. HHSRS refers to hazards from 'high humidity' (p. 22).
2.	Excess cold – exposure to low temperatures.	–	'healthy indoor temperature . . . 21°C . . . Below 16°C . . . serious health risks for the elderly' (p. 23).
3.	Excess heat – exposure to high temperatures.	–	'smaller dwellings . . . more prone . . . south facing glazing' (p. 25).
	Pollutants (non-microbial)		
4.	Asbestos and MMF – exposure to asbestos fibres or manufactured mineral fibres.	Asbestos and other deleterious materials.	'rockwool . . . glass fibre . . . during maintenance' (p. 27–28).
5.	Biocides – exposure to chemicals used to treat timber and mould growth.	–	'*chemicals* . . . to treat *timber and/or mould growth*' (p. 28).
6.	Carbon monoxide and fuel combustion products – exposure to – (a) carbon monoxide; (b) nitrogen dioxide; (c) sulphur dioxide and smoke.	Gas leaks and carbon monoxide poisoning.	HSS hazard also appears against item 9. 'May impair foetal growth . . . can cause . . . death . . . (v) entilated lobby between integral garage and living accommodation' (p. 30).
7.	Lead – the ingestion of lead.	Lead water pipes and lead paint.	'exhaust fumes . . . mental retardation . . . children . . . particularly vulnerable' (p. 31).
8.	Radiation – exposure to radiation.	High radon levels. Overhead power lines, which may cause issues relating to electromagnetic fields (EMFs).	'five percent of lung cancers could be traced to residential radon' (p. 31).
9.	Uncombusted fuel gas – exposure to uncombusted fuel gas.	Gas leaks and carbon monoxide poisoning.	'*asphyxiation* resulting from . . . escape of fuel gas' (p. 31).
10.	Volatile organic compounds – exposure to volatile organic compounds.	–	'*Formaldehyde* is included in this hazard' – see Chapter? (environment) (p. 32).

Table 13.2 Comparison of HHSRS and HSS hazards, 'physiological requirements – space, security, light and noise' and 'protection against infection'

No.*	HHSRS hazard	HSS closest equivalent	Notes
Physiological requirements – space, security, light and noise			
11.	Crowding and space – a lack of adequate space for living and sleeping.	Inappropriate use of accommodation (for example, non-conforming roof space conversion and bedrooms in damp basements).	'linked to *psychological distress . . . mental disorders . . . contagious disease*' (p. 33).
12.	Entry by intruders – difficulties in keeping the dwelling or HMO secure against unauthorised entry.	–	'emotional impact after burglary affects more than 75 per cent of victims . . . local area . . . crime' (p. 33).
13.	Lighting – a lack of adequate lighting.	–	'*depression* . . . lack of window with a view . . . artificial external lighting . . . *convulsive reactions*' (p. 35).
14.	Noise – exposure to noise.	–	'lack of sufficient sound insulation . . . road traffic . . . *stress* . . . (p. 35).
Protection against infection **Hygiene, sanitation and water supply**			
15.	Domestic hygiene, pests and refuse (1) Poor design, layout or construction such that the dwelling or HMO cannot readily be kept clean. (2) Exposure to pests. (3) An inadequate provision for the hygienic storage and disposal of household waste.	Animals and vermin (bird droppings, rats, dog waste, etc.)	'*gastro-intestinal disease* . . . *infections* . . . HMOs particularly vulnerable' (p. 37).
16.	Food safety – an inadequate provision of facilities for the storage, preparation and cooking of food.	–	'Fifty per cent of food poisoning . . . in the home' (p. 40–41).
17.	Personal hygiene, sanitation and drainage – an inadequate provision of – (a) facilities for maintaining good personal hygiene; (b) sanitation and drainage.	–	'*skin infections* . . . *dysentery* (between 2,000 and 20,000 notified cases per annum) . . . cracks/ chips . . . to . . . facilities' (p. 41).
18.	Water supply – an inadequate supply of water free from contamination for drinking and other domestic purposes.	Legionnaire's disease.	'legionella . . . 10–15 percent of cases . . . fatal' (p. 43).

Table 13.3 Comparison of HHSRS and HSS hazards: 'protection against accidents, falls, electric shocks, fires, burns and scalds, collisions, cuts and strains'.

No.*	HHSRS hazard	HSS closest equivalent	Notes
	Protection against accidents Falls		
19.	Falls associated with baths etc. – falls associated with toilets, baths, showers or other washing facilities.	–	'Possible death weeks/ months after' (p. 45).
20.	Falling on level surfaces etc. – falling on any level surface or falling between surfaces where the change in level is less than 300 mm.	–	'floors/yards/paths . . . Following a fall . . . elderly person may deteriorate' (p. 45).
21.	Falling on stairs etc – falling on stairs, steps or ramps where the change in level is 300 mm or more.	Falls from height, lack of safety rails, steep stairs, and serious and significant tripping hazards.	'Internal . . . (e)xternal steps or ramps . . . falls on stairs account for around 25 per cent of all home falls' (p. 45–46).
22.	Falling between levels – falling between levels where the difference in levels is 300 mm or more.	–	'balconies . . . landing balustrades . . . retaining walls . . . *head/brain/spinal injuries*' (p. 47–48).
	Electric shocks, fires, burns and scalds		
23.	Electrical hazards – exposure to electricity.	Dangerous electrics.	'shocks and burns' (p. 49).
24.	Fire – exposure to uncontrolled fire and associated smoke.	Lack of emergency escape, inadequate fire precautions and fire protection measures. The implications of external wall systems and their combustibility/incorrect fixing. Referral should be made to competent persons where issues of combustibility and/or incorrect fixing appear to exist.	'*burns, deaths* . . . cookers . . . (f)ire stops to cavities . . . (d)etectors/smoke alarms . . . means of escape' (p. 50–51). HSS reference is clearly to the Grenfell Tower fire – (MHCLG 2018).
25.	Flames, hot surfaces etc – contact with (a) controlled fire or flames; (b) hot objects, liquid or vapours.	–	'over 200 people . . . die from burn and scald injuries', half are children (p. 52).
	Collisions, cuts and strains		
26.	Collision and entrapment – collision with, or entrapment of body parts in, doors, windows or other architectural features.	Absence of safety glass to openings and outbuildings. Automatic gates.	'window injuries . . . worse . . . piercing by glass' (p. 54).

(Continued)

Table 13.3 (Continued)

No.*	HHSRS hazard	HSS closest equivalent	Notes
27.	Explosions – an explosion at the dwelling or HMO.	Absence of test certificates for services/appliances/private water supply.	'Incidence . . . low . . . *(p)ossible scalding* if a hot water appliance is involved' (p. 54).
28.	Position and operability of amenities etc. – the position, location and operability of amenities, fittings and equipment.	–	'*physical strain* associated with functional space' (p. 54).
29.	Structural collapse and falling elements – the collapse of the whole or part of the dwelling or HMO.	Unstable parts of the building, especially at high level. Unsecured fireplace surrounds.	'objects falling . . . extremely rare' (p. 55).
–	–	Unprotected garden ponds and swimming pools.	This HSS hazard is not in the HHSRS.

reported on, so long as it is reasonable to do so and the hazard is not too remote. The process for hazards that meet these criteria should be:

1. Every such hazard should be included in the report; but
2. In some cases, due to the surveyor's risk assessment process, some of those hazards may warrant action such as remedial works and/or improvements to be taken, with allocation of either a CR2 or, more likely, a CR3 – see later for how to decide this.

The HSS states a residential survey is not a 'formal assessment of statutory health and safety risks (for example a Housing Health and Safety Rating System)', but confirms hazards that present 'a safety risk to occupants **must** be described in the report' (RICS 2019, p. 17). The word in bold confirms this as a mandatory requirement.

The guidance establishes a clear difference in how surveyors should report hazards at survey levels two and three:

- Level two – identify and list the risks, explaining the nature of the hazard;
- Level three – likewise and explain how to resolve or reduce the risks.

The HSS further differentiates between a survey for a homeowner and one for a client buying to let the property to a tenant, confirming the surveyor should 'adjust the scope of the service so the client can be properly advised on statutory risks and hazards to health and safety of occupants', i.e. tenants (RICS 2019, p. 17). This advice reflects practice by many surveyors, who have differentiated between these survey types for many years, emphasising hazard implications in 'buy-to-let' reports. This is because of the potential duty of care owed by a surveyor's client, the landlord, to the tenant for injuries and/or losses suffered due to a hazard. The landlord may, in turn, take similar action against the surveyor for failing in their duty of care to highlight issues arising from any hazard.

The HSS therefore confirms a greater duty of care on surveyors preparing survey reports for potential landlords. The surveyor must not only identify and report the hazard but also explain the legal implications of actions that might be taken against the landlord arising from the presence of hazards, such as enforcement action.

13.2.4 Housing Health and Safety Rating System (HHSRS)

Surveyors preparing reports for 'buy to let' properties must therefore be familiar with the twenty-nine 'matters and circumstances' that can justify enforcement action against landlords by a local authority under the HHSRS Regulations. These are set out in column 2 of Tables 13.1, 13.2 and 13.3 (HMSO 2005, pp. 5–7).

Under the HHSRS, a local authority inspector considers hazards arising from defects (as surveyors do for HBRs); *and* harm tenants might suffer from 'any *deficiency* that can give rise to a hazard' – 'deficiency' is defined as a 'failing of some kind . . . an element . . . not . . . to an acceptable standard' (DCLG, p. 5 & 19). Surveyors preparing reports for potential landlords must therefore consider hazards arising from defects *and* deficiencies to help their client landlords avoid enforcement action, including:

- A prohibition order – no use allowed of part or all of the property;
- Emergency action – an imminent risk of serious harm, so the local authority can do the work and charge the landlord;
- A 'hazard awareness notice' – drawing the landlord's attention to the matter; and
- Demolition order or clearance – extreme action but allowed under the Regulations (ODPM 2006b).

We believe surveyors who prepare reports for owner-occupied property should include all relevant hazards as long as it is reasonable and not too remote. This is one of the reasons we have included consideration of the HHSRS in this chapter.

13.2.5 Other information sources

13.2.5.1 Introduction

There are many other references surveyors will find helpful, including information from their local authority – 'knowledge of locality'. Some of these are as follows.

13.2.5.2 Building Regulations Approved Documents (BRADs)

BRADs provide excellent advice about hazards in and around buildings. Some of the 'benchmarks' or 'rules' are referred to throughout this book. The particularly helpful BRADs include:

- 'B1' – means of warning in a fire, and escape for single-family properties and flats (OPSI 2019);
- 'C' – dampness, radon and other ground-based gases (OPSI 2013);
- 'H' – household waste storage and disposal;

- 'J' – safe positions of boiler flues and vents, oil storage (OPSI 2013a);
- 'K' – numerous benchmarks for safe staircases and landings, balconies, safety glass (OPSI 2013b);
- 'M1' – disabled access requirements, e.g., outside ramps and paths; and (OPSI 2019);
- 'Q' – unauthorised access and security (OPSI 2015);

13.2.5.3 LACORS

Useful guidance on fire safety in housing. It is particularly helpful for basement accommodation, HMOs and flats, including conversions (LACORS 2008).

13.2.5.4 RICS Asbestos Guidance Note

Provides surveyors with a 'balanced and pragmatic appreciation of the various issues surrounding asbestos' including most likely locations and advice regarding common areas in flats (Winstone et al. 2011, p. 4). For properties built after 2000, it suggests 'it is reasonable to conclude . . . it does not contain' PACMs, unless evidence suggests otherwise (Winstone et al. 2011, p. 25).

13.3 Risk assessment

13.3.1 Introduction

In order to condition-rate hazards, surveyors preparing survey and/or valuation reports must be generally aware of how to carry out a risk assessment process. This is necessary for owner occupied property so that the owner is aware when action must be taken to remove or reduce any hazard. For let property, the landlord needs to be aware whether it is possible or likely enforcement action might be carried out by the local authority on behalf of a tenant. The process we suggest is not as complicated or as 'accurate' as that carried out by a local authority inspector but is sufficient to provide necessary guidance to a client.

Some surveyors might argue they haven't been trained in risk assessment. However, we all carry out risk assessments in our lives – visualise the difference between:

- A pedestrian crossing a busy main road with a lorry approaching; and
- Crossing that same road with no traffic visible.

13.3.2 Guidance on risk assessment

The Health and Safety Executive (HSE) issues good advice on how to complete risk assessments. Surveyors should consider any assessment within the framework of the HSE's 'Five steps':

1.   Identify the hazards;
2.   Decide who might be harmed and how;
3.   Evaluate risks and decide on precautions;

4. Record findings and implement them; and
5. Review the assessment and update if necessary.

More particularly, a surveyor must appreciate there is a difference between the terms 'risk' and 'hazard'. Thus:

- 'A **hazard** is anything that may cause harm, such as chemicals, electricity, working from ladders, an open drawer etc;
- The **risk** is the chance, high or low, that somebody could be harmed by these and other hazards, together with an indication of how serious the harm could be'.

(HSE 2006, p. 2)

Residential buy-to-let property hazards include any activity it is reasonable to expect an occupier to be doing, such as using the stairs, or a defect or deficiency in the building making it more likely harm could result, such as a loose slate or non-safety glass in a door. The surveyor's role should be to anticipate an HHSRS inspector's risk assessment approach in considering *likelihood* and *severity*; e.g. 'how likely is a fire to break out, what will happen if one does?' (ODPM 2006, p. 3). This should not involve an equivalent HHSRS calculation but should adopt the approach suggested by HSE, to put the client 'in the right ballpark', as some of the writers advise in training sessions. One method of doing this is to use a risk estimator matrix, as in Figure 13.1.

13.3.3 Risk assessments for surveys

A local authority HHSRS inspector judges the *severity* of the risk by considering, for 'each hazard which is obviously worse than average for that type and age of property . . .:

i) The likelihood of an occurrence over the next twelve months; and
ii) The probable spread of harms which could result from such an occurrence' (ODPMa, p. 24).

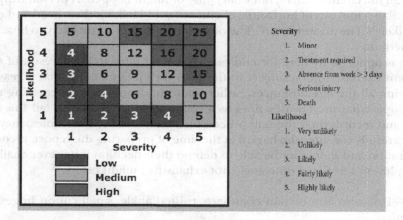

Figure 13.1 A risk estimator matrix (Russen et al. 2010, p. 13).

Hazard Matrix	'Likelihood' of hazard occurring – increasing ⟹		
'Severity of outcome' – severity increasing		Low	Medium	High
	Low	CR1	CR1	CR2
	Medium	CR1	CR2	CR3
	High	CR2	CR3	CR3

Figure 13.2 Simple risk estimator matrix for survey hazards.

Surveyors may wish to use the risk estimator matrix at Figure 13.1, or their own version. We suggest initially using the simpler matrix at Figure 13.2, used by some surveyors in site notes to record their thought process.

The 'likelihood of hazard occurring' should be based on a 'reasonable' assessment. Lack of a fire-resisting partition in the roof space on the line of the party wall between two semi-detached homes <u>could</u> allow a fire to spread from the adjacent home into the property. However, is there likely to be such a fire? The answer will be based on the circumstances the surveyor records in their site notes, but in most cases will be a resounding 'no'; the likelihood of this hazard occurring is therefore recorded in the site notes as 'low'.

The 'severity of outcome' should be based on a reasonable assessment of what is likely in all the circumstances, not what <u>could</u> or might happen. A person tripping on a paving slab 5 mm out of horizontal alignment <u>could</u> fall, hit their head and die. But is that 'likely'? The answer is 'no'. The outcome is possibly more likely to be a broken arm ('medium').

There is opportunity here for endless debate and ultimately the result of the risk assessment exercise will be subject to the surveyor's judgment of what is <u>reasonable</u> considering all the circumstances, rather than what <u>could</u> occur. Providing the surveyor adopts what the law describes as a 'reasonable' approach in 'all the circumstances' and records their thought process on a matrix in their site notes, they will be able to carefully consider the hazard at the time of preparing the report, record such consideration and thereafter be able to defend their decision if it is ever challenged.

Examples of 'severity of outcome' (not exhaustive) might be:

Low – Bruise to arm or skin elsewhere, twisted ankle, small cut on finger, leg or similar;

Medium – Partial loss of mobility, broken arm or other limb, lost eye;

High – Significant loss of mobility or some other faculty, amputation of limb, death.

13.4 Case studies and methodologies

13.4.1 Introduction

In this section we demonstrate how surveyors can practically assess hazards for owner occupiers and buy-to-let landlords. It is emphasized the methodologies are neither a replacement, nor equivalent to, the sophisticated HHSRS risk assessment process.

13.4.2 Case study 1 – fire escape from basement accommodation

13.4.2.1 Background

The surveyor inspects for a level two pre-purchase survey on a property with accommodation at first floor and attic conversion level. There are no smoke detectors. The client tells the surveyor they intend to let individual rooms, including the basement room, to local college students. This case study only considers the basement room and does not consider the radon risk.

13.4.2.2 Considerations

The RICS HSS requires the surveyor to adjust the scope of service for this potential buy-to-let HMO property. It is therefore appropriate to consider carrying out a risk assessment exercise; not as complex as a HHSRS exercise but considering *severity* and *likelihood*.

The basement is under the front part of the property

An entrance lobby has been added outside the original rear entrance door

The property was built in the 1890s and has a former coal cellar converted into a basement room.

Figure 13.3 Case study 1, front and rear external views.

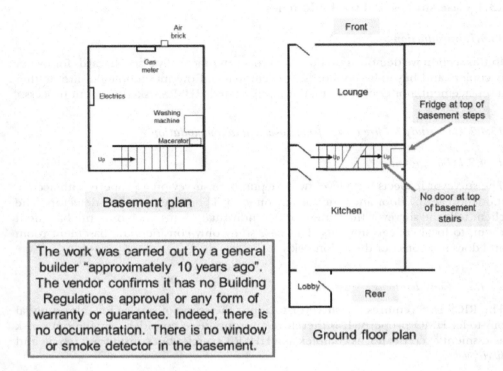

Figure 13.4 Case study 1, basement and ground floor layouts (from rear).

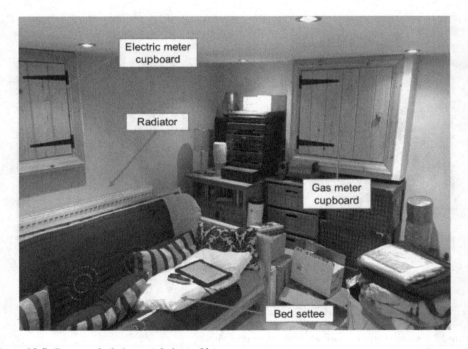

Figure 13.5 Case study 1, internal view of basement room

Matters to consider might include:

- The total number of annual accidental fires in dwellings in the UK fell between 2010 and 2019, from 31,856 to 26,610; of those figures, 'misuse of equipment or appliances' was responsible for 9.025 incidents, followed by 'faulty appliances and leads' at 3,961 (HO 2019a). By comparison, the total number of fatalities in England from the same cause over the same period fell from 335 to 251, with a significant spike caused by the deaths in Grenfell Tower (HO 2019). Indeed, 'fire-related fatalities had been on a downward trend since the 1980s, but have plateaued . . . a very small proportion of fires resulted in a fire-related fatality' (HO 2019b);
- Several *deficiencies* including:
  - BRAD 'B' requires 'basement storeys containing habitable rooms' should have either an emergency escape window or a protected stairway giving access to a final exit – neither are present (OPSI 2019, p. 16);
  - There is no means of warning in the event of a fire and the means of escape is through the room where the fire is most likely to occur – the kitchen;
  - Presence of the meter installations and other household equipment in the basement and along the escape route are important issues – especially in view of HO warnings about equipment and appliances above.

In view of all the above, a surveyor might conclude:

- *Likelihood* of hazard occurring – 'medium', not many fires occur in homes, but there are several items of fixed and moveable equipment that could become faulty and cause a fire; and
- *Severity* of outcome – 'high', a tenant trapped in, or trying to escape from, the basement could suffer severe harm, possibly death.

The decision process is subject to the surveyor's judgment and could be questioned later, in the event of audit, complaint, fire, or enforcement action. However, providing the surveyor records the process, they will at least be able to demonstrate they

Hazard Matrix	'Likelihood' of hazard occurring – increasing ⟹		
'Severity of outcome' – severity increasing ⬇	Low	Medium	High
Low	CR1	CR1	CR2
Medium	CR1	CR2	CR3
High	CR2	CR3	CR3

Figure 13.6 Case study 1 – risk estimator matrix assessment – CR3.

followed a reasonable protocol system to arrive at their final condition rating. They might be shown as being 'wrong' later, but that doesn't mean they are 'negligent' – so long as they can show they did what a 'reasonably competent' surveyor must do. That record-keeping process could include:

- Drawing a circle in their site notes around one of the 'low', 'medium' and high' indicators on each axis, and around the final condition rating;
- Adding comments confirming why each indicator has been chosen, such as the preceding notes supporting the decisions on *likelihood* and *severity*.

In this case it is likely the condition rating, **CR3**, might include a recommendation for further investigation or a recommendation the internal layout of the property is likely to require fundamental alteration, together with advice the basement room should not be used for habitable purposes until the matter has been satisfactorily resolved; and/or in the case of a 'buy-to-let 'enquiries by the legal adviser regarding works the local authority might require;

13.4.3 Case study 2 – glazed internal doors without safety glass

13.4.3.1 Background

This is for a level two pre-purchase survey for a private homeowner. The case study concerns two glazed doors the surveyor estimates have been present for around 50 years. There are no significant trip hazards close to the doors.

Figure 13.7 Case study 2 – detached house, 1886, front elevation.

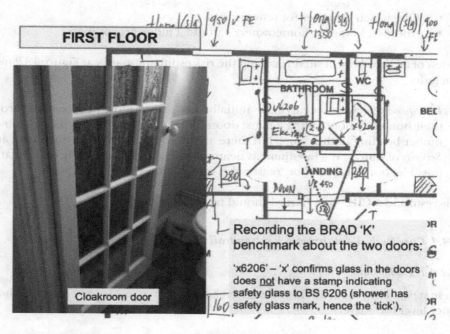

Figure 13.8 Case study 2 – site notes, picture of one of the doors and part of first floor plan.

13.4.3.2 Considerations

Following guidance in the HBR PS, a possible protocol might be:

Table 13.4 Hazard assessment process suggested in RICS HBR PS

Step	Question	Answer	Notes
1.	Is this a safety hazard?	Yes.	The risk estimator matrix at Figure 13.2 could help to answer the question.
2.	Is it a 'direct threat'	Yes.	The PS stipulates hazards must not be too remote.
3.	Does the hazard affect the condition rating?	No.	If the hazard has existed for some time and there is no defect, the PS stipulates the hazard does not affect the condition rating.
4.	Condition rating therefore?	1.	The hazard is reported in the 'element' section of the report and again in the summary section.

Alternatively, some surveyors consider:

- The definition of the term 'defect' includes the description 'deficiency' in relation to hazards (and for a buy-to-let property the local authority inspector considers any hazard that is a 'deficiency' or a 'defect'); or
- They want to emphasise hazards; and, anyway

- Why apply two safety standards:

    - One which is higher for tenants; and
    - Another, lower for homeowners – would a judge accept that?

In view of all the above, a surveyor using the risk estimator matrix at Figure 13.2 might conclude:

- *Likelihood* of hazard occurring – initially '**low**', since most people move around their homes without falling against doors and some protection is provided by the timber beading in the doors; but there are two doors, so choose '**medium**'; and
- *Severity* of outcome – '**medium**' as death or loss of limb is unlikely (yes, 'death' is possible, but is it 'likely' or 'reasonably foreseeable'?).

This results in a **CR2** – the hazard should be remedied soon.

13.4.4 Case study 3 – stairs with low handrail

13.4.4.1 Background

These stairs are in the same 1886 detached house. The report is a pre-purchase home-owner level two survey.

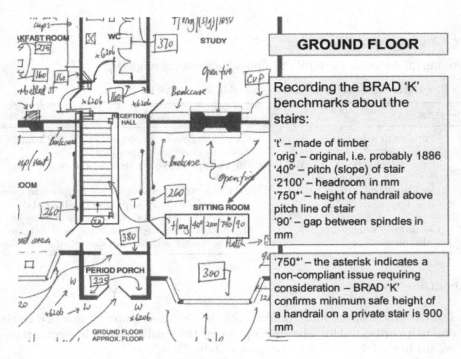

Figure 13.9 Case study 3 – site note details of stairs.

Figure 13.10 Case study 3– view from first floor landing on left.

13.4.4.2 Considerations

The first protocol used for the glazed internal doors suggests a CR1, as there is no defect and the hazard has existed for some time.

A risk assessment process might conclude:

- *Likelihood* of hazard occurring – 'low', since most people use their stairs on a very regular basis without falling over the handrail and the hazard (handrail height) is only 150 mm less than the current BRAD 'K' benchmark; and
- *Severity* of outcome – 'medium' as broken arm is most likely.

Using the hazard matrix at Figure 13.2, this results in a **CR1** – no action is required, but the hazard is flagged up in the element section and the summary of hazards section.

13.4.5 Case study 4– water tank in loft

13.4.5.1 Background

The surveyor is preparing a pre-purchase level two report on a much-extended 1920s detached bungalow to be let for single-family occupation.

The property has been empty for six months over a warm summer and the water system has not been drained down.

Figure 13.11 Case study 4, front elevation.

Figure 13.12 Case study 4, inside of water tank.

13.4.5.2 Considerations

Matters to consider include:

The HSE reports Legionnaires' disease is a potentially fatal type of pneumonia. Some people are at higher risk, including:

- 'People over 45 years of age . . . ;
- People suffering from chronic respiratory or kidney disease; and
- Anyone with an impaired immune system'.

(HSE 2012, p. 1)

It is not known who the tenant will be, but bungalows are perhaps more likely to be occupied by older people. There are clear indications of hazards in the water tanks and the system has been stagnant for a considerable time.

The protocol is likely to suggest a CR3 (see Table 13.4), as there is a hazard which is a defect and a 'direct threat'.

A risk assessment process might conclude:

- *Likelihood* of hazard occurring – 'high', given circumstances described in the water system generally and the water tank specifically; and
- *Severity* of outcome – 'medium', Table 13.2, HHSRS hazard 18 suggests a 15% possibility of death.

Using the hazard matrix at Figure 13.2, this results in a **CR3** – action is required; the hazard is flagged up for remedial work in the element section and the summary of hazards section. Given, in any event, the fact that annual legionella testing of water installations is required for let properties, a recommendation for such a test would also be flagged up for the legal adviser, the surveyor effectively acting as their 'eyes and ears'.

For surveyors who have been practicing for some time and are unskilled in risk assessments, this suggested methodology may appear complicated and sometimes unnecessary. We have some sympathy for this reaction. However, with careful practice it will be found that using this approach becomes as normal as their usual consideration of defects in other building elements.

13.5 'Services that kill'

13.5.1 Introduction

In residential surveying, 'services that kill' are defined as those services which, generally because they involve carbon-based fuels and/or electricity or another hazard defined by the HSE or by law, require regular (usually annual, but not always) testing to confirm their safety by a 'competent person' within the meaning of the Building Regulations. Table 13.5 sets out those requirements. Such testing with necessary servicing also, coincidentally, helps maintain the efficiency of the installation, thereby helping to reduce running costs and prolong the life of the system. For more detailed information, see volume 3 of 'Residential Property Appraisal'.

Table 13.5 Regularity of testing of 'service installations that kill' and comment

Item	Service	Regularity of testing	Comments
1.	Electricity	Such testing should usually be every 5 years for a property let to a tenant, or 10 years for 'owner-occupied' property, or if there is a 'material change' (significant alteration) in the electrical system in the property, or upon 'change of ownership', i.e., when the property is sold.	In practice, the 'change in ownership' requirement usually means a CR3 is almost always allocated in a survey report to the electricity installation, since a surveyor is almost never presented with a current electricity test certificate. The surveyor should beware of accepting a certificate that relates only to part of the system, such as a part 'P' Building Regulations certificate, which usually applies to an alteration such as an extension or new shower room. The surveyor should bear in mind electricity installations are used in many other service and other installations, e.g., outbuildings, sewage treatment plants, electrically operated vehicle entrance gates.
2.	Gas	Annually.	The test should include the entire system. Thus, the gas installation might include distribution pipes, gas boiler, gas hob and a gas fire in the living room. The surveyor must check to ensure the test certificate includes all those parts of the installation – it is often the case that parts of the system are omitted, for many different reasons, usually relating to cost.
3.	Oil	Annually.	The test should include comment on the oil storage facility (tank, including tank support), pipe, valves and boiler.
4.	Water	Annually, to some parts.	In all property types, an annual test is required for pressurised hot water cylinders (which can be likened to an unexploded bomb if not properly serviced; whilst legionella testing is required for all let properties.
5.	Heating	Annually.	Heating facilities include any boiler (if not tested as part of the service installation), and parts such as wood or multi-fuel burning stoves, open fires, electric convector heaters, storage heaters, and independent heaters.

13.5.2 Discussion

The surveyor's inspection procedure is two-fold in relation to such services. Firstly, enquiries are made of the vendor regarding the availability of current certificates, defined above. Verbal confirmation that the service has been tested is insufficient – the actual certificate must be seen and interrogated for details. If such certificates are not available, the presumption must be that the installation is potentially hazardous; even if there are no visual indications of defects, deficiencies and/or hazards – see later. This is because surveyors are not trained and/or skilled in acting as 'competent persons' for the particular service installation and cannot therefore practically, ethically or legally sign off the installation as safe.

In practice, as indicated in Table 13.5 above, this means that in relation to test certificates, most 'services that kill' are reported as CR3. The reasons for this fact are:

- Most vendors do not have their service installations tested as regularly as is recommended, although many boilers <u>are</u> tested on a regular basis. Unless the property has been let and the letting agent has arranged for testing as a matter of course (or, as above, the boiler has been tested), such certificates usually only exist if the entire installation or part of it has been replaced or refurbished; or
- Even if the installation has been recently tested, by the time the surveyor carries out the inspection, all the necessary documentation has been 'sent to the legal adviser' and is not available to consider.

Secondly, the surveyor must carefully inspect the installation for visual evidence of defects, deficiencies and/or hazards. Typical examples of such issues might include:

- The electricity fuse-board (consumer unit) is not contained in a metal enclosure or box to help reduce the risk of a fire spreading in the property if a fault occurs in the fuse-board. This is a relatively recent modern requirement for all new electricity installations due to an increase in fires arising from the fact most consumer units are made of plastic (which burns) and/or nor not contained in such a metal enclosure;
- LPG cylinders not secured to the adjoining wall. This can allow the cylinders to fall over, resulting in an escape of gas, with a consequent risk of fire or explosion;
- The flue to the central heating boiler is too close to an opening into the property, e.g., the kitchen window. This conclusion will be based on either the benchmarks in BRAD 'J' regarding proximity of flues to openings and/or the manufacturer's information for the particular boiler. Proximity of the flue to the opening could allow dangerous flue gases to enter the property.

Any such visual indication of a defect, deficiency and/or hazard in a service that kills should prompt the surveyor to confirm a CR3, because of the risks of death; even if there is a current test certificate. The surveyor should not be shy in doing this; since even competent service engineers are prone to make mistakes, just like surveyors. The maxim should be 'better to be safe than sorry' when it comes to the safety of clients.

In this regard, it is incumbent on the surveyor to maintain their knowledge of safety and other improvements through their 'life-long learning' (LLL), or 'continuing professional development' (CPD). One of the main reasons for this requirement is that improvements are often incorporated into the BRADs or the competent persons' schemes for reasons of occupant safety. Thus, the installation may have been 'safe' 5 years ago, but if a modern safety requirement is not included, it no longer complies with current good practice and this deficiency must be flagged up as a hazard.

The surveyor must ensure all of the above is reported to the legal adviser in the report, so the client can be advised about any necessary action as appropriate.

To summarise, for a 'service that kills' to be reported as CR1, the surveyor must:

1. See a current, satisfactory, service test certificate signed off by an apparently (see later) 'competent person'; and
2. There must be no visual indications of defects, deficiencies and/or hazards.

In practice, this is seldom so. There may be a certificate, but there are very often indications of defects, deficiencies or other hazards. Or, there are no visual indications of defects, deficiencies or hazards; but there is no certificate. That is why most 'services that kill' will in practice be reported as CR3.

13.6 Summary

Surveyors and valuers should tailor their service to reflect clients' intentions for the property they are buying so far as possible. Hazard reporting comprises a significant proportion of a survey and must be carefully considered for a valuation report. Key requirements for hazards include:

- In all properties, the surveyor must try to confirm existence of, and interrogate, certificates prepared by a 'competent person' trained in testing and servicing for the particular installation;
- The surveyor must note defects, deficiencies and/or other hazards that constitute safety threats;
- A hazard risk assessment should be carried out where a hazard is noted, to decide whether appropriate action is required (i.e., CR2 or CR3);
- For a 'service that kills' to achieve a CR1 there must be:

  - A current and satisfactory test certificate prepared by a 'competent person', and
  - No visual indications of defects, deficiencies and/or other hazards.

- **Where the property is, or will be, a 'buy to let', surveyors must note all hazards (defects *and* deficiencies), condition rate them as appropriate and warn the client of the legislative regime (principally the HHSRS) and implications;**
- The surveyor should confirm the position regarding existence or otherwise of certification to the legal adviser in the report.

Homeowners and tenants have become much more aware of risks and threats to their personal safety; most recently arising from the fire and subsequent reporting of the awful events at Grenfell Tower in London. It is therefore incumbent on surveyors and valuers who prepare advice on such threats and risks to inspect and report accordingly.

References

Department for Communities and Local Government (DCLG) (2006). *Housing Health and Safety Rating System (HHSRS) – Guidance for Landlords and Property Related Professionals*. Department for Communities and Local Government, Eland House, Bressenden Place, London SW1E 5DU.

Health and Safety Executive (HSE) (2006). *Five Steps to Risk Assessment*. HSE Books, PO Box 1999, Sudbury, Suffolk CO10 2WA.

Health and Safety Executive (HSE) (2012). *Legionnaires Disease, A Brief Guide for Dutyholders*. HSE Books, PO Box 1999, Sudbury, Suffolk CO10 2WA.

HMSO (2005). *The Housing Health and Safety Rating System (England) Regulations 2005, SI 2005 No. 3208*. The Stationery Office Ltd, London.

Home Office (HO) (2019). *Fire Statistics Table 0501: Fatalities and Non-Fatal Casualties by Population and Nation*. Home Office, 2 Marsham Street, London SW1P 4DF.

Home Office (HO) (2019a). *Fire Statistics Table 0601: Primary Fires in Dwellings and Other Buildings, by Cause of Fire, England.* Home Office, 2 Marsham Street, London SW1P 4DF.

Home Office (HO) (2019b). *Fire and Rescue Incident Statistics, England, Year Ending March 2019.* Home Office, 2 Marsham Street, London SW1P 4DF.

Local Authority Coordinators of Regulatory Services (LACORS) (2008). *Housing – Fire Safety, Guidance on Fire Safety Provisions for Certain Types of Existing Housing.* LACORS, Local Government House, Smith Square, London SW1P 3HZ.

Ministry of Housing, Communities and Local Government (MHCLG) (2018). *Housing Health and Safety Rating System, Operating Guidance, Addendum for the Profile for the Hazard of Fire and in Relation to Cladding Systems on High Rise Residential Buildings.* Ministry of Housing, Communities and Local Government, Fry Building, 2 Marsham Street, London SW1P 4DF.

Office of the Deputy Prime Minister (ODPM) (2006). *Reducing the Risks – The Housing Health and Safety Rating System.* ODPM, Eland House, Bressenden Place, London SW1E 5DU.

Office of the Deputy Prime Minister (ODPM) (2006b). *Housing Health and Safety Rating System, Enforcement Guidance.* ODPM, Eland House, Bressenden Place, London SW1E 5DU.

Office of Public Sector Information (OPSI) (2013). Building Regulations approved document 'C'. *Site Preparation and Resistance to Contaminants and Moisture.* OPSI, Richmond, Surrey.

Office of Public Sector Information (OPSI) (2013a). Building Regulations approved document 'J'. *Combustion Appliances and Fuel Storage Systems.* OPSI, Richmond, Surrey.

Office of Public Sector Information (OPSI) (2013b). Building Regulations approved document 'K'. *Protection from Falling, Collision and Impact.* OPSI, Richmond, Surrey.

Office of Public Sector Information (OPSI) (2015). Building Regulations approved document 'Q'. *Security – Dwellings.* OPSI, Richmond, Surrey.

Office of Public Sector Information (OPSI) (2019). Building Regulations approved document 'B'. *Fire safety, Volume 1 – Dwellings.* OPSI, Richmond, Surrey.

RICS (2016). *RICS HomeBuyer Report – Survey, 1st edition, June 2016, RICS Professional Statement (PS).* Royal Institution of Chartered Surveyors, Parliament Square, London SW1P 3AD.

RICS (2019). *RICS Professional Statement, Home Survey Standard,* 1st edition. (HSS), Royal Institution of Chartered Surveyors, Great George Street, London SW1P 3AD.

Royal Institution of Chartered Surveyors & Incorporated Society of Valuers and Auctioneers (RICS & ISVA) (1997). *The RICS/ISVA Homebuyer Survey & Valuation HSV Practice Notes.* RICS business Services Ltd, 12 Great George Street, London SW1P 3AD.

Russen, Larry, Rees, Simon and Neale, Stephen (2010). *Commercial Energy Assessor's Handbook.* Royal Institution of Chartered Surveyors, Surveyor Court, Westwood Business Park, Coventry CV4 8JE.

Winstone, Paul, et al. (2011). *Asbestos and its Implications for Surveyors and Their Clients,* 3rd edition, Guidance Note (GN 38/2011). Royal Institution of Chartered Surveyors, |Surveyor Court, Westwood Business Park, Coventry CV4 8JE.

14 Writing the report

Contents

DOI: 10.1201/9781003253105-14

14.1 Introduction

Writing the report is the final stage of the survey process and its purpose is to communicate to the client the relevant outcomes from all the work that has been undertaken during the inspection process. Failure to concentrate on this final element can undo any good work that has gone on before as the client will act on your recommendations and this is likely to represent a substantial investment.

Differing criteria will apply depending up on the type of report required. The main differences will be the scope of what has been agreed in the terms of engagement. If an inspection is limited then you are entitled to make assumptions about those areas that could not be seen. These assumptions may not always be correct but that is the risk the client must accept if the level of inspection was clearly communicated and agreed. However, how those limitations and assumptions are reported to the client is very important.

Survey reports usually follow a format and because the different levels are tiered, you need to make sure that the scope of the report does not exceed the terms of engagement agreed with the client.

For those residential surveyors who are members of RICS then the Home Survey Standard (or HSS for short) is considered the benchmark that should be adopted (RICS 2019). As with all benchmarks they are there as a starting point but if the circumstances of any particular instruction vary from this benchmark, then it will almost certainly need reporting to the client.

To complement the standard guidance, this section identifies other key features that you should consider when producing a survey report:

- For RICS members, you should comply with the Professional and Ethical standards as laid down by RICS in the Rules of Conduct specifically relating to Objectivity and Independence. Although this latter reference relates to valuation it is a good benchmark upon which to act for all surveys reports;
- Comply with the legal precedents and guidance as laid down in case law;
- Follow the Professional Statement as set down for the Home Survey Standard;
- Satisfy your client requirements as set down in the terms of engagement unless these conflict with the above points;
- Use common sense in your reporting style and do not blindly use standard paragraphs and phrases unless they are 'property specific' (RICS 2019, p. 14).

14.2 Report writing principles – Introduction

Before we dive into the details of this chapter, we think a quick review of RICS standard report formats over the last 20 years or so will help set the context of the topic.

Since the first edition of this book was published, standard report formats have changed considerably. Back in 2001, the book was written to suit the then 'Homebuyers Survey and Valuation' as published by the RICS. This was roughly equivalent to a level two service today.

In 2008, the Scottish Government introduced the Home Report and this package of documents included the compulsory Single Survey that introduced the concept of condition rating for the first time in the UK. After a period of reflection, by 2010 RICS had reviewed their own standard formats and launched the RICS 'Home Surveys' brand. This consisted of the RICS Condition Report, RICS HomeBuyer Report and the RICS Building Survey Report. These were equivalent to the current levels one, two and three services respectively. All three were supported by detailed Practice Statements that provided mandatory standards to which RICS members had to conform. These products further developed and used the red/amber/green condition rating system.

Although these formats were amended from time to time, by 2018 the unprecedented rate of social, economic, political and technological change had left the Home Surveys brand tired and outdated (just like the authors of this book). To resolve this problem and following a consultation process, RICS published the Home Survey Standard in 2019. This set out a series of concise mandatory requirements or 'benchmarks' around which RICS members could design and deliver services that met their clients' needs in a changing environment.

The key characteristic of the new Home Survey Standard (HSS) is that unlike its predecessors, it is non prescriptive. As long as RICS members conform to the overarching benchmarks of the HSS then they have the freedom to offer their clients the type of services they think are appropriate. For RICS members who did not want to develop their own product, RICS published a series of standard report templates titled 'Home Survey' at levels one, two and three that can be used by RICS members under license and for a fee. However, these new templates have limited functionality, are not linked to any sample phrases and are not supported by detailed guidance notes.

Because of the variation of products in the marketplace, this chapter will not be about 'how to write an RICS report', instead it will focus on report writing principles that we think match those RICS benchmarks described in the Home Survey Standards and may help you develop your own style and products.

14.3 The current RICS formats

As mentioned above, the RICS Home Survey includes standard format at levels one, two and three. Level three is equivalent to the former building survey and so is beyond the scope of this book because of the deeper level of technical knowledge required to provide that level of service.

14.3.1 The Home Survey Standards – report writing benchmarks

Although the Home Survey Standards are less prescriptive than previous RICS guidance, it still lays down some important principles (RICS 2019, Section 4.1, p14):

The report **must** be property specific and:

- Be clearly presented and follow a logical structure so clients can quickly find the required information;
- Be factual and unambiguous, and clearly separate fact and the RICS member's opinion;
- Use non-technical terms throughout – if technical words are occasionally used, the client will find a layperson's explanation helpful; and
- Provide a balanced perspective of the condition of the property.

Although we don't want to provide a line-by-line commentary on this document there are some important principles that underpin good report writing. We have highlighted a small number of these as follows:

'The report must be property specific' – This may sound like simplistic and elementary advice but we have reviewed a large number of reports where the content is so general it could be applied to any property of that type. This is a clear reminder to keep the report focused on the dwelling that has been inspected and the client requirements.

'Factual and unambiguous, and clearly separate fact and the RICS member's opinion'. It is important that the reader can separate 'fact' (for example, *'to the rear roof slope near the chimney, several roof tiles are missing'*) from opinion (for example *'In my opinion, this has occurred because the fixing nails have corroded. More tiles could become loose especially in stormy weather'*). This separation will help you provide the client with a *'. . . balanced perspective of the condition of the property'.*

'Use non-technical terms throughout'. Will your client know the difference between the purlin and the dragon tie? Do they know the difference between a TRV and a tundish? Probably not. We will return to this important topic later in this chapter.

These principles provide a starting point and provide the parameters within which you can build your own preferred report writing style.

14.3.2 Level specific reporting requirements

In Chapter 2 we looked at selecting the right level of service depending on the client's requirements, the surveyor's knowledge and experience and the nature of the property. In this section we will assume the choice of service has already been made and so our focus will be on reporting requirements at level one and two.

14.3.3 Level one reports

For level one service, the HSS states that for each element of the building a surveyor should (RICS 2019, p14):

- Describe the part or element in enough detail so it can be properly identified by the client;
- Describe the condition of the part or element that justifies the RICS member's judgement; and
- Provide a clear and concise expression of the RICS member's professional assessment of each part or element.

This assessment should help the client gain an objective view of the condition of the property that helps them make a purchase decision. Additionally, if the client buys the property, the report could help them establish appropriate repair/improvement priorities.

As this is a level one service, it will be an objective report that states the facts and does not include any advice because a level one service is better suited to '. . . conventionally built, modern dwellings in a satisfactory condition' (RICS 2019, p. 22). By advice, we adopt a dictionary definition of 'guidance or recommendations offered with regard to prudent future action'.

Therefore, at level one, the report should be concise, objective and focused. It should describe the following issues:

- The subject element and its location in the property;
- Any faults, defects, or other problems (if there are any) that affect the particular element; and
- State the appropriate Condition Rating.

Although we have used the term 'condition rating', these are not mandatory in the HSS. The standard states members should provide an assessment of each element and the use of a '. . . *condition rating system is one way of achieving this'*. The use of condition ratings is discussed in more detail later in this chapter.

14.3.4 Level two reports

In addition to those matters included in level one reports, the HSS describes some additional characteristics that level two reports should address. At this level, we think concise 'advice' should be included and the relevant sections are discussed here (RICS 2019, p14):

- *Comments where the design or materials used in the construction of a building element may result in more frequent and/or more costly maintenance and repairs than would normally be expected.* Classic examples of this include a warning that flat roofs can have a short life or that listed buildings will be more expensive to maintain and repair when compared to non-listed buildings.
- *The likely remedial work should be broadly outlined and what needs to be done by whom and by when should be identified.* This does not mean a detailed description of the repair works but a concise outline that will help the client understand the likely scope of any repair work that may be required. In other words, we think you should try and put the client in the right 'ball-park'. A condition rating system will help the client understand the *'by when'* advice.
- *Concise explanations of the implications of not addressing the identified problems should be given.* The word 'concise' helps remind us the report should be 'short and to the point' but include sufficient information so the client can understand what will happen if the problem isn't resolved. A typical phrase might include: 'If the roof tiles are not repaired soon rain could get into the building and damage the roof structure and the ceilings below'.
- *Cross-references to the RICS member's overall assessment should be included.* We think this is the most important part of any report. This 'overall assessment' normally consists of a summary of conditions ratings that sits alongside a written opinion.

Therefore, at level two, the report builds on those points identified for level one and in addition includes the following.
  The report should describe:

- Description of the part of the building and its location on or in the subject property;
- Any faults, defects, or other problems (if there are any) and their likely consequences and/or possible risks;
- An appropriate Condition Rating;
- The action the client needs to take. This should include an outline of the likely remedial work, who should carry this out and by when;
- Supplementary information that will help the client better understand both the nature of the problem and what they need to do about the issue. This may include features that give rise to more costly maintenance and repairs than would normally be expected or matters that could affect future saleability and so on;
- Link to the overall opinion where appropriate.

Before we delve into the detail of report writing techniques, it will be useful to review the condition rating system because, taken together with the text of the report, we think a method of prioritising repairs is an important part of any reporting format.

14.4 Revisiting the condition rating system

14.4.1 Why condition ratings work

As described above, the professional guidance that supported the former RICS Home Surveys range of products used between 2011 and 2019 was prescriptive in an attempt to increase the level of consistency across the sector. Arguably, one of the most important features of these reports was the use of a condition rating system. In our view, the ratings helped emphasise the issues raised in the written part of the report and enabled clients to quickly focus on the factors that were likely to affect their purchase decision.

The former RICS Practice Statements that supported the Home Buyer report defined a range of criteria that would help surveyors select the appropriate condition rating. Using these defined criteria, in 2010, we developed a more sophisticated 'condition rating protocol' for one of our earlier publications (Parnham 2011, p. 46). The protocol aimed to make the selection of the appropriate condition rating a straightforward process. We never saw the protocol as a precise science that automatically resulted in the 'right answer'. Instead, it was seen as a decision-making tool that objectively put surveyors in the right 'ballpark' with enough flexibility to allow them to apply their own professional judgement in marginal cases.

Although we like to think the condition rating system was successful, not all surveyors agreed. Some considered the approach too prescriptive and so chose to offer their own bespoke reports tailored to the needs of their local market. These often included their own prioritisation methodology.

Consequently, the new Home Survey Standards adopted a more flexible approach. Although it called for reports to establish 'appropriate/improvement priorities', it stopped short of making condition rating mandatory. For example, when describing prioritisation methodologies for survey level one reports, the HSS states, '. . . A condition rating system is one way of achieving this, although RICS members may use their own methodology. Whatever the choice, any system must be clearly defined in the information given to the client' (RICS 2019, p. 14).

As a consequence, we have adopted the following approach:

- We have continued to promote the use of the three condition ratings because we think that, after over a decade of use, these are well understood by both the public, residential surveyors and the broader residential sector;
- We have simplified the selection criteria to make the rating system easier to use; and
- We have built in greater scope for residential surveyors to make professional judgements.

This revised condition rating definitions and condition rating protocol are described in the next section.

14.4.2 Condition rating definitions (the 2021 version)

For any condition rating system to work consistently, individual ratings have to be clearly defined. Our refreshed definitions are as follows, and although not exactly the same, they are based on those included in the original RICS 2011 publications.

Condition rating definition	Explanation
Condition rating one No repair is currently required.	This rating is appropriate where there are no indications of present or suspected defects that need repairing or replacing. The test for condition rating one is: • *Does the element need to be repaired or replaced?* • *If any work is required, is it more than normal maintenance?* If the answer to both statements is **NO**, then condition rating one will apply. For the purposes of this assessment, normal maintenance activities are not usually treated as a repair.
Condition rating two A repair or replacement is required but is not considered to be either serious or urgent.	This rating is appropriate where repairs or replacements are required but the defect is not considered to be serious or urgent. The test for condition rating two is: • *is the problem a 'serious defect' which compromises the structural/ functional integrity of the element?* • *Is the problem an 'urgent defect' where the repair or replacement has to be done as soon as possible?* If the answer to both statements is **NO,** then condition rating two applies.
Condition rating three A repair or replacement is required and it is considered to be serious and/or urgent or needs further investigation.	This rating is appropriate where the defect is of a serious nature; or where a repair or replacement is required as soon as practically possible; or where the defect has resulted in a serious safety hazard; or where the surveyor feels unable to reach the necessary conclusion with reasonable confidence and recommends a further investigation. The test for a condition rating three is: • Does the defect compromise the structural/functional integrity of the element? • If the defect is not repaired or replaced as soon as possible, will it cause serious defects in other building elements and/ or result in a significant threat to the health and safety of the building users? • Does the defect require further investigation? If the answer to any of these statements is **YES,** then the condition rating applies.
Not inspected	This rating should be applied when it is not possible to inspect parts of the property usually included in a standard level of service.

Figure 14.1 Condition ratings definition (the 2021 version).

14.4.3 *Understanding condition rating definitions*

Keeping within the spirit of the Home Survey Standards, we do not want to include pages of nuanced and abstract discussion that attempts to explore the detailed descriptions of the condition ratings and their differences. However, any type of prioritisation methodology does need to be understood by the surveyor using it, so a concise explanation of some of the definitions has been included.

Condition rating one

The most important distinction here is the difference between a repair and/or replacement and the type of activities that would be covered by 'normal maintenance'. In our previous publication, we quoted a former RICS definition of normal maintenance:

> '. . . *work or a reoccurring nature which certain building elements routinely require in order to preserve their integrity and functionality'.*
>
> (Parnham 2011, p. 39).

We went on to suggest this might typically include annual clearing out of gutters and drainage gullies, yearly servicing of boilers and fires, repainting of external wood and metal surfaces, removal of sludge from septic tanks and so on.

Although this is reasonably clear-cut, two examples may illustrate the cross-over between 'normal maintenance' and a 'repair':

- Consider the clearing out of rainwater gutters. In dwellings near trees and other types of vegetation, it will be normal to see an amount of debris in gutters, especially around autumn time. In these circumstances, a reasonable property owner would organise gutter and gully cleaning after leaf fall or just before the spring and this would be clearly considered as 'normal maintenance'. Although you might want to advise your client of the importance of this maintenance activity, it should not affect the judgement to apply a condition rating of one.

  Now imagine the same property where the owner had fallen behind on property maintenance and the gutters had not been cleaned out for a few years. The gutters may be full of leaves and other debris, out of which other plants are growing. Although this might not yet be causing noticeable moisture problems below (especially if it is dry during your inspection), you may consider the removal of such a lot of debris that could soon result in rainwater spilling down the walls below as something more than 'normal maintenance'. In this case, allocating a condition rating two might be more appropriate;
- The same approach can be applied to the painting of timber window frames. During normal maintenance cycles, the paint on timber surfaces can begin to flake off, but as long as the windows are redecorated every four to six years, then the timber should remain sound. However, if the window frames had not been decorated for some time and the paint film was in a very poor condition, wood rot could soon develop if not attended to in the near future. This could be another example of where activities usually described as 'normal maintenance' develop into a 'repair' justifying a condition rating two assessment.

These two straightforward examples show how difficult it is to use a condition rating methodology that produces clear and objective decisions every time – you will always have to use your professional judgement.

Condition rating two

This rating assumes the surveyor has made the decision that a defect exists and a repair is now required. The transition from a condition rating two to a condition rating three is focused on whether the repair or replacement is 'serious or urgent'.

Using the examples of the blocked gutter and poorly painted windows previously described, consider the following:

- Blocked gutters – if the blocked gutters are causing the occasional overflowing of water that has not yet adversely affected the building below, then a condition rating two would be appropriate as the required action is not yet 'serious or urgent'. However, if the blockage has clearly caused regular overflowing of water that has saturated the walls below and resulted in moisture problems internally, then a condition rating three would be appropriate because the effects of the problem are 'serious' and need to be urgently resolved.
- Poorly painted windows – as previously described, if the paint is flaking off and only small and isolated areas of wood rot are noted then the repair and repainting of the window frames could be done 'soon' and so justify a condition rating two. In other words, it would not be serious or urgent. However, if the window frames were in such poor condition that they were letting water get into the building and/or parts were loose and unsafe (especially during stormy weather), then a condition rating three would be required.

Condition rating three

The decision to allocate a condition rating three (CR3) to an element is a more complex process. This is because one of our main duties to a client is to provide balanced advice from which they can make a well-informed purchase decision. On the one hand, too many condition rating threes (CR3) may result in the client not going ahead with the purchase, but on the other if the surveyor allocates a CR2 to a 'serious or urgent' problem, the client may be faced with an unexpected level of repair once they move in.

14.4.4 Condition rating tests

To support an objective decision, the following initial tests should be applied:

Does the defect compromise the structural/functional integrity of the element?

In our previous book, we defined **'compromising structural integrity'** as a defect that has a pronounced effect on an element, leaving part of it potentially unstable. This does not mean it is a dangerous structure, but does indicate that if action is not taken soon, the element could become unstable in the not-too-distant future. Examples would typically include:

- Part of a wall that is cracked and distorted (say category three damage as described in Chapter 5);
- A ceiling that is cracked, bowing downwards and parts of it moves under light finger pressure; or
- Roofing components that have deflected or split resulting in uneven roof slopes.

If the defect has **'compromised the functional integrity'** of part of a building, then the element usually no longer fulfils one of its primary functions. The focus here is on the primary function, as most elements have a number of less important secondary functions as well. Examples might include:

- Roofs that are leaking allowing water to affect ceilings and habitable spaces below;
- Suspended floors that deflect excessively under normal loadings so the ceilings below have cracked; and
- Service systems that are unsafe to use.

If the defect is not repaired or replaced as soon as possible, will it cause serious defects in other building elements and/or result in a significant threat to the health and safety of the building users?

This relates to the criteria of 'urgency' and the roof leak described above is a typical example. If the leak is bad enough, then it could affect the bedroom ceiling below and cause it to become unstable and, in some cases, collapse. Staying with this example, if the ceiling collapses, it could injure an occupant. Therefore, it should be considered a serious safety hazard. Such a defect would tick both boxes in this test.

However, if the roof leak was less serious, resulting in only light water staining on the ceiling that is otherwise intact, then a condition rating two would be appropriate. To keep a sense of balance, it is important that we do not become too speculative about how such a problem could develop. Yes, roof leaks do get worse, especially during stormy weather, but we must be led by evidence that we see on the day of our inspection rather than the worst-case scenario an overactive imagination might conjure up.

Does the defect/problem require a further investigation?

The need for further investigations has been explicitly highlighted in the Home Survey Standards (RICS 2019, p. 19). This part of the standard acknowledges that an RICS member may suspect that a visible defect may affect other concealed building elements and if this is the case, they must recommend that a further investigation is undertaken. Conversely, the HSS goes on to say that members should exercise professional judgement and must not call for further investigations only to cover themselves '. . . against future liabilities'.

These contrasting statements represent the dilemma posed by further investigations. A non-destructive, point-in-time inspection will never be able to diagnose the full extent of every defect you come across and to make sure the client is properly advised, further investigation by an appropriately qualified person will be necessary from time to time. However, we have seen a culture of 'defensive surveying' develop amongst residential surveyors, especially as clients have become increasingly litigious. This has resulted in an increasing trend of recommending further investigations in place of making objective and balanced professional judgements based on what can be seen at the time of inspection.

Although this is further discussed in Section 14.6.1 of this chapter, at this stage, the following statements may help you decide where further investigations are required:

- **Further investigations should NOT be recommended just because an element could not be inspected.** Furnishings, floor finishes and a lack of access panels can prevent the inspection of parts of a property, but unless there is clear evidence that these concealed areas may be affected by a hidden defect, then they should be rated as 'Not inspected' and the client advised of the implications.

- **Further investigations SHOULD be recommended where there is a suspicion that a visible defect may have affected other concealed elements.** A typical example would be a carpeted, suspended timber floor that vibrates excessively during a heel drop test, has high moisture meter readings to the adjacent skirtings and does not have sufficient ventilation to the underfloor area (see Chapter six for a more detailed case study). In this case, a further investigation should be recommended, as there is a clear suspicion and high risk that the floor structure may be affected by concealed defects.
- **Are you technically competent to make a diagnosis and give advice on this type of defect?** Most residential surveyors have a broad knowledge of building construction and pathology, but usually are not 'specialists' in any one particular area or discipline. In these technological times, dwellings are becoming increasingly specialised and sophisticated with such features as air source heat pumps, whole house ventilation systems and walls formed with structurally insulated panels (SIPs), to name but a few. These add to the already growing list of services about which a residential surveyor can say little. Although we might be able to give the client an idea about the likely scope of the problem/solution, if they want to be sure, then they will have to appoint a specialist to carry out further investigations.

In the first instance, you should use these questions to help you decide whether a further investigation is required. We will return to this topic to discuss how such matters should be reported.

14.4.5 The condition rating protocol and how to use it

In the 'Surveyor's guide to RICS Home Surveys' (Parnham 2011), we created the 'condition rating protocol'. This was an easy-to-use methodology designed to help residential surveyors develop a consistent approach to condition rating using RICS's original condition rating definitions. As described above, we have refreshed the rating definitions and consequently revised the protocol.

Before we described the protocol in detail, it is important to understand two important principles:

- The protocol is not a precise science. It is not a quantitative process that automatically gives the right answer. Instead, it is a decision-making tool that gets you in the vicinity of an appropriate rating;
- Although many condition rating decisions will be clear and can be quickly taken, some choices will be more marginal. The condition rating protocol will help, but in these cases, it is you who will have to apply a professional judgement and make the final choice.

Using the protocol

The protocol is split into two stages:

Stage one

IS THE BUILDING ELEMENT IN A SATISFACTORY CONDITION?

If the answer is yes, there are no indications of present or suspected defects that require the undertaking of a specific repair, and it requires normal maintenance only, then it is a condition rating one. If the building element needs to be repaired, replaced or investigated, then you should move on to the questions in Stage Two.

Stage two

If you have decided that some form of intervention is required, then the purpose of this second stage is to decide whether a condition rating two or three is appropriate.

To help you do this, we have deconstructed the 'tests' in the condition rating definitions to produce six questions that will help identify the critical features of the problem. You should apply each question to the identified issue, ticking each 'yes' or 'no' box as appropriate. Once you have responded to each question, you should look at the results as a whole. The following advice may help:

- If you have ticked all the 'no' boxes, a condition rating two may be appropriate;
- If you have ticked one or more 'yes' boxes, a condition rating three may be applicable.

Stage One

Is the building element in a satisfactory condition?

☐ **Yes:** There are no indications of present or suspected defects that require the undertaking of a specific repair. It requires normal maintenance only.

☐ **No:** The building element needs to be repaired, replaced or investigated.

Stage Two

	No	Yes
Question A Does the problem compromise the functional integrity of the building element?	☐	☐
Question B Has the defect caused structural failure or serious defects in other building elements?	☐	☐
Question C Has the defect compromised the structural integrity of the element?	☐	☐
Question D Does the defect seriously and directly threaten the safety of the building users?	☐	☐
Question E Are urgent repairs or replacements needed now?	☐	☐
Question F Does the problem require further investigation?	☐	☐

Figure 14.2 Condition rating protocol (the 2021 version).

This is not a precise science and it does not give the 'right answer'. Instead, you should use it as a decision-making tool that gets you in the right 'ballpark'. Although it encourages an objective approach, there is enough flexibility for you to apply your own professional judgment and discretion in marginal cases.

Condition rating case studies

Although we think the protocol is clear, any methodology remains abstract until it is applied to realistic examples. We have included three typical roof covering examples.

Condition rating case study A

This semi-detached house was built in the late 1950s. The roof covering is original to the house and has a sarking felt below the tiles (see Figures 14.3 and 14.4). Although the roof covering is over 70 years old, it is still performing satisfactorily and there are no roof leaks. Based on a visual inspection using binoculars, it was noted that the pointing to a small number of ridge tiles was either loose or missing (see 14.5).

Figure 14.3 The front view of Condition rating case study A property.

Figure 14.4 Rear view of Condition rating case study B property.

Figure 14.5 Closer view of the ridge to Condition rating case study A property.

Using this information, the following condition-rating protocol was produced:

Stage One

Is the building element in a satisfactory condition?

☐ **Yes:** There are no indications of present or suspected defects that require the undertaking of a specific repair. It requires normal maintenance only.

☑ **No:** The building element needs to be repaired, replaced or investigated.

Stage Two

	No	Yes
Question A Does the problem compromise the functional integrity of the building element?	☑	☐
Question B Has the defect caused structural failure or serious defects in other building elements?	☑	☐
Question C Has the defect compromised the structural integrity of the element?	☑	☐
Question D Does the defect seriously and directly threaten the safety of the building users?	☑	☐
Question E Are urgent repairs or replacements needed now?	☑	☐
Question F Does the problem require further investigation?	☑	☐

Figure 14.6 Condition rating protocol for Condition rating case study A.

Explanation for case study A

In terms of the protocol's standard set of questions, we felt that the roof of Case Study A did not undermine the functional integrity of the roof (in other words, it wasn't leaking). No other building element was affected by the problem and the structural integrity of the roof covering itself was unaffected at the time of inspection. Although loose ridge tiles can present a danger, at this moment, we felt the safety of the building users was not yet affected. Although the ridge tiles needed to be repaired at some point, this did not have to be done urgently and the roof covering did not require any further investigation, as the 'reasonable' residential practitioner should be able to come to a view on this matter. Consequently, the checked boxes under the 'no' column strongly suggest that a condition rating two rating would be appropriate.

Condition rating case study B

This detached house was built in 1913 and is covered with a natural slate roof that was original to the house (see Figure 14.7). The covering does not have a sarking felt beneath and the back pointing (or torching) was in a poor condition and dropping away in a number of areas (see Figure 14.8). There are a small number of cracked or

Figure 14.7 View of roof slope to Case Study B.

Figure 14.8 View of the underside of roof covering of Case Study B.

Figure 14.9 Broken slate to Case Study B.

Figure 14.10 Leak to bedroom ceiling of Case Study B.

missing slates and on the rear slope and it could be seen from the dormer window that one slate has cracked and come loose (see 14.9). This resulted in a roof leak that has stained the bedroom ceiling below (see 14.10). The loose slate was resting in the adjacent gutter.

Using this information, the following condition-rating protocol was produced:

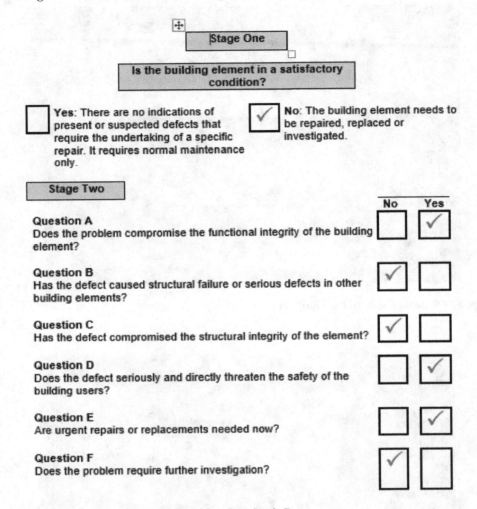

Figure 14.11 Condition rating protocol for Case Study B.

Explanation for case study B

In terms of the protocol's standard set of questions, we felt that the functional integrity of the roof was undermined by the active leak. Although the leak affected the ceiling below, it had not yet caused structural failure in the ceiling or resulted in a serious defect. The overall structural integrity of the roof covering at the time of inspection had not been undermined, but the loose slate that was wedged in the adjacent gutter did threaten the safety of the building users. It would not be too speculative to suggest that stormy weather could easily dislodge the slate. Consequently, because of the active leak and the safety issue, this problem is urgent and has to be resolved as soon as possible. As with Case Study A, as long as a residential practitioner could carry out this type of visual inspection, then they should be able to come to a view without referring the matter for further investigation.

This case study shows how making an objective assessment of the condition of an element can be challenging. For example, take question B of the protocol: 'Has the defect caused . . . serious defects in other building elements?' Although we responded 'no' to this question, many practitioners would point out that although the leak may be relatively minor now, the next storm could result in a dramatic increase in the level of damage. We wouldn't disagree with that, but it is important not to be too speculative and base your judgement on what you can see at the time of your inspection.

Whatever your own judgements, as a result of the higher number of checked boxes under the 'yes' column, in our opinion a condition rating three rating is clearly justified for a roof covering in this condition.

Condition rating case study C

This stone-built terraced property is likely to be at least 200 years old (see Figure 14.12). The roof slopes are covered with heavy stone slates typical of the region (Figure 14.13). There are no missing or cracked stone slates or roof leaks and an internal inspection revealed the covering has a relatively modern roofing felt beneath with a number of newer purlins and rafters. This suggests that the roof has been extensively repaired and recovered in the recent past (see Figure 14.14).

Figure 14.12 The front view of Case Study C property.

Figure 14.13 Closer view of the stone slates to Case Study C.

Figure 14.14 View of the roof structure of Case Study C.

IMPORTANT NOTE: This is an older property that was built using local materials and construction techniques particular to this area. In practice, before offering services on a property like this, you should be confident that you have the appropriate knowledge and experience (RICS 2019, HSS, p. 8).

Using this information, the following condition-rating protocol was produced:

| Stage One |

| Is the building element in a satisfactory condition? |

☑ **Yes**: There are no indications of present or suspected defects that require the undertaking of a specific repair. It requires normal maintenance only.

☐ **No**: The building element needs to be repaired, replaced or investigated.

| Stage Two |

	No	Yes
Question A Does the problem compromise the functional integrity of the building element?	☐	☐
Question B Has the defect caused structural failure or serious defects in other building elements?	☐	☐
Question C Has the defect compromised the structural integrity of the element?	☐	☐
Question D Does the defect seriously and directly threaten the safety of the building users?	☐	☐
Question E Are urgent repairs or replacements needed now?	☐	☐
Question F Does the problem require further investigation?	☐	☐

Figure 14.15 Condition rating protocol for Case Study C.

Explanation for case study C

This is a straightforward case. In our view, there were '. . . no indications of present or suspected defects that require the undertaking of a specific repair'. The roof was not leaking, so it had not caused problems in other building elements. The roof covering looked to be in satisfactory condition and there were no safety hazards. No repairs were required and there were no 'trails of suspicion' that required further investigation. Therefore, a condition rating one would be appropriate.

Coming to this view does not mean that the whole roof has a clean bill of health. For example, we would like to know who did the repair/reroof of the property and when was it done? Did it have appropriate building regulation approval? Also, with roof coverings that are as heavy as stone slates, it is important to carefully assess the nature of the supporting roof structure. However, in an elemental report, none of these issues will usually affect the assessment of the roof covering element itself. These important but broader matters will be considered in the next section – report writing.

14.5   Report writing – practical examples

14.5.1   Adding depth and breadth

The previous section explained how the different elements of a building can be objectively assessed and an appropriate condition rating applied. Although this helps the reader of the report identify the repair priorities, in our view, this is only half the story. Revisiting these three case studies, consider the following:

- **Case study A (condition rating two).** Although the ridge tiles of this roof covering were not yet loose enough to present a safety hazard, the increase in weather extremes could result in further damage during stormy weather. Additionally, although this problem could be resolved by rebedding two or three ridge tiles on a mortar bed, a number of other factors should be taken into account:

  - Modern roofing standards now recommend that mechanically fixed ridge tile systems are used in an effort to reduce safety hazards often posed by those bedded in cement mortar; and
  - Although rebedding two or three ridge tiles may not take very long, the provision of safe access equipment (in other words a scaffold) will add considerably to the cost of the repair.

- **Case study B (condition rating three).** As previously described, at the time of the inspection, only one slate was loose. However:

  - This is an older roof so the likelihood of increasing numbers of slates working loose over the next few years is high;
  - The lack of a sarking felt would make this roof vulnerable to wind-blown snow and rain especially during stormy weather.

- **Case study C (condition rating one):** This is a traditional roof covering to an older building located in a national park:

  - Although the covering has been rated CR1, if future repairs are required then only appropriately contractors experienced with this type of covering should be used as unsuitable repairs could result in additional problems; and
  - The property is in a National Park so it is likely there will be local planning controls on the type of building materials used.

All of these issues are important and will affect the complexity and costs of any repairs/maintenance required. It is important that clients are aware of these matters when they are coming to a purchase decision and no matter how sophisticated a condition rating system is, it will not be able to highlight these more subtle considerations. This is where the written part of the report becomes very important.

Criteria for level two reports	*Proposed text*
Describe the part of the building and its location on or in the subject property.	**The sloping roof of the property is covered with clay pantiles over a secondary waterproof barrier (roofing felt) that are the same age as the house.**
Outline any faults, defects, or other problems (if there are any) and their likely consequences and/or possible risks.	**A small number of the tiles along the top of the roof (the ridge tiles) are poorly secured. If these are not repaired soon, they may become dislodged during stormy weather and become a safety hazard.**
Allocate an appropriate Condition Rating.	**Condition rating two**
State the action the client needs to take. This should include an outline of the likely remedial work, who should carry this out and by when.	**Although small numbers of ridge tiles can be repaired, the work may dislodge or damage the adjacent tiles. It may be economical to replace all the other ridge tiles with a mechanically fixed system. You should ask an appropriately experienced roofing contractor to carry out this work soon.**
Include concise supplementary information that will help the client better understand both the nature of the problem and what they need to do about the issue.	**To carry out this repair safely without damaging the roof covering, the contractor will have to use appropriate access equipment (for example scaffolding, hydraulic platforms and so on) and this will add to the cost of the work.**

Figure 14.16 Deconstructing the text. This table shows how we built up the text for Case Study A using the criteria laid down in the Home Survey Standard.

14.5.2 Constructing the text

Using the principles described in Section 14.3.1 and the details of the case studies described above, we have included text extracts we would include in a level two report for each of the roof covering case studies. Rather than simply presenting our preferred text, we have shown how the report can be built up using the criteria listed in 14.3.1 for Case Study A. These criteria are presented in the left-hand column of Figure 14.16 and the suggested text for the report included in the right-hand column.

Rather than trying to make sense of this in a tabular form, we have re-presented the text in the form that it would appear in the report.

Roof coverings

The sloping roof of the property is covered with clay pantiles over a secondary waterproof barrier (roofing felt) that is the same age as the house. A small number of the tiles along the top of the roof (the ridge tiles) are poorly secured (see photo X). If these are not repaired soon, they may become dislodged during stormy weather and become a safety hazard.

Condition rating two

Although small numbers of ridge tiles can be repaired, the work may dislodge or damage adjacent tiles. It may be economical to replace all the other ridge tiles with a mechanically fixed system at the same time. You should ask an appropriately experienced roofing contractor to carry out this work soon.

To carry out this repair safely without damaging the roof covering, the contractor will have to use appropriate access equipment (for example scaffolding, hydraulic platforms and so on) and this will add to the cost of the work.

Commentary on case study A

Although residential surveyors should always follow good report-writing practice, the overall style will vary between surveyors. To help you reflect on this effort, we have included a number of points:

- **Concise** – This is a level two report so we have been as concise as is practically possible;
- **Structure** – We try to follow a consistent structure. The information above the 'condition rating line' is factual, while the text below the condition rating aims to interpret the condition of the roof and provide the client with broader and useful advice;
- **Terminology** – We prefer to use the layperson's word first ('secondary waterproof barrier') and the technical term second ('roofing felt').
- **Prioritisation** – We try to communicate the relative urgency of any repairs in two ways. First, by the condition rating. In the definition for condition rating two, it states the repair is neither 'serious nor urgent'. Second, by the use of the word 'soon'. We accept that this can be vague, but the alternative is being too precise. Some surveyors use phrases like 'medium term', 'before the next winter season' or even 'within the next two years'. All options come with their own pros and cons and we prefer a looser phrase that can be discussed with the client after they have read the report. For a condition rated three element, we would use the term 'now' to indicate an appropriate level of urgency;
- **Supplementary information** – Level two services should outline remedial work without going into technical detail. But in this case, it is important to let the client know that all the ridge tiles are likely to be in a similar condition, making it difficult to carry out a limited repair. Additionally, the British Standards for new roof coverings were changed a number of years ago, making dry fixed ridge systems compulsory as mortar bedding alone was no longer considered sufficient (BS 5544 British Standard for slating and tiling 2014). We think this should be reflected even in a level two report.

To carry out this amount of repair work, the operatives will have to spend a reasonable amount of time working at a high level. To do this safely, some form of scaffold access will have to be provided and this can be costly. Although precise figures should never be used, the level of description used will help the client understand the implications.

It is important to point out that this reflects our approach to report writing that will not suit everyone. It is important to develop your own style that reflects your own knowledge, experience and approach while at the same time meeting the requirements of the Home Survey Standards and those of your client.

Reports for case study B and C

We have included our report extracts for the other case studies as follows:

Case study B

ROOF COVERINGS MAIN ROOF

The sloping roof of the property is covered with natural slates and does not have a secondary waterproof barrier beneath (roofing felt). The roof covering is likely to be the same age as the property. A small number of slates are cracked, several others have been repaired and one slate to the rear slope has broken away and is sitting in the adjacent gutter causing a leak that is staining the ceiling below. If these problems are not repaired now, further leaks could occur and the loose slate could become a safety hazard, especially during stormy weather.

CONDITION RATING THREE

It is normal for properties of this age and type not to have roofing felt. Without this, water may get into the roof space, especially during stormy weather. The lack of roofing felt and the other repairs indicates the covering will have a limited life and you should plan to recover the roof soon.

If you did choose to renew the roof covering, the work may need building regulation approval from the local authority and to meet current standards, you may have to carry out additional work (for example strengthening of the roof structure, improving thermal insulation and so on). This can increase the amount and cost of work required.

To carry out repairs safely without damaging the roof covering, the contractor will have to use appropriate access equipment (for example scaffolding, hydraulic platforms and so on) and this will add to the cost of the work.

Case study C

MAIN ROOF

The sloping roof of the property is covered with thick sandstone slates and has a secondary waterproof barrier beneath (roofing felt). The presence of the roofing felt indicates that the roof has been recovered in the relatively recent past.

CONDITION RATING ONE

Although the roof covering is in a satisfactory condition, any future repairs should be carried out by appropriate contractors experienced with this type of traditional roof covering. Additionally, because the property is in a National Park, you should contact the local planning department before any future repairs are carried out.

You should ask your legal adviser to investigate whether the recovering of the roof received approval under the building regulations and whether it is covered by any guarantee or warranty (see the legal section).

We hope you find these report examples useful.

14.5.3 Handling technical terms

It is important to use simple language when dealing with clients. Few members of the public have much technical knowledge and lose patience if presented with a barrage of technical jargon. The following example from a real report shows how comments can be misleading:

> *Internally, one or two floors flexed substantially under load and these should be checked to ensure that adequate support is given.*

Is this the load of furniture, humans or what? Who should check them and what should they look for? The fact that they 'flex' means that adequate support is obviously not 'given' at the present time. Does this mean that the floors are going to collapse and pose a danger to the occupants of the building? All this may be obvious to a technical fraternity, but to intelligent and proactive lay people who pay your fees, it will just seem like a poor service. Also, if 'additional support' was required, it could be disruptive and expensive to the occupier.

The importance of handling technical terms properly is emphasised in the Home Survey standards. The HSS states that the report '. . . must use non-technical terms throughout – if technical terms are occasionally used, the client will find a layperson's explanation useful' (RICS 2019, p. 14).

Despite this, we still see numerous reports of all types that seem to have been written for other residential surveyors rather than the typical client.

There are a number of ways of dealing with technical jargon:

- Similar to our own approach, use a layperson's phrase first with the technical term in brackets straight after. One drawback with this approach is that the layperson alternative can sometimes be tortuous. For example, consider these examples:

  - **Purlin** – The large timber beam that supports the roof structure;
  - **Flaunching** – The cement bedding around the base of the chimney pot;
  - **Benching** – The shaped cement surfacing at the bottom of the inspection chamber.

  Although we like this approach, describing technical terms in this way can literally add 'weight' to a concise report and so it might not be to everyone's taste.
- Another method is using a glossary that provides the layperson's explanation for a range of technical terms in a tabular format. This can be placed at the end of the report and provide the reader with more complete explanations that don't clutter the main body of the report. A possible disadvantage with this approach is that readers have to do all the work – every time they come across an unfamiliar term, they have to find the glossary and scroll through the document to find the relevant word or phrase. Additionally, there is a danger that the surveyor may over-rely on the glossary and fail to accommodate the needs of the lay reader;
- The other method is to use images instead of words. This could range from the commonly used 'cut-away' house diagram used by RICS in their standard formats or something that makes full use of recent technological developments. In digital documents, this could range from 'pop up' explanations of technical terms that

can be activated within the text to hyperlinked images and other illustrations of the particular features.

The flexibility of the HSS and the developing technologies provide the capacity for innovation for those with the appropriate skills and motivations.

14.6 Other report-writing issues

The previous section introduced core report-writing skills. This part of the chapter will deal with a range of specific and important report-writing issues.

14.6.1 Reporting on further investigations

For many clients, one of the most disappointing features of any condition report is the high number of referrals for further investigations that are sometimes recommended by the surveyor. When a person has paid what they feel is a lot of money for a professional to inspect and report on a property they are interested in buying, they don't expect to have to go to a number of other professionals, specialists and contractors before they can get enough information to enable them to make a purchase decision. We have seen many reports where residential surveyors have referred their clients to six or more 'specialists' – in our view, this is totally unacceptable.

In our experience, there are usually a number of reasons for this:

- The wrong level of service has been instructed. For example, carrying out a level two service on an older property that has been altered and extended a number of times will inevitably result in numerous further investigations. This is because a level two service was never meant to suit that type of property (see Chapter 2.5 for further discussion on this point);
- The residential surveyor may not have sufficient knowledge and/or experience to provide that level of service on the property. Although a surveyor may be a member of RICS or any other equivalent professional organisation, this does not mean that they can automatically offer any level of service on any property. Assessing your own competence and limitations is a vital part of offering a professional service (see Chapter 2 for more information);
- Surveyors sometimes lack confidence in their own abilities, maybe because of a recent complaint or claim. They then engage in what is known as 'defensive' reporting in an effort to minimise their own liability. We understand why this attitude may arise, as both of us have had claims against us. However, although defensive reporting may give comfort for a short time, client dissatisfaction will not be good for business and such a negative attitude is not good for the profession.

The HSS acknowledges the need for further investigations and states, '. . . a RICS member's knowledge will, at times, lead to a suspicion that a visible defect may affect other concealed building elements'. In these circumstances, the RICS member '. . . must recommend that a further investigation is undertaken' (RICS 2019, p. 19). However, the HSS goes on to say that members should exercise professional judgement and must not call for further investigations '. . . only to cover him or herself against future liabilities'.

Getting the right balance between giving a clear professional judgement on a matter and calling for further investigations is a delicate one. Based on the HSS and our own experience, here are a few tips.

Limitation of inspection or further investigations?

Imagine a situation in which you were inspecting the pitched roof of a terraced house. Although you could see all the front slope from the road, the small backyard prevented you from seeing any of the rear roof slope. However, you gained access to the roof space from where you inspected the underside of the rear slope and found no evidence of a defect. In these circumstances, you should be able to come to a professional judgement without the need for further investigations. Additionally, it is vital that you inform the client of the restriction and explain the implications.

Consider a different set of circumstances. You may be inspecting a similar terraced property where you could see both roof slopes to the front and rear from ground level, but you couldn't get access to the roof space. You also notice watery brown stains on the bedroom ceilings that suggest a leaking roof. In this case, because there is a 'trail of suspicion', a referral for a further investigation would be justified.

Making a further investigation recommendation

Where a further investigation is recommended, the RICS suggest the following information should be included in the client's report:

* A description of the affected element and why a further investigation is required;
* When the further investigation should be carried out; and
* A broad indication of who should carry out the further investigation (for example their qualifications, membership of a trade body, competent person scheme and so on).

We welcome this clarity because if we are sending the client off to get a further report from a 'specialist', it is important they have as much information as possible so they can properly instruct the specialist and hopefully properly interpret the resulting report. In addition to this information, we would add two additional points:

* **Who should carry out the further investigation**. In some cases, this will be obvious and straightforward, so we can be specific about their qualifications. Typical examples would include a structural engineer or building surveyor to investigate building movement or a Gas Safe engineer to inspect a suspect gas appliance. For other areas, it may not be so simple. Consider electrical work. Traditionally, surveyors have referred their clients to an electrician registered with the National Inspection Council for Electrical Installation Contracting (NICEIC). Although this is acceptable, the government's Competent Person Scheme has given a number of organisations the right to self-certify their own work. At the time of writing, the following run their own schemes: BESCA, Blue Flame Certification, Certsure, NAPIT, OFTEC, Stroma. 'Certsure' are a relatively new scheme that now incorporates both of the schemes run by NICEIC and Elecsa. The reason for bombarding you with these meaningless acronyms is that giving a client a 'broad

indication' of who should carry out the further investigation is not as easy as it used to be.

Consequently, where there are multiple organisations that deliver the same specialist service, you might consider using a phrase like 'suitably or appropriately qualified and experienced person' and providing a hyperlink to the government's competent person schemes website.

- **Scope of the further investigation**. One aspect we find frustrating is that when we refer a matter for further investigation, the specialist will often be limited by the same restrictions we faced during the inspection. To illustrate this point, we were inspecting a property during a level two service. The surface of one of the suspended timber floors at ground floor was covered with a laminate covering and we noted the following:

  - This floor felt very bouncy after a 'heel drop' test;
  - High moisture meter readings were recorded to a number of skirtings; and
  - There was not enough ventilation to the underfloor area from the outside.

We clearly suspected a possible wood rot/moisture problem and because of the limitations of the inspection, we recommended a further inspection by 'an appropriately qualified and experienced person'. One was instructed, but because they were not allowed to lift the laminate flooring, they found themselves in the same position as we did – all they could do was speculate about the condition of the floor. Consequently, in addition to the information recommended by RICS outlined above, we now add the phrase 'To properly assess the condition of the (name element), parts of the construction may have to be removed and/or opened up and this can be disruptive. You should discuss this with the vendor'. This can help the client understand the complexity of some of the further investigations that may be required.

14.6.2 Legal matters: regulations, guarantees and other legal matters

Although the condition of the property is one of the main influences on the client's purchase decision, the legal implications of owning that particular property can be just as important. The HSS acknowledges this and clarifies the relationship between the role of the residential surveyor and the legal adviser (RICS 2019, 4.6, p. 17). The following points may help you understand this relationship:

- Although the legal adviser is responsible for checking the relevant documents and issues, they will not be familiar with the property;
- Consequently, residential surveyors will be the 'eyes and ears' of the legal adviser. In this role, we have to clearly identify apparent and specific items and features that have possible legal implications so they can be brought to the attention of the legal adviser.;
- It is unlikely the legal adviser will read the whole report. So it is vital residential surveyors must clearly highlight the relevant legal matters, assemble them in a separate legal section and remind the client they should bring these matters to the attention of their legal adviser;
- Although we don't want to do our legal colleagues a disservice, they may not look at the long list of carefully itemised and described legal issues until a few days/

hours before the exchange of contracts. When they notice the serious implications of the issues raised, they will frantically try to resolve matters quickly so they do not delay the sale. Sometimes this can result in clients being 'urged' to accept less-than-optimal solutions as a result.

The other problem we come across in survey reports is that surveyors use vague and non-specific phrases such as '*legal adviser to check all approvals and permissions are in place*'. Faced with such a meaningless sentence, the legal adviser will make their usual standard enquiries and submit the outcomes to their client. Consequently, they will never know about the two-storey extension that doesn't have building regulation approval, the established right of way across the middle of the rear garden, or the noise from the illegal car repair business next door. In other words, the very things the 'eyes and ears' of the surveyor should have picked up. If any problems result from missed features like these, then it will be the residential surveyor who will be liable and not the legal adviser.

The scope of the legal section

The HSS includes a list of matters that should be included in the legal section and these are split between:

- **Regulations:** including planning-related matters; competent person schemes; building regulations and the like;
- **Guarantees:** including structural work, timber and damp treatment works and Japanese knotweed management plans to name but a few; and
- **Other matters**: other features and issues that may have an impact on the property and require further investigation by the legal adviser. This will include a broad range of issues noted during the visual inspection or through the surveyor's knowledge of the locality.

We have not included a full listing of the relevant matters because they have been satisfactorily summarised in the HSS, to which you should refer for more information (RICS 2019, 4.6, p. 18). Instead, we have highlighted a number of topics we think will help you use this section more effectively:

Relationship between the main body of the report and the legal section: In a similar way to the Risks section (see Chapter 13), it is important you link the information within the main body of the report and the summary information in the legal section. If not, confusion and repetition can be the result. In our view, you should locate the primary description of the legal issue and its implication under the element/section of the report it affects together with a cross-reference to the legal section. You should include only a brief summary of the issues in the separate legal section itself;

Include a 'what if not' clause: If you suspect there could be a legal issue associated with a particular feature of a property and properly highlight this in your report and your client promptly brings it to the attention of their legal adviser, the matter could still remain unresolved for several weeks. This could be for a variety of reasons, including that formal enquiries of a third party could take some time to process, the equivalent of the legal adviser's 'in-tray' may have been very full, or they run a 'just in time' service where matters are only processed by a deadline date. Whatever the cause, if the legal enquiries identify a significant issue during the latter stages of the transaction, the sale as a whole could be threatened. Whereas if the client had known about the impediment earlier in the process, they might have had time to find other ways of resolving the matter.

One way of helping the client manage these challenges is to try and give them an early 'heads-up' about the problem. We call it 'what if not' advice that can be best illustrated by an example.

Imagine you are providing a level two service for a 1980s semi-detached property where a single-storey extension extends across the rear elevation and wraps around part of the gable wall. Three openings have been formed through the outside walls into the extension and although the extension is in satisfactory condition, there are some hairline cracks around the head of the through openings. The property is empty and the selling agent does not know anything about the dwelling's history.

There is no way of knowing whether the extensions and structural openings have been properly constructed and one of the best ways of getting some reassurance is by determining whether the work has received building control approval. The following clause could be added to the end of the section on walls or maybe even the 'other' element:

> You should ask your legal adviser to confirm whether rear and side extensions, the structural openings between the original kitchen and kitchen extension, the kitchen and dining area and sitting room and dining area have received building regulation approval (including the issuing of a final completion certificate) from the local council and advise on the implications (see legal section).
>
> If this has not been granted, retrospective building regulation approval will have to be obtained and this could be disruptive and costly. For example, parts of the plasterwork and flat roof over the extensions will have to be removed so the beam support can be inspected.

The second paragraph of this example is the 'what if not' phrase and can help the client understand what 'retrospective approval' will involve. If they have an early insight into the nature of the issue, they will be able to proactively manage the problem themselves. Better to know now rather than at the last minute.

14.6.3 *Risks to the occupants*

In Chapter 13, we described how safety hazards and risks can be identified and assessed. In this section, we will briefly describe how these can be reported to the client.

The most relevant advice is contained in the section of the HSS that describes how the report should be formulated (RICS 2019, p. 17). To summarise, you should:

* Describe the risk in the appropriate part of the report; and

- Create a separate section in the report where the risks can be 'concisely' listed. These should be cross-referenced to the appropriate element within the main body of the report.

The HSS states that the range of identified matters will be the same for each level of service, but what will vary is the explanation:

- A level one report will identify and list the risks and give no further explanation;
- A level two report will identify and list the risks and explain the nature of these problems;
- A level three report will do all this and explain how the client may resolve or reduce the risk.

Using the outcomes from case study one in Chapter 13, we have outlined how this requirement could be met for each of the different levels:

- **Level one report:** *The room in the basement does not have a safe escape in the event of a fire (see section titled 'Risks to occupants');*
- **Level two report:** *The room in the basement does not have a safe escape route in the event of a fire. Consequently, this space should not be used for habitable purposes (see section titled 'Risks to occupants');*
- **Level three report:** *The room in the basement does not have a safe escape route in the event of a fire. Consequently, this space should not be used for habitable purposes. Further investigations are required by an appropriately qualified person and although the implications will not be known until a report is received, it is likely that an alternative form of escape from the basement may be required (see section titled 'Risks to occupants').*

In the 'risk to occupants' section itself, there should be a concise cross reference back to the appropriate part of the main report. For example:

Lack of a safe escape route from the basement in the event of a fire (see Section D9 Basement).

This would be the same for each level of service. The extra detail should be included in the main body of the report.

As you can see, additional information and details are added at each respective level. The usefulness of the risks section is that the client can quickly see the range of hazards associated with the property and, depending on the level of service, get an insight into the nature of the problems and how they could be resolved.

Although the HSS's approach is effective, it is important that the relationship between the risks section and the main body of the report is clear. If not, as with the legal section, confusion and repetition could result. Consequently, you should locate the primary description of the potential hazard under the element/section of the report it affects, together with a cross reference to the risks section. You should include only a brief summary of the issue in the separate risks section itself.

14.6.4 *Overall opinion*

This section has been left until the end of this chapter because, like an overall opinion in a real report, it should be the last part you write. This is because you will need to reflect on your inspection, allocate the condition ratings and write the report. You will then be in a position to construct your overall opinion.

According to the HSS, the overall summary should have the following characteristics (RICS 2019, p. 16):

- The summary should provide a brief, simple and clear overview;
- It should be as concise as possible, be property specific and not repeat descriptive details;
- It should express the RICS member's view of the main positive and negative features of the property and highlight areas of concern.

The HSS also points out that condition ratings or other prioritisation methods will help place such assessments in context and give a balanced view of the property.

A well-written summary that properly reflects the true nature of the property can help orient the client. Despite this, for years this part of the report has been dominated by a small number of bland and meaningless paragraphs that were once formally sanctioned in RICS guidance. Here is a typical example of a non-valuation version:

> *This property is considered to be a reasonable proposition for purchase, provided that you are prepared to accept the cost and inconvenience of dealing with the various repair/improvement works reported. These deficiencies are common in properties of this age and type. Provided that the necessary works are carried out to a satisfactory standard, I see no reason why there should be any special difficulty on resale in normal market conditions.*

If we had a pound for every time we have seen this on the front of a report, we would not need to supplement our income by writing books like this. This is clearly not 'property specific' and gives very little information. For example, the '. . . *cost and inconvenience of dealing with the various repair/improvement works reported'* may mean different things to different people. The client is left to plough their way through the whole report in an effort to get a clearer picture.

You might find this a more suitable approach:

- Once you have written your report, review the balance of the condition ratings. For a dwelling of this type and age, does this look like a typical distribution? We do not suggest a statistical analysis, as this is not only about the ratio of condition ratings threes to condition rating ones. Watch out for those properties that have a large proportion of condition rating twos – although any one item may not be serious or urgent, the cumulative effect of the condition rated two repairs could be considerable;
- Based on this review, identify three positive aspects of the property and three of the most troubling features. This will help you construct a balanced view of the dwelling;
- Did the client have any particular requests when the service was set up? Did they ask you to look at any particular features? If yes, these can be appropriately incorporated into the overall opinion, as it shows you have made this property specific.

The HSS also reminds us that the overall opinion should be:

- Concise as possible;
- Property specific; and
- Not repeat the descriptive detail of the main report.

A typical example of an overall opinion is as follows.

Overall opinion of the property

This semi-detached property is located on the hillside, with picturesque views over the Styx valley. The layout of the property and the size of the rooms are satisfactory and it benefits from a kitchen extension to the eastern side.

This house is approaching 50 years old and although it is in a similar condition to other houses of its age and type in the neighbourhood, a number of repairs are now required. The significant number of condition rating twos indicate that while each individual problem is not serious or urgent, taken together they represent a significant amount of repair work. I would particularly draw your attention to the following matters:

- *Like many houses of this age, a number of components may contain asbestos and you will need to employ a specialist to make these safe;*
- *Although the owner has extended the property and made sure this work has been done in accordance with the appropriate regulations, other parts are unlikely to meet the current standards. The electrical system is a typical example;*
- *A number of legal checks are required to make sure the repairing liabilities are clearly known especially for the large retaining walls to the front and side of the property.*

To place this summary into context, it is important you should read the whole of this report.

14.6.5 The lure of standard paragraphs

The prized possession of many residential surveyors is their collection of standard report paragraphs and phrases. Whether they were written by themselves, obtained through a licence with a commercial provider or 'borrowed' from a previous employer, these paragraphs can save a lot of time when assembling the report and they can help you produce technically consistent and reliable information. As with most things, there are two sides to every coin and the overuse of standard phrases can also have negative effects. For example:

- Rather than choosing only relevant phrases and adjusting them to suit the specific property, some surveyors will throw every standard phrase in their armoury against any particular element or problem. Many of these paragraphs focus on broad generalities rather than the particular nature and condition of the element under consideration;
- We use the term 'armoury' in the previous point on purpose. This is because many surveyors will often use these phrases in an attempt to protect themselves from challenge – a tactic that will usually unravel. This approach is often adopted

by surveyors who have been subject to a previous complaint or those who lack the technical knowledge and experience to properly advise on that type of property;
- The over-use of standard phrases usually results in over-long reports. Several general and meaningless paragraphs can quickly increase the length of what otherwise would be a concise report;
- As a consequence, clients soon realise that much of the report is non-specific and does not help them make that all important decision. This will often result in impatience and disappointment in the service – the essential ingredients for a future complaint.

Additionally, the technical and regulatory content of standard phrases must be continually reviewed and amended; otherwise, the client may receive out-of-date advice. One common mistake we still see in survey reports is the name of the trade association for damp-proofing and timber treatment. Several surveyors still use the 'British Wood Preserving and Damp Course Associations (BWPDA)' despite their name change to the Property Care Association in 2003. Clear evidence that the surveyor has not taken the time to update their standard phrases. Other inappropriate uses of standard phrases include:

- A level three report on a timber-framed property built in the late 1700s that included the phrase 'we have not tested any of the concrete components for high alumina cement';
- A warning in a level two report on a property in the west midlands that the dwelling may be subjected to coastal flooding.

Effective use of sample phrases

A more appropriate use of standard paragraphs was discussed by the RICS in their book titled 'Survey Writer: sample phrases for the RICS HomeBuyer Report' (Parnham 2011). Although this book contained hundreds of 'sample phrases', it recommends the reader should '. . . adjust the style of the text to suit your preference . . . and situation'. This publication emphasised two other important principles:

- It used the term 'sample phrases' rather than 'standard paragraphs' to highlight the flexible nature of the text. The terms 'standard' or sometimes 'preferred' paragraphs suggest a rigid and unalterable text which would be inappropriate; and
- Both in the first and second editions of the Survey writer book, the aim was to provide a phrase that provided readers with 80–90% of the text they required. The remaining 10 to 20% would be the part that made the phrase 'property specific'.

This approach is reinforced in the Home Survey Standard (RICS 2019, p. 19) that clearly states the report must be '. . . property specific . . .'.
    In conclusion, sample phrases should:

- Be relevant, up to date and regularly reviewed;
- Match your own preferred style of writing so they fit seamlessly into the report as a whole.

14.6.6 Use of photographs in reports

When the first edition of this book was written in 2001, digital cameras had only been around for a few years and the first commercial 'camera phone' became available in 2000 and could store up to 20 photos. In Chapter 3 of the first edition under the heading of survey equipment, we suggested a '. . . suitable camera (with flash) and spare film' would be a useful addition to a surveyor's kit. How times have changed.

At the time of writing, technological developments have provided residential surveyors with a range of options:

- Separate digital cameras some of which can be attached to camera poles;
- Camera in smartphones;
- Cameras built into tablet devices;
- Those built into drones and other more specialist inspection equipment such as borescopes;
- 360-degree cameras on poles;
- Thermal imaging cameras.

Most of these different types can take both photos and videos and their storage capacity has increased exponentially. Whether it is through storage capacity on the device itself or through a cloud-based application, surveyors are able to take and store many hundreds of photographs.

The use of digital images is, to some extent, a matter of choice. We do not want to try to describe best practice for two reasons: technological development is so rapid that anything we do set down will be out of date before this book is published; furthermore, we are not best placed to write on these matters (we hardly know our AIs from our VRs). However, you may find the following comments useful.

Use of images during the site inspection

Although technology allows us to take many hundreds of high-quality images and many minutes of video during inspections, you should carefully consider what purpose these images will serve. Consider the following:

- Although useful, in our opinion images will never be a complete substitute for written site notes or sketches. For example, it is difficult to photograph a bulge in a wall and to show a crack in context. Sunny days can be the photographer's nightmare, and the resulting contrast can hide a lot of important detail;
- Hundreds of images can be very useful but make sure you have got the facilities to properly catalogue and store them all. Most digital images can be quickly sorted using appropriate software packages and you should adopt the approach that best suits your own organisation; and
- Like all other digital information, make sure you back up the images for at least six years.

Use of images in the report

Photographs are a very useful way of highlighting points in the report, although they need to be of good quality, explain what is being shown and they need to be in context. There is nothing worse than a report that includes a large number of unexplained images in an appendix. There are a number of different approaches:

- Each image should have a clear purpose. For example, is it showing something that is difficult to describe in words or something the client specially wanted to see? Do not use images simply to 'pad out' the report;
- Every image should be properly referenced and preferably annotated. This will enable the reader to link the text of the report to the appropriate image;
- Consider where you want to locate the image. Do you want to place it as near to the text as possible? If yes, then make sure the software can accommodate this without those frustrating page breaks. The other option is to put all the images in an appendix but to link them to the main text with hyperlinks. This will avoid the reader from having to scroll backwards and forwards all the time.

The other matter to consider is how the use of developing technologies can affect your terms of engagement. For example, the use of camera poles and drones may affect the scope of the inspections. You may want to reflect on the usual limitation that the inspection uses the vantage point of a ladder 3 m off the ground when the images in the report show the flaunching detail that only an eight-metre-high camera pole could take.

14.7 Discussing the findings of the report with the client

In our view, one of the most important aims of the Home Survey Standards is to improve the quality of liaison between residential surveyors and their clients both before the terms of engagement are agreed upon (see Chapter 2.3) and after the report has been delivered. In our view, this will help reduce misunderstandings and the likelihood of complaints later on.

In relation to post-report delivery liaison, RICS make it clear that adequate time must be set aside '. . . to discuss the findings of the report' (RICS 2019, p. 20). The nature of this liaison will vary between surveyors and the nature of the service provided but might typically include:

- Telephone discussions;
- Video consultations over the internet; and
- Contact through social media.

We are aware of other methods, including:

- Meeting the client at the property at the end of the inspection so the most important issues can be pointed out;
- Face to face meetings in the surveyor's office, coffee bar or even the client's home.

Whatever your chosen method, it is important that this stage is built into your fee for the service; otherwise, there will be a tendency to resent any post report contact. We have met a considerable number of surveyors who, once the report has been sent to the client, would like nothing more than never hearing from the client ever again. This is clearly not in line with RICS standards.

14.7.1 The pitfalls of post report liaison

Although most commentators agree that working closely with the client is a very important part of the process, there are some risks that have to be handled carefully. The main ones are highlighted in the Home Survey Standard and usually occur when the residential surveyor qualifies and/or expands on the delivered report during these discussions (RICS 2019, p. 20). This can result in both extending liability and confusing the client – outcomes this process is meant to avoid.

It is difficult to offer clear advice on how to approach this part of the service, as it relies on achieving the right balance. Based on RICS's guidance in the HSS, you should consider the following:

- Clearly explain the status of the post report discussion/exchanges at all stages. This might typically include an explanation in your general literature describing your service, in the terms of engagement and particularly during the direct discussions with the client;
- During these discussions, do not go beyond the scope of the service. This may prove challenging, especially for level two services on properties with a number of problems. Clients will often ask: 'How much do you think that will cost?', or 'What do you think the specialists will find out?', and even 'Do you really think that is a condition rating three or do you have to say that because of your insurance?' Although you may well have thoughts on all of these questions, it is important that you keep within your terms of engagement without being unduly abrupt with your clients;
- Keep a record of any discussions/exchanges. This will depend on the nature of the exchange but could typically include:
  - If it was a telephone call or a face-to-face meeting, then consider either recording the call or making a written note during and/or straight after the discussions. It is important to confirm the issues discussed and this could be done by sending an email/text/message to the client confirming the time and date of the discussion together with a summary of the topics discussed. If your client wants clarification on your summary, they will contact you to discuss – the ball will be in their court;
  - If it was a video-based consultation then recording the exchange would be relatively straightforward as long as the client agrees;
  - Like with all digital records, it is important to conform to the appropriate regulations and legislation.
- Many residential surveyors report that although offers are made to discuss the outcomes of the report, a number of clients do not take up the offer. This can be very frustrating, especially if complaints are received in the future. This is the client's prerogative, but in such cases, it is important to keep a file copy of all communications where such an offer was made.

14.8 Conclusions

When putting together your report, you must:

- Conform to RICS's Home Survey Standards and the terms of engagement agreed with the client;

- Be clearly presented and follow a logical structure making appropriate use of visual images;
- Use of an appropriate prioritisation method (such as condition ratings) and an overall summary;
- Use non-technical terms throughout;
- Provide a balanced perspective on the condition of the property; and
- Properly account for further investigations, legal matters, and risk to occupants.

If you follow these principles, then the client will feel much better served and less likely to be disappointed.

References

RICS (2019). *RICS Professional Statement, Home Survey Standard,* 1st edition. Royal Institution of Chartered Surveyors, Great George Street, London.
Parnham, P. (2011). *A Surveyor's Guide to RICS Home Surveys.* Royal Institution of Chartered Surveyors, Great George Street, London.

15 Service completion, documentation storage and retention

Contents

15.1 Introduction

Once you have delivered the report and discussed any issues with your client, in most busy offices, there is a temptation to consider the job complete and transfer your attention to other clients who think their needs are more important. However, this is a tricky time. Although administration methods will vary between surveyors, at this stage, many of the service files could be a cacophony of emails, background research, and certification and associated documentation, photographs, video and audio files. Some files may be in the 'cloud', others on the hard drive with some paper copies languishing in the rarely checked physical in-tray.

At this stage, it is important to properly close any project file because:

- It will enable you to quickly track down information that may be required to answer any subsequent queries from the client. These may be perfectly straightforward questions that arise once they move into a home that is new to them. A well-organised file will help you do this quickly and efficiently;
- A properly closed file will help you meet most professional organisations audit requirements and if you are an RICS member it will show compliance with the Home Survey Standards (RICS 2019, p. 20);
- If you do find yourself in the position of having respond to a complaint or even a legal challenge, if you can quickly provide a well organised and clearly presented project file it will help build a profile of competence;
- It will protect business critical records and improve business resilience. For example, what would happen if your office burnt down or was flooded?
- An effectively organised and closed project file will minimise storage requirements and reduce costs.

DOI: 10.1201/9781003253105-15

In busy offices where individual surveyors may be carrying out one or two condition-related inspections a day, file closing can be low on priority lists. If this job is left unattended for a couple of weeks or more, memories will dim and the job of bringing sense to chaotic project files even more daunting.

15.2 Organising and assembling project files

Many surveyors will have the valuable support of non-technical staff and an effective quality assurance system that helps organise matters as the service is delivered. Whether this be an office manager, administrators, virtual assistants or the traditional secretary, this support can keep project files in cost-effective order using an agreed office-wide system. Using specific folders and even prearranged digital file names, material can be slotted in as they are generated – a facility that is very useful if the surveyor is ill while the service is being delivered. A locum will welcome a well-organised filing system.

For many sole traders, this level of support will not usually be available, although the principles of project administration will be exactly the same. Computer/cloud-based project management software can help plug this gap as long as they are understood and effectively used by the surveyor. General advice on general aspects of data management can be obtained from a variety of academic websites, including the 'Information Management' section of the National Archive website (National Archive 2022) and the Research data section of the University of Cambridge website (University of Cambridge 2022).

Whatever your set-up, closing a file will normally involve assembling and updating all the relevant information and communications. Rather than describing how this can be done in detail (which will soon be outdated by technological and regulatory changes), we have identified a number of important areas that will need resolving.

15.3 Client liaison

As previously discussed in Chapter 2.3 and 14.7, appropriate client liaison both before the terms of engagement are agreed and after the report has been delivered are vital. This can help ensure their requirements are met and, if there is a query or complaint, help clarify the nature of the contract as well as broader expectations. When closing the file, you may have to consider the following:

- How do you store the outcomes of the discussions? Depending on your systems, this may include such diverse communication methods as paper-based written comments, telephone conversations, social media messages (Messenger, WhatsApp), internet video apps and so on;
- Making sure all queries and/or questions are properly resolved or answered. Did you promise to get back to your client with some additional information, but never closed the loop?
- If your client did not take advantage of the offer to discuss the outcomes of the service, make sure this is clearly identified and recorded in the project file. This may prove useful if the client raises an issue at a later date.

15.4 Assembling information from different formats

In addition to those file formats used to record client discussion, other research may present a variety of different sources, such as:

- Images from Google Street View, satellite and maps; environmental information such as radon, flooding and transport noise maps;
- PDF documents of certificates, published guidance, manufacturer's information and so on;
- Photographic and video images. When surveyors took 10 to 20 real photographs of a property, cataloguing and storing was not a problem. With modern digital cameras and smartphones, many hundreds of images can be recorded and stored and each one could help resolve a future query. Organising this amount of files will be a balance between accessibility and time spent.

15.5 Cleaning up files

Even if you have a very effective document management system, you may have created and obtained a range of files or information that are no longer required and do not form part of the service. A typical example would be old internal drafts that were not shared with the client.

15.7 Retaining copies of documentation

Although the regulations may change over time, the project file must be securely stored for an appropriate period of time. The Home Survey Standards states that a legal liability may extend up to a maximum period of 15 years in England and Wales (RICS 2019, p. 20), although a six-year retention period is often employed across the residential sector. In our opinion, we recommend that a 15-year period be used.

References

National Archive (2022). 'How to manage your information'. Available at www.nationalarchives. gov.uk/information-management/manage-information/. Accessed 9 January 2022.

RICS (2019). *Home Survey Standard. RICS Professional Statement.* RICS, London.

University of Cambridge (2022). 'Organising your data'. Available at www.data.cam.ac.uk/data-management-guide/organising-your-data. Accessed 9 January 2022.

Index

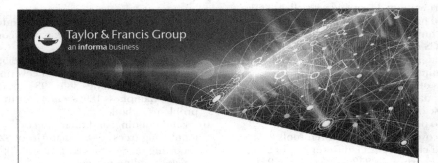

Printed in the United States
by Baker & Taylor Publisher Services